Introduction

The question of how the geometry of a projective variety is determined by its hyperplane sections has been an attractive area of algebraic geometry for at least a century. A century ago Picard's study of hyperplane sections led him to his famous theorem on the 'regularity of the adjoint '. This result, which is the Kodaira vanishing theorem in the special case of very ample line bundles on smooth surfaces, has led to many developments to this day. Castelnuovo and Enriques related the first Betti number of a variety and its hyperplane section. This and Picard's work led to the Lefschetz hyperplane section theorem and the modern work on ampleness and connectivity. A large part of the study of hyperplane sections has always been connected with the classification of projective varieties by projective invariants. Recent new methods, such as the adjunction mappings developed to study hyperplane sections, have led to beautiful general results in this classification. The papers in this proceedings of the L'Aquila Conference capture this lively diversity. They will give the reader a good picture of the currently active parts of the field.The papers can only hint at the friendly 'give and take' that punctuated many talks and at the mathematics actively discussed during the conference.

The success of this conference was in large part due to the Scientific and Organizing Committe: Professor Mauro Beltrametti (Genova), Professor Aldo Biancofiore (L'Aquila), Professor Antonio Lanteri (Milano), and Professor Elvira Laura Livorni (L'Aquila). The publication of this proceedings would not have been possible except for the efforts of Professor E.L.Livorni.

Andrew J. Sommese

Table of contents

INFINITESIMAL DEFORMATIONS OF NEGATIVE WEIGHTS AND HYPERPLANE SECTIONS

Lucian Bădescu

Introduction

Consider the following:

Problem. Let (Y,L) be a normal polarized variety over an algebraically closed field k, i.e. a normal projective variety Y over k together with an ample line bundle L on Y. Then one may ask under which conditions the following statement holds:

(#) Every normal projective variety X containing Y as an ample Cartier divisor such that the normal bundle of Y in X is L, is isomorphic to the projective cone over (Y,L), and Y is embedded in X as the infinite section.

Recall that the projective cone over (Y,L) is by definition the projective variety $C(Y,L) = \mathrm{Proj}(S[T])$, where S is the graded k-algebra $S(Y,L) = \bigoplus_{i=o}^{\infty} H^{\circ}(Y,L^{i})$ associated to (Y,L), and the polynomial S-algebra $S[T]$ (with T an indeterminate) is graded by $\deg(sT^{i}) = \deg(s)+i$ whenever $s \in S$ is homogeneous. The infinite section of $C(Y,L)$ is by definition the subvariety $V_{+}(T)$, and it is isomorphic to Y.

This problem has classical roots (see [3] for some historical hints). In [1], [2], [3] and [4], among other things, we produced several examples of polarized varieties (Y,L) satisfying (#). If Y is smooth of dimension $\geqslant 2$, and if T_Y is the tangent bundle of Y, Fujita subsequently proved in [6] the following general criterion: (Y,L) satisfies (#) if $H^{1}(Y,T_{Y}\otimes L^{i}) = o$ for every $i < o$.

In this paper we prove two main results. The first one (which is in the spirit of [4]) considers the case where Y has singularities, and is a criterion for (Y,L) to satisfy (#). This criterion (see theorem 1 in §1) improves a result of [4] and involves the space of first order infinitesimal deformations of the k-algebra $S(Y,L)$. In §2 we apply it to check that the singular Kummer varieties of dimension $\geqslant 3$ and the symmetric products of certain varieties satisfy (#) with respect to any ample line bundle. In §3 we make a few remarks when Y is smooth and state an open question. It should be noted that in the first two sections the Schlessinger's deformation theory (see [18], [19]) plays an essential role.

The second main result (see theorem 6 in §4) shows that if Y is a P^n-bundle ($n \geq 1$) over a smooth projective curve B of positive genus, and if X is a normal singular projective variety containing Y as an ample Cartier divisor, then X is isomorphic to the cone C(Y,L). The case $B = P^1$ was discussed in [3], while the case when X is smooth (and B of arbitrary genus), in [1] and [2]. Putting these results together, we get a complete description of all normal projective varieties containing a P^n-bundle over a curve as an ample Cartier divisor (see theorem 7 in §4).

Unless otherwise specified, the terminology and the notations used are standard.

§1. The first main result

In the set-up and notations of the above problem, the graded k-algebra $S = S(Y,L)$ is finitely generated because L is ample (see e.g. [8], chap. III). Let $a_1,\dots,a_n$ be a minimal system of homogeneous generators of S/k, and denote by $k[T_1,\dots,T_n]$ the polynomial k-algebra in n indeterminates $T_1,\dots,T_n$, graded by the conditions that $\deg(T_i) = \deg(a_i) = q_i$ for every $i = 1,\dots,n$. Then S is isomorphic (as a graded k-algebra) to $k[T_1,\dots,T_n]/I$ in such a way that a_i corresponds to $T_i \bmod I$ for every $i = 1,\dots,n$ (where I is the kernel of the homomorphism mapping T_i to a_i). Let $f_1,\dots,f_r$ be a minimal system of homogeneous generators of I, and set:

$$(1) \qquad d = \max(d_1,\dots,d_r), \text{ where } d_i = \deg(f_i).$$

Theorem 1. In the above notations assume the following:

i) $H^1(Y,L^i) = o$ for every $i \in \mathbb{Z}$, or equivalently, $\operatorname{depth}(S_{S_+}) \geq 3$, where S_+ is the irrelevant maximal ideal of S.

ii) $T^1_S(-i) = o$ for every $1 \leq i \leq d$, where d is given by (1), $T^1_S = T^1(S/k,S)$ is the space of first order infinitesimal deformations of the k-algebra S, and $T^1_S = \bigoplus_{i\in\mathbb{Z}} T^1_S(i)$ is the decomposition arising from the G_m-action of the graded k-algebra S (see [18], [17]).

Then the property (#) holds for (Y,L).

Proof. Let X be a normal projective variety containing Y as an ample Cartier divisor such that $O_X(Y)\otimes O_Y \cong L$. Let $t \in H^\circ(X,O_X(Y))$ be a global equation of Y in X, i.e. $\operatorname{div}_X(t) = Y$. Denote by S' the graded k-algebra $S(X,O_X(Y)) = \bigoplus_{i=o}^{\infty} H^\circ(X,O_X(iY))$. Then using the standard exact sequence

$$o \longrightarrow O_X((i-1)Y) \xrightarrow{t} O_X(iY) \longrightarrow L^i \longrightarrow o,$$

the hypothesis i), and a theorem of Severi-Zariski-Serre saying that $H^1(X,O_X(iY)) = o$ for every $i \ll o$, one immediately sees that $S'/tS' \cong S$ (isomorphism of graded k-algebras, where $\deg(t) = 1$).

Then choose $b_1,\ldots,b_n \in S'$ homogeneous elements of degrees $q_1,\ldots,q_n$ respectively, such that $b_i \bmod tS' = a_i$, $i = 1,\ldots,n$. Then $S' = k[b_1,\ldots,b_n,t]$. Denote by P the polynomial k-algebra $k[T_1,\ldots,T_n,T]$ in n+1 indeterminates $T_1,\ldots,T_n,T$, graded by $\deg(T_i) = q_i$, $i = 1,\ldots,n$, and $\deg(T) = 1$. For every $m \geq 1$ set $S^m = S'/t^mS'$, and consider the surjective homomorphism $\beta_m: P \longrightarrow S^m$ such that $\beta_m(T_i) = b_i'$, $i = 1,\ldots,n$, and $\beta_m(T) = t'$, where for every $b \in S'$ we have denoted by b' the element $b \bmod t^mS'$. Let $F_1,\ldots,F_s$ be a system of homogeneous generators of the ideal $J = \mathrm{Ker}(\beta_m)$, and put $e_i = \deg(F_i)$, $i = 1,\ldots,s$.

Now, according to [18], §1 (or also [14]), we can consider:

- The S^m-module $Ex(S^m/k,S)$ of all isomorphism classes of extentions of S^m over k by the S^m-module $S = S^m/t'S^m$. Recall that an extention of S^m/k by S is a k-algebra E together with a surjective homomorphism of k-algebras $E \longrightarrow S^m$ whose kernel is a square-zero ideal of E, isomorphic as an S^m-module to S.

- The S^m-module $T^1(S^m/k,S)$ defined by the following exact sequence

$$(2) \quad \mathrm{Der}_k(P,S) \xrightarrow{u} \mathrm{Hom}_{S^m}(J/J^2,S) \longrightarrow T^1(S^m/k,S) \longrightarrow o,$$

where $\mathrm{Der}_k(P,S)$ is the S^m-module of all k-derivations of P in S, and u is defined in the following way: if $D \in \mathrm{Der}_k(P,S)$ then $u(D)$ is the element of $\mathrm{Hom}_{S^m}(J/J^2,S)$ defined by the restriction D/J (which necessarily vanishes on J^2). It turns out that $T^1(S^m/k,S)$ is independent of the choice of the presentation P/J of S^m.

Now, the point is that there is a canonical isomorphism of S^m-modules (see [18], theorem 1, page 12, or also [14], page 41o):

$$(3) \qquad \mu: Ex(S^m/k,S) \xrightarrow{\sim} T^1(S^m/k,S).$$

Since S^m is a graded k-algebra, $T^1(S^m/k,S)$ has a natural gradation $T^1(S^m/k,S) = \bigoplus_{i\in\mathbb{Z}} T^1(S^m/k,S)(i)$ arising from the G_m-action of S^m (see [17], page 19).

Coming back to our situation, consider the element of $Ex(S^m/k,S)$ given by the exact sequence

$$(a_m) \qquad o \longrightarrow S \cong t^m S'/t^{m+1}S' \longrightarrow S^{m+1} \longrightarrow S^m \longrightarrow o.$$

We need to compute $\mu(a_m) \in T^1(S^m/k,S)$ explicitly. By the definition of the isomorphism μ (see [18]), we need to consider the commutative diagram with exact rows

$$\begin{array}{ccccccccc}
o & \longrightarrow & J/J^2 & \longrightarrow & P/J^2 & \longrightarrow & P/J & \longrightarrow & o \\
 & & \downarrow v & & \downarrow v' & & \downarrow & & \\
o & \longrightarrow & t^m S'/t^{m+1}S' \overset{w}{\cong} S & \longrightarrow & S^{m+1} & \longrightarrow & S^m & \longrightarrow & o
\end{array}$$

where v' is the map deduced from β_m. Thus $v(F_i \bmod J^2) = t^m G_i(b_1,\dots,b_n,t) \bmod t^{m+1}S'$, with $G_i(b_1,\dots,b_n,t) \in S'_{e_i-m}$ homogeneous of degree e_i-m. Then $w \circ v \in \mathrm{Hom}_{S^m}(J/J^2,S)$ corresponds to the vector $(G'_1,\dots,G'_s)$, with $G'_i = w(t^m G_i(b_1,\dots,b_n,t) \bmod t^{m+1}S') = G_i(a_1,\dots,a_n,o)$, and recalling the exact sequence (2) we have $\mu(a_m)$ = class of $w \circ v \in T^1(S^m/k,S)$.

According to the explicit description of the gradation of $T^1(S^m/k,S)$ given in [17], page 19, the elements of $T^1(S^m/k,S)(j)$ of degree j correspond to those elements of $\mathrm{Hom}_{S^m}(J/J^2,S)$ given by vectors $(h_1,\dots,h_s)$ with $h_i \in S_{e_i+j}$ homogeneous of degree e_i+j, $i = 1,\dots,s$. Since $\deg(G'_i) = e_i-m$, the foregoing discussion implies:

$$(4) \qquad \mu(a_m) \in T^1(S^m/k,S)(-m) \quad \text{for every } m \geq 1.$$

Now take $m = 1$. Since $S^1 = S$, it follows that $\mu(a_1) = o$ by hypothesis ii). But the trivial extension of $\mathrm{Ex}(S/k,S)$ is

$$o \longrightarrow S \cong TS[T]/(T^2) \longrightarrow S[T]/(T^2) \longrightarrow S[T]/(T) \cong S \longrightarrow o,$$

and therefore there is an isomorphism of extensions

$$\begin{array}{ccccccccc}
o & \longrightarrow & S \cong TS[T]/(T^2) & \longrightarrow & S[T]/(T^2) & \longrightarrow & S[T]/(T) \cong S & \longrightarrow & o \\
 & & \| & & \downarrow \wr & & \downarrow \wr & & \\
o & \text{———} & S \cong t'S^2 & \text{———} & S^2 & \text{———} & S & \text{———} & o
\end{array}$$

such that the vertical isomorphism in the middle maps $T \bmod (T^2)$ into t'.

Assume now that we know that for some m, $2 \leq m \leq d$, there is an isomorphism $S[T]/(T^i) \cong S^i$ for every $1 \leq i \leq m$, which maps $T \bmod (T^i)$ into $t' = t \bmod t^i S'$. Then recall that there is a general exact sequence (see [18])

$$T^1(S^m/S,S) \longrightarrow T^1(S^m/k,S) \longrightarrow T^1(S/k,S),$$

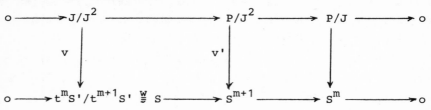

where the maps are homogeneous and the second map corresponds to the inclusion $S \hookrightarrow S^m$ obtained by composing the natural inclusion $S \hookrightarrow S[T]/(T^m)$ with the isomorphism $S[T]/(T^m) \cong S^m$. Using this and hypothesis ii) we infer that the map $T^1(S^m/S,S)(-m) \longrightarrow T^1(S^m/k,S)(-m)$ is surjective, wich together with (4) implies that the extension (a_m) comes from $\mathrm{Ex}(S^m/S,S) \cong T^1(S^m/S,S)$. In other words, S^{m+1} is an S-algebra and the canonical surjective map $S^{m+1} \longrightarrow S^m$ is a map of S-algebras. Then we can easily define an isomorphism of extensions

$$\begin{array}{ccccccccc}
o & \longrightarrow & S \cong T^m S[T]/(T^{m+1}) & \longrightarrow & S[T]/(T^{m+1}) & \longrightarrow & S[T]/(T^m) & \longrightarrow & o \\
 & & \| & & \downarrow & & \downarrow \wr & & \\
o & \longrightarrow & S = t^m S'/t^{m+1}S' & \longrightarrow & S^{m+1} & \longrightarrow & S^m & \longrightarrow & o
\end{array}$$

where the middle vertical isomorphism is the homomorphism of S-algebras mapping $T \bmod (T^{m+1})$ into $t' = t \bmod t^{m+1}S'$.

Summing up, we have proved by induction on m that there is an isomorphism of graded k-algebras $S[T]/(T^{d+1}) \cong S^{d+1}$ such that $T \bmod (T^{d+1})$ corresponds to $t \bmod t^{d+1}S'$. In particular, there is a commutative diagram

$$\begin{array}{ccc}
S & \xrightarrow{\quad h \quad} & S^{d+1} = S'/t^{d+1}S' \\
 & \text{id} \searrow \quad \swarrow \text{canonical surjection} & \\
 & S = S^1 &
\end{array}$$

Choose homogeneous elements $c_i \in S'_{q_i}$ such that $h(a_i) = c_i \bmod t^{d+1}S'$, $i = 1,\dots,n$. Then we claim that

$$(5) \qquad f_i(c_1,\dots,c_n) = o \quad \text{for every } i = 1,\dots,r.$$

Indeed, since $f_i(c_1,\dots,c_n) \bmod t^{d+1}S' = f_i(h(a_1),\dots,h(a_n)) = h(o) = o$, it follows that $f_i(c_1,\dots,c_n) \in t^{d+1}S'$ for every $i = 1,\dots,r$. If for some i we would have $f_i(c_1,\dots,c_n) \neq o$, it would follow that $d_i = \deg(f_i(c_1,\dots,c_n)) \geqslant d+1$, a contradiction because $d = \max(d_1,\dots,d_r)$.

Finally, using (5) we can construct a homomorphism of graded k-algebras $f: S \longrightarrow S'$ by putting $f(a_i) = c_i$, $i = 1,\dots,n$. The equations (5) show that this definition is correct. Then we get a unique homomorphism of graded k-algebras $g: S[T] \longrightarrow S'$ such that $g/S = f$ and $g(T) = t$. Then it is clear that g is surjective, and hence an isomorphism, because both $S[T]$ and S' are domains of the same dimension. In other words, we have proved that $X \cong C(Y,L)$. Q.E.D.

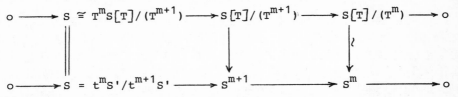

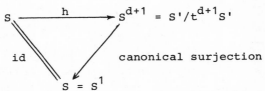

Remarks. 1) Theorem 1 had been proved in [4] in the stronger hypothesis that $T^1_S = o$, where we had in mind an application to weighted projective spaces.

2) Unfortunately, the hypothesis i) of theorem 1 is quite restrictive. We do not know whether theorem 1 remains still valid if one drops hypothesis i), even if one assumes for example that char(k) = o and $T^1(-i) = o$ for every $i \geqslant 1$.

Corollary 1. In the notations of theorem 1, assume that ii) holds. Let X be a normal projective variety containing Y as an ample Cartier divisor such that the normal bundle of Y in X is L and $H^1(X, O_X(iY)) = o$ for every $i \geqslant o$. Then X is isomorphic to the projective cone C(Y,L) and Y is embedded in X as the infinite section.

Indeed, the exact sequence from the beginning of the proof of theorem 1 together with the hypothesis that $H^1(X, O_X(iY)) = o$ for every $i \geqslant o$ imply that $S'/tS' \cong S$ (in the proof of theorem 1 the hypothesis i) was used only to deduce this isomorphism).

Another immediate consequence of the proof of theorem 1 is the following purely algebraic result:

Corollary 2. Let $S = k[T_1, \dots, T_n]/I$ be an $\mathbb{N}$-graded k-algebra, where the polynomial k-algebra $k[T_1, \dots, T_n]$ in the indeterminates $T_1, \dots, T_n$ is graded by $\deg(T_i) = q_i > o$, $i = 1, \dots, n$, for some fixed system of weights $(q_1, \dots, q_n)$, and I is the ideal generated by some homogeneous polynomials $f_1, \dots, f_r$ of positive degrees. Let S' be an $\mathbb{N}$-graded k-algebra such that S'/tS' is isomorphic to S as a graded k-algebra for some non-zero divisor $t \in S'$ of degree 1. If $T^1_S(-i) = o$ for every $1 \leqslant i \leqslant \max(\deg(f_1), \dots, \deg(f_r))$, then S' is isomorphic (as a graded k-algebra) to the polynomial S-algebra S[T] in such a way that t is mapped into T.

§2. Applications of theorem 1

The tools for verifying hypotheses of type ii) of theorem 1 have been developed by Schlessinger in [19]. The lemma 1 below (which is essentially due to Schlessinger) provides examples of singular normal polarized varieties (Y,L) satisfying the condition ii) of theorem 1.

Start with a smooth projective variety V and a finite group G acting on V. Denote by Y the quotient variety V/G and by $f: V \longrightarrow Y$ the canonical morphism. Let L be an ample line bundle on Y and set $M = f^*(L)$. Since f is a finite morphism, M is also ample. Let S = S(Y,L) and $A = S(V,M)$ be the graded k-algebras associated to (Y,L) and (V,M) respectively.

Lemma 1. *In the above notations assume the following:*

i) *$\mathrm{Dim}(V) \geqslant 3$ and $\mathrm{char}(k)$ is either zero, or prime to the order $/G/$ of G.*

ii) *G acts freely on V ouside some closed G-invariant subset of V of codimension $\geqslant 3$.*

iii) *$H^1(V,M^{-i}) = o$ for every $i \geqslant 1$ (in characteristic zero this is always fulfilled by Kodaira's vanishing theorem).*

iv) $H^1(V,T_V \otimes M^{-i}) = o$ *for every $i \geqslant 1$, where T_V is the tangent bundle of V.*

Then $T^1_S(-i) = o$ *for every $i \geqslant 1$.*

Proof. Since lemma 1 is not given in [19] in this form, we include its proof for the convenience of the reader. From ii) we infer that the singular locus of Y, Sing(Y), is of codimension $\geqslant 3$, and that f is étale outside Sing(Y). Using this, the normality of Y and [16], §7, it follows that $f_*(M^i)^G = L^i$ for every $i \geqslant o$. This shows that G acts on A by automorphisms of graded k-algebras and that the k-algebra of invariants A^G coincides with S. Consider the cartesian diagram

$$\begin{array}{ccc} \mathrm{Spec}(A)-(A_+) = W & \xrightarrow{\ g\ } & U = \mathrm{Spec}(S)-(S_+) = W/G \\ \Big\downarrow q & & \Big\downarrow p \\ V & \xrightarrow{\ f\ } & Y = V/G \end{array}$$

with q and p the canonical projections of the G_m-bundles W and U respectively (see [8], chap. II, §8). If F is the ramification locus of f, then $q^{-1}(F)$ is the ramification locus of g, and hence g acts freely on W ouside a closed G-invariant subset of W. In particular, the singular locus Z of U is of codimension $\geqslant 3$ in U. Then by [19] and [2o] we get that $T_U = g_*(T_W)^G$, where T_U is the tangent sheaf of U. Taking into account of hypothesis i) we infer that T_U is a direct summand of $g_*(T_W)$, and in particular

(6) $\quad H^1(U,T_U)$ is a direct summand of $H^1(U,g_*(T_W)) \cong H^1(W,T_W)$.

On the other hand, it is well known that there is a canonical exact sequence (see e.g. [14] or [21])

$$o \longrightarrow O_W \longrightarrow T_W \longrightarrow q^*(T_V) \longrightarrow o$$

which yields the exact sequence

(7) $\quad H^1(W,O_W) \longrightarrow H^1(W,T_W) \longrightarrow H^1(W,q^*(T_V))$

One has the natural isomorphisms $H^1(W,O_W) \cong \bigoplus_{i\in\mathbb{Z}} H^1(V,M^i)$ and $H^1(W,q^*(T_V)) \cong \bigoplus_{i\in\mathbb{Z}} H^1(V,T_V\otimes L^i)$, which give natural gradings on $H^1(W,O_W)$ and on $H^1(W,q^*(T_V))$ respectively. On the other hand, the middle space in (7) has also a natural gradation $H^1(W,T_W) = \bigoplus_{i\in\mathbb{Z}} H^1(W,T_W)(i)$ arising from the G_m-action on W, and all these three gradations are compatible with the maps in (7). Therefore, using hypotheses iii) and iv) we get that $H^1(W,T_W)(i) = o$ for every $i<o$. There is also a natural gradation $H^1(U,T_U) = \bigoplus_{i\in\mathbb{Z}} H^1(U,T_U)(i)$ arising from the G_m-action on U, and this gradation is compatible via (6) with the gradation of $H^1(W,T_W)$. Consequently we get:

$$(8) \qquad H^1(U,T_U)(i) = o \quad \text{for every } i<o$$

Since U has only quotient singularities in codimension ≥ 3, by [19] and [2o] we infer that all the singularities of U are rigid, and in particular, $\text{depth}_Z(T_U)\geq 3$. Then the exact sequence of local cohomology shows that the restriction map $H^1(U,T_U) \longrightarrow H^1(U-Z,T_U)$ is an isomorphism.

Finally, since U has only quotient (and hence Cohen-Macaulay) singularities and $\text{codim}_U(Z)\geq 3$, by [19] and [2o] we get $T^1_S \cong H^1(U-Z,T_U)$. Recalling (8) and the isomorphism $H^1(U-Z,T_U) \cong H^1(U,T_U)$ we get the conclusion of lemma 1. Q.E.D.

Now we illustrate how theorem 1 can be applied -via lemma 1 - on some examples. First we apply theorem 1 to the singular Kummer varieties of dimension ≥ 3. Recall that a singular Kummer variety Y is a variety of the form V/G, where V is an abelian variety of dimension $d\geq 2$ and $G\subset \text{Aut}(V)$ is the subgroup of order 2 generated by the involution $u:V \longrightarrow V$ defined by $u(x) = -x$ for every $x\in V$ (where $-x$ is the inverse of x in the group-law of V). If $\text{char}(k) \neq 2$, there are exactly 2^{2d} points of order 2 on V (see [16]), and hence Y = V/G has exactly 2^{2d} isolated singularities (which are all quotient singularities). Now we have:

<u>Theorem 2. Let Y be a singular Kummer variety of dimension $d\geq 3$ and let L be an arbitrary ample line bundle on Y. If char(k) $\neq$ 2 then the property (#) holds for (Y,L).</u>

<u>Proof.</u> We first show that lemma 1 implies that $T^1_S(-i) = o$ for every $i\geq 1$, with S = S(Y,L). Indeed, the hypotheses i) and ii) of lemma 1 are clearly satisfied, while iii) and iv) follow using the fact that the tangent bundle of an abelian variety is trivial, together with the fact that the Kodaira's vanishing theorem holds for an abelian variety

in arbitrary characteristic (see [16], §16).

It remains to check that $H^1(Y,L^i) = o$ for every $i \in \mathbb{Z}$ (which is the first hypothesis of theorem 1). If $f:V \longrightarrow Y$ is the canonical morphism, then by [19], L^i is a direct summand of $f_*f^*(L^i)$ because $\mathrm{char}(k) \neq 2 = /G/$, and hence $H^1(Y,L^i)$ is a direct summand of $H^1(Y,f_*f^*(L^i)) \cong H^1(V,f^*(L^i))$. By [16], §16 the latter space is zero for every $i \neq o$ because $f^*(L)$ is ample. On the other hand, if $i = o$, according to Schlessinger [19], page 24, we infer that $H^1(Y,O_Y) = H^1(V,O_V)^G$, and G acts on $H^1(V,O_V)$ by $t \longrightarrow -t$. It follows that $H^1(Y,O_Y) = o$. Applying theorem 1 we get the conclusion. Q.E.D.

Further examples of singular normal varieties satisfying (#) with respect to any ample line bundle are the symmetric products of certain smooth projective varieties. Let Z be a smooth projective variety of dimension $d \geqslant 3$, and let Y be the symmetric product $Z^{(n)} = V/G$, where: $n \geqslant 2$ is a fixed integer, $V = Z^n$ (the direct product of Z with itself n times), and G is the symmetric group of degree n acting on V by $g\cdot(z_1,\dots,z_n) = (z_{g(1)},\dots,z_{g(n)})$ for every $g \in G$ and $(z_1,\dots,z_n) \in V$. Then the ramification locus of the canonical morphism $f:V \longrightarrow Y$ has codimension $d = \dim(V) \geqslant 3$ in V.

<u>Theorem 3. Let Z be a smooth projective variety of dimension $d \geqslant 3$ such that $H^1(Z,M) = o$ for every line bundle M on Z, and let $n \geqslant 2$ be an integer such that either $\mathrm{char}(k) = o$, or $n < \mathrm{char}(k)$ if $\mathrm{char}(k) > o$. Then for every ample line bundle L on $Y = Z^{(n)}$ the property (#) holds for (Y,L).</u>

<u>Note.</u> The simplest examples of varieties Z satisfying the hypotheses of theorem 3 are all smooth hypersurfaces in P^{d+1} with $d \geqslant 3$.

<u>Proof of theorem 3.</u> The hypotheses imply in particular that $H^1(Z,O_Z) = o$, and then the see-saw principle (see [16],§5) immediately implies that $f^*(L) \cong p_1^*(L_1) \otimes \dots \otimes p_n^*(L_n)$, with $L_1,\dots,L_n \in \mathrm{Pic}(Z)$ and $p_i:V \longrightarrow Z$ the projection of V onto the i-th factor. Since L is ample on Y and f is finite, $f^*(L)$ is ample on V, and hence L_i is ample on Z for every $i = 1,\dots,n$. As in the proof of theorem 2, it will be sufficient to check the following:

$$H^1(V,f^*(L^i)) = o \quad \text{for every } i \in \mathbb{Z}, \text{ and}$$

$$H^1(V,T_V \otimes f^*(L^i)) = o \quad \text{for every } i < o,$$

in order to deduce (via lemma 1) that the hypotheses of theorem 1 are satisfied. But these vanishings are easily checked using the Künneth's formulae, the fact that $T_V = p_1^*(T_Z) \oplus \dots \oplus p_n^*(T_Z)$, the hypotheses of the theorem, and the fact that L_i is ample for $i = 1,\dots,n$ (which implies that $H^{\circ}(Z,L_i^j) = o$ for every $j < o$ and $i = 1,\dots,n$). Then the conclusion of the theorem follows from theorem 1. Q.E.D.

§3. A few remarks when Y is smooth

In this section we shall assume that Y is smooth and char(k) = o. Then it is known that the space $T_S^1(i)$ can be computed in the following way (see [23], page 337 and theorem 3.7). First, there is an exact sequence of vector bundles

$$o \longrightarrow O_Y \longrightarrow M \longrightarrow T_Y \longrightarrow o,$$

which is the dual of the exact sequence

$$o \longrightarrow \Omega_Y^1 \longrightarrow F \longrightarrow O_Y \longrightarrow o$$

corresponding to the image of L in $H^1(Y,\Omega_Y^1)$ via the canonical map $H^1(Y,O_Y^*) \cong \mathrm{Pic}(Y) \longrightarrow H^1(Y,\Omega_Y^1)$ induced by the map $O_Y^* \longrightarrow \Omega_Y^1$ given by $f \longrightarrow df/f$. Then it is proved in loc. cit. that

$$(9) \quad T_S^1(i) = \mathrm{Ker}(H^1(Y,M\otimes L^i) \longrightarrow H^1(Y,\bigoplus_{j=1}^{n} L^{i+q_j})) \text{ for every } i\in\mathbb{Z},$$

where $S = S(Y,L)$ and $q_1,\dots,q_n$ have the same meaning as at the beginning of §1.

Using (9), the first exact sequence and the Kodaira's vanishing theorem, it follows that the condition "$T_S^1(-i) = o$ for every $i \geq 1$" is a consequence of the condition "$H^1(Y,T_Y\otimes L^{-i}) = o$ for every $i \geq 1$". If Y is smooth and char(k) = o, one can get rid of the unpleasant hypothesis i) of theorem 1 because of the following:

<u>Theorem 4</u> (See [6]). <u>Let (Y,L) be a smooth polarized variety of dimension ≥ 2 such that</u> $H^1(Y,T_Y\otimes L^{-i}) = o$ <u>for every $i \geq 1$ and char(k) = o. Then the property (#) holds for (Y,L).</u>

Theorem 4 is proved in [6]; via a quick argument, it is also a consequence of theorem 2 in [22]. Using theorem 4 and the main result of [22] we prove the following:

<u>Theorem 5. Let (Y,L) be a smooth polarized variety such that: char(k) = o, dim(Y) $\geq$ 2,</u> $H^1(Y,T_Y\otimes L^{-i}) = o$ <u>for i = 1 and i = 2, and the linear system |L| contains a smooth divisor. Then the property (#) holds for (Y,L).</u>

<u>Proof.</u> By theorem 4 it will be sufficient to show that $H^1(Y,T_Y\otimes L^{-i}) = o$ for every $i \geq 1$. Let $H \in |L|$ be a smooth divisor of |L|. Since dim(Y) $\geq$ 2, H is also connected. If we denote by L_H the restriction $L\otimes O_H$ and by T_H the tangent bundle of H, we have the canonical

exact sequence

$$o \longrightarrow T_H \otimes L_H^{-i} \longrightarrow (T_Y \otimes L^{-i}) \otimes O_H \longrightarrow L_H^{1-i} \longrightarrow o,$$

which yields the exact sequence

$$(10_i) \quad H^\circ(H, T_H \otimes L_H^{-i}) \longrightarrow H^\circ(H, (T_Y \otimes L^{-i}) \otimes O_H) \longrightarrow H^\circ(H, L_H^{1-i}).$$

For every $i \geq 2$ the last space is zero. On the other hand, by the main result of [22] (which extends a theorem of Mori-Sumihiro), the first space could be $\neq o$ only if $(H, L_H) \cong (P^1, O(1))$ (and then $i = 2$), in which case it follows easily that $(Y, L) \cong (P^2, O(1))$, and hence the property (#) holds for (Y,L) in this case. Thus we may assume that $H^\circ(H, T_H \otimes L_H^{-i}) = o$ for every $i \geq 2$. Then by (10_i) we get that the space in the middle is zero for every $i \geq 2$. Finally, using this and the exact sequence

$$(11_i) \quad o \longrightarrow T_Y \otimes L^{-i-1} \longrightarrow T_Y \otimes L^{-i} \longrightarrow (T_Y \otimes L^{-i}) \otimes O_H \longrightarrow o,$$

we infer that the map $H^1(Y, T_Y \otimes L^{-i-1}) \longrightarrow H^1(Y, T_Y \otimes L^{-i})$ is injective for every $i \geq 2$. Therefore $H^1(Y, T_Y \otimes L^{-i}) = o$ for every $i \geq 1$. Q.E.D.

<u>Corollary.</u> <u>Let (Y,L) be a smooth polarized variety of dimension $d \geq 2$ such that there is a smooth divisor $H \in |L|$ for which the exact sequence</u>

$$(12) \quad o \longrightarrow T_H \longrightarrow T_Y \otimes O_H \longrightarrow L_H \longrightarrow o$$

<u>is not split (in particular,</u> $H^1(H, T_H \otimes L_H^{-1}) \neq o$). <u>Assume moreover that $\mathrm{char}(k) = o$ and</u> $H^1(Y, T_Y \otimes L^{-1}) = o$. <u>Then the property (#) holds for (Y,L).</u>

<u>Proof.</u> According to the proof of theorem 5, the exact sequence (11_1) shows that it is sufficient to prove that $H^\circ(H, (T_Y \otimes L^{-1}) \otimes O_H) = o$. The exact sequence (10_1) yields the exact sequence

$$(13) \quad H^\circ(H, T_H \otimes L_H^{-1}) \longrightarrow H^\circ(H, (T_Y \otimes L^{-1}) \otimes O_H) \longrightarrow H^\circ(H, O_H) \xrightarrow{\partial} H^1(H, T_H \otimes L_H^{-1}).$$

By [22], the first space could be $\neq o$ only in one of the following cases: either $(H, L_H) \cong (P^{d-1}, O(1))$, or $(H, L_H) \cong (P^1, O(2))$. In the first case $(Y,L) \cong (P^d, O(1))$, and hence (Y,L) has the property (#) in this case; the second case is ruled out because then $H^1(H, T_H \otimes L_H^{-1}) = o$, and hence (12) splits. Therefore we may assume $H^\circ(H, T_H \otimes L_H^{-1}) = o$, and then (13) shows that $H^\circ(H, (T_Y \otimes L^{-1}) \otimes O_H) = o$ if and only if $\partial(1) \neq o$. Since $\partial(1)$ is the obstruction in $H^1(H, T_H \otimes L_H^{-1})$ such that (12) be split, we get the result. Q.E.D.

<u>Remark.</u> In a more special situation, L'vovskii proved in [15] a better result than theorem 5 and its corollary. More precisely, assume that $Y \subset P^n$ is a smooth non-degenerate projective subvariety of P^n of

dimension $\geqslant 2$ and degree $\geqslant 3$, such that $H^1(Y,T_Y(-1)) = o$ and char(k)= = o. Let $X \subset P^{n+1}$ be an irreducible subvariety of P^{n+1} such that $X \cap P^n =$ = Y, and X is smooth along Y and transversal to P^n, where P^n is embedded in P^{n+1} as a hyperplane. Then X is a cone over Y. In fact, L'vovskii has an even weaker assumption than $H^1(Y,T_Y(-1)) = o$ (loc. cit.). His proof uses completely different techniques.

Coming back to the above corollary, we may ask the following:

Question. Let (Y,L) be a smooth polarized variety of dimension $d \geqslant 2$ such that L is generated by its global sections. Find sufficient conditions ensuring that there is a smooth divisor $H \in |L|$ such that the corresponding exact sequence (12) is not split. Or, enumerate the situations when (12) is split for $H \in |L|$ general.

A necessary condition such that this question has a positive answer is that $H^1(H,T_H \otimes L_H^{-1}) \neq o$ for $H \in |L|$ general. Is it also sufficient? In the case of surfaces, the pairs (Y,L) for which $H^1(H,T_H \otimes L_H^{-1}) = o$ for $H \in |L|$ general, can be easily enumerated. Indeed, by duality and Riemann-Roch on the curve H one gets that this happens if and only if $(H,L_H) \cong (P^1,O(i))$ with i=1, 2, or 3. And by a well-known classical result, (Y,L) is isomorphic to one of the following: $(P^2,O(1))$, $(P^1 x P^1,O(1,1))$, or any smooth hyperplane section of $P^1 x P^2 \subset \subset P^5$ via the Segre embedding (the latter surfaces are all isomorphic to the projective plane blown up at a point).

§4. P^n-bundles over an irrational curve as hyperplane sections

Let B be a smooth projective curve, and let E be a vector bundle of rank n+1 on B, with $n \geqslant 1$. Denote by Y = P(E) the projective bundle associated to E, and by $p: Y \longrightarrow B$ the canonical projection. The main result of this section is the following:

Theorem 6. In the above notations, assume that the genus of B is positive and char(k) = o. Let X be a singular normal projective variety containing Y = P(E) as an ample Cartier divisor. Then X is isomorphic to the projective cone C(Y,L) and Y is embedded in X as the infinite, where L is the normal bundle of Y in X.

The motivation of theorem 6 lies in the fact that, combining it with some results from [1], [2], and [3], we get the following complete description of all normal projective varieties whose hyperplane sections are P^n-bundles over a curve:

Theorem 7. Assume that B is a smooth projective curve of arbitrary genus, and let $Y = P(E)$ be a P^n-bundle over B ($n \geqslant 1$). Assume furthermore that char(k) = o. Let X be a normal projective variety containing Y as an ample Cartier divisor. Then one has one of the following possibilities:

a) $X \cong P^3$, $Y \cong P^1 x P^1$, and Y is embedded in X as a quadric.

b) X is isomorphic to a smooth hyperquadric in P^4, $Y \cong P^1 x P^1$, and Y is embedded in X as the intersection of X with a hyperplane of P^4.

c) There is an exact sequence of vector bundles of B of the form

$$o \longrightarrow O_B \longrightarrow F \xrightarrow{\ \mu\ } E' \longrightarrow o$$

such that F is an ample vector bundle in the sense of [1o], $E' = E \otimes L'$ for some $L' \in \mathrm{Pic}(B)$, $X \cong P(F)$, and $Y \cong P(E')$ is embedded in X via μ.

d) X is isomorphic to the projective cone C(Y,L), where L is the normal bundle of Y in X, such that Y is embedded in X as the infinite section.

Remarks. 1) In certain (but not all) cases theorem 6 was proved in [3], theorem 6.

2) Theorem 7 is obtained as the result of a long case-by-case discussion (see [1], theorem 5, [2], theorems 1, 2 and 3, and [3], theorems 3, 4 and 5, and theorem 6 above). The most difficult case is when Y is a surface, i.e. E is a rank two vector bundle. Note that the proof of the result in case $Y = P(O_{P^1} \oplus O_{P^1}(-1))$ and X is smooth is completely given in our short note, L. Bădescu, The projective plane blown up at a point as an ample divisor, Atti Accad. Ligure Sci. Lettere, 38 (1981), 3-7 (cf. also lemma 2 in [3] and its proof, in the case X is singular). Another proof of theorem 7 in case X is smooth was subsequently given by P. Ionescu in [12], as an application of the general adjunction mapping, using Mori's theory of extremal rays and Kawamata-Shokurov contraction theorem.

Proof of theorem 6. According to [2] and [3], the Lefschetz theorem and the Albanese mapping yield the commutative diagram

$$\begin{array}{ccc} Y & \longrightarrow & U \subset X \\ & p \searrow \quad \swarrow q & \\ & B & \end{array}$$

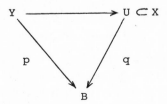

where U is an open neighbourhood of Y in X (in fact, we can take U = $= X_{reg}$). Then X has finitely many singularities, and by Hironaka [11], there is a desingularization $f:X'' \longrightarrow X$ with the following properties: f induces an isomorphism $f^{-1}(U) \cong U$, the rational map $q'' = q \circ f: X'' \longrightarrow B$ is in fact a morphism, and the exceptional fibres of f are divisors of normal crossings (i.e. with smooth components of codimension one intersecting transversely). Then the normal bundle of Y in X" is L, and since L is ample, L is in particular p-ample. One of the main point in the proof of theorem 6 is the following lemma, which is essentially the relativization of theorem 4.2, chap. III of [9].

Lemma 2. Let $q'':X'' \longrightarrow B$ be a surjective morphism between the normal projective varieties X" and B, and let Y be an effective Cartier divisor on X" such that the restriction $p:Y \longrightarrow B$ of q" is surjective. Assume that the normal bundle L of Y in X" is p-ample. Then there is a canonical commutative diagram

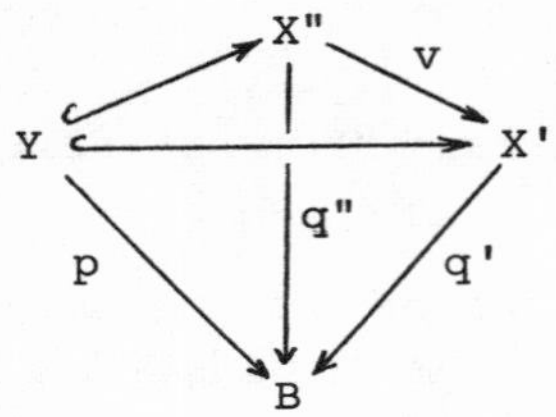

with X' a normal projective variety, $q':X' \longrightarrow B$ a morphism, v a birational morphism such that v is an isomorphism in a neighbourhood of Y, and v(Y) is a q'-ample effective Cartier divisor on X'.

Proof of lemma 2. First we are going to show that for $i \gg o$ the following three conditions are satisfied:

i) L^i is p-very ample.

ii) The canonical map $q''^{*}q''_{*}(O_{X''}(iY)) \longrightarrow O_{X''}(iY)$ is surjective.

iii) The canonical map $q''_{*}(O_{X''}(iY)) \longrightarrow p_{*}(L^i)$ is surjective.

Indeed, since L is p-ample, i) holds. Now we prove iii). Consider the exact sequence $(i \geq 1)$

$$q''_{*}(O_{X''}(iY)) \longrightarrow p_{*}(L^i) \longrightarrow R^1q''_{*}(O_{X''}((i-1)Y)) \xrightarrow{\mu_i} R^1q''_{*}(O_{X''}(iY)) \longrightarrow R^1p_{*}(L^i)$$

induced by $o \longrightarrow O_{X''}((i-1)Y) \longrightarrow O_{X''}(iY) \longrightarrow L^i \longrightarrow o.$

The last sheaf is zero for $i \gg o$ because L is p-ample (see [8], chap.III, (2.2.1)). Hence the map μ_i is surjective for every $i \geq j$ for some $j > o$. Since q" is a projective morphism, $R^1q''_{*}(O_{X''}(jY))$ is coherent on B, and therefore μ_i becomes an isomorphism for $i \gg o$, i.e. iii) holds.

To prove ii), observe that by [8], chap. II, (3.4.7), ii) is equivalent to the fact that for every affine open subset D = Spec(A) of B, the sheaf $O_{X''}(iY)/q''^{-1}(D)$ is generated by its global sections for $i \gg o$. But by iii), the natural map $H^{\circ}(D, q''_{*}(O_{X''}(iY))) \cong H^{\circ}(q''^{-1}(D), O_{X''}(iY)) \longrightarrow H^{\circ}(D, p_{*}(L^{i})) \cong H^{\circ}(p^{-1}(D), L^{i})$ is surjective for $i \gg o$ because D is affine. Using the fact that L^{i} is p-very ample, it follows that $L^{i}/p^{-1}(D)$ is generated by its global sections, and hence (by the above surjectivity), $O_{X''}(iY)/q''^{-1}(D)$ is generated by its global sections.

Now we fix an $i \gg o$ such that i), ii) and iii) are fulfilled. From ii) it follows that there is a unique B-morphism $v_1 : X \longrightarrow P :=$ $= P(q''_{*}(O_{X''}(iY)))$ such that $v_1^{*}(O_P(1)) \cong O_{X''}(iY)$. Set $X_1 = v_1(X'')$ and $Y_1 =$ $= v_1(Y)$. Since L^{i} is p-very ample, we know that $v_1/Y : Y \longrightarrow Y_1$ is an isomorphism and that iY_1 is a B-very ample Cartier divisor on X_1. Furthermore $Y = v_1^{-1}(Y_1)$ because a global equation of the effective Cartier divisor iY on X" separates points x and x' such that $x \in Y$ and $x' \in X''-Y$. Then consider the Stein factorization of v_1

$$
\begin{array}{ccc}
X'' & \xrightarrow{\quad v \quad} & X' := \mathrm{Spec}(v_{1*}(O_{X''})) \\
 & {\scriptstyle v_1} \searrow \quad \swarrow {\scriptstyle w} & \\
 & X_1 &
\end{array}
$$

Since $v_{*}(O_{X''}) \cong O_{X'}$ and X" is normal, X' is also normal. Notice that $v/Y : Y \longrightarrow Y' = v(Y)$ is an isomorphism and $Y = v^{-1}(Y')$, so by Zariski's main theorem (see [8], chap. III, (4.4.1)), v is an isomorphism in neoghbourhood of Y in X". Since w is a finite morphism and Y_1 is B-ample, $Y' = w^{*}(Y_1)$ is q'-ample on X', where q' is the composition $X' \xrightarrow{w} X_1 \hookrightarrow P \longrightarrow B$. Lemma 2 is proved.

<u>Note.</u> The above proof of lemma 2 is an adaptation of the proof of theorem 4.2, chap. III in [9] to the relative case.

<u>Proof of theorem 6, continued.</u> We apply lemma 2 to the desingularization X" of X such that $q'' = q \circ f$ is a morphism, and get the normal projective variety X' with all properties stated in lemma 2 (in particular, Y is an effective q'-ample Cartier divisor on X'). Notice that v blows down to points only subvarieties of X" that are contained in the exceptional locus of f, and since X' is normal, by [8], chap. II, (8.11.1) we infer that there is a unique morphism $u : X' \longrightarrow X$ such that $q \circ u = q'$ and $f = u \circ v$. Notice also that the construction of u and X' is canonical and depends only on X, Y and the rational map q, and not on the choice of the desingularization $f : X'' \longrightarrow X$.

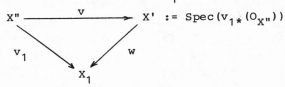

With this construction in hand, we can proceed further. Since $Y = P(E)$ and L is a p-ample line bundle, there is an $M \in \mathrm{Pic}(B)$ and a positive integer $s \geq 1$ such that

$$(14) \qquad L \cong O_Y(s) \otimes p^*(M^{-1}), \text{ where } O_Y(s) = O_{P(E)}(s).$$

Replacing E by $E \otimes N$, with $N \in \mathrm{Pic}(B)$ of sufficiently high degree, we get that $L = O_{P(E\otimes N)}(s) \otimes p^*(N^{-s} \otimes M^{-1})$. In other words, we may assume that in (14) M has a sufficiently high degree.

According to the Lefschetz theorem, there is an $F \in \mathrm{Pic}(U)$ such that $F \otimes O_Y \cong O_Y(1)$ (cf. e.g. [2], proof of theorem 2). Set $U'' = f^{-1}(U)$ and $U' = u^{-1}(U)$. Since $U'' \cong U \cong U'$, we may consider the sheaf F on U", and since X" is smooth, F extends (non-uniquely) to a line bundle on X", still denoted by F. Since the map $\mathrm{Pic}(U) \longrightarrow \mathrm{Pic}(Y)$ is injective, (14) can be translated into $F^s/U'' \cong (O_{X''}(Y) \otimes q''^*(M))/U''$. Therefore there is a divisor D supported by the exceptional fibres of f, such that $F^s \cong O_{X''}(Y) \otimes q''^*(M) \otimes O_{X''}(D)$. If $D = D_+ - D_-$, with D_+ and $D_- \geq o$, after replacing F by $F \otimes O_{X''}(D_-)$ (which still has the restriction $O_Y(1)$ to Y), we may assume that $D \geq o$. Furthermore, since M is of sufficiently high degree, for a general divisor $b_1 + \ldots + b_m \in |M|$ (with $b_i \neq b_j$ for $i \neq j$), the fibres $X''_i = q''^{-1}(b_i)$ are all smooth and transverse to all components of D as well as to all their possible intersections. Thus, replacing D by $D'' = D + D'$, with $D' = X''_1 + \ldots + X''_m$, we get

$$(15) \qquad F^s \cong O_{X''}(Y) \otimes O_{X''}(D''),$$

where D" is a normal crossing divisor on X" such that $D'' = D + D'$, with $D \geq o$ and Supp(D) contained in the exceptional fibres of f, and D' a sum of distinct fibres of q" (and hence D' reduced). Then for every $i \in \mathbb{Z}$ put $F^{(i)} = F^i \otimes O_{X''}([-iD''/s])$, where if $\Delta = \sum_j a_j \Delta_j$ is a $\mathbb{Q}$-divisor on X" (with Δ_i irreducible and reduced, and $\Delta_i \neq \Delta_j$ for every $i \neq j$), $[\Delta]$ denotes the integral divisor $\sum_j [a_j] \Delta_j$, where for every real number a, $[a]$ denotes the largest integer $\leq a$. Notice that if $i = js + r$ is an arbitrary integer such that $o \leq r \leq s-1$, then by (15) we get:

$$(16) \qquad F^{(i)} \cong O_{X''}(jY) \otimes F^{(r)}.$$

Now, the second main point in the proof of theorem 6 is the following:

<u>Lemma 3.</u> <u>$R^b q''_*(F^{(-i)}) = o$ for every $i \geq 1$ and $b = o,1$.</u>

<u>Proof of lemma 3.</u> The proof follows the well known philosophy consisting in using global vanishing theorems in order to get relative (i.e. local) ones (see e.g. [5], appendix 1).

Let N be a sufficiently ample line bundle on B such that $N\otimes R^b q''_*(F^{(-i)})$ is generated by its global sections and such that $H^a(B,N\otimes R^b q''_*(F^{(-i)}) = o$ for $a\geq 1$, $b = o,1$ and $i\geq 1$ (i fixed). Consider the Leray spectral sequence

$$E_2^{a,b} = H^a(B,N\otimes R^b q''_*(F^{(-i)}) \Longrightarrow H^{a+b}(X'',q''^*(N)\otimes F^{(-i)}).$$

By the choice of N, we have $E_2^{a,b} = o$ for $a>o$, which implies that $H^\circ(B,N\otimes R^b q''_*(F^{(-i)}) \cong H^b(X'',q''^*(N)\otimes F^{(-i)})$. Since $N\otimes R^b q''_*(F^{(-i)})$ is generated by its global sections, it is sufficient to show that the left-hand side is zero, or by the above isomorphism, that the right-hand-side is zero. To this end, using the fact that N is sufficiently ample, by Bertini we can choose a divisor $c_1+\ldots+c_e\in |N|$ (with $c_i \neq c_j$ if $i\neq j$) such that $X''_i = q''^{-1}(c_i)$ is smooth, not included in Supp(D"), and transverse to D. Then we have the exact sequence

$$H^b(X'',F^{(-i)}) \longrightarrow H^b(X'',q''^*(N)\otimes F^{(-i)}) \longrightarrow H^b(Z,F^{(-i)}),$$

with $Z = X''_1+\ldots+X''_e$. Notice that since $f^*(O_X(Y)) \cong O_{X''}(Y)$ and Y is ample on X, $O_{X''}(iY)$ is generated by its global sections for $i\gg o$, and the self-intersection number $(O_{X''}(Y)^{\cdot\dim(X'')})$ is strictly positive (in particular, the divisor Y is nef and big in the terminology of [21]). Therefore, recalling (15) and the definition of $F^{(i)}$, the first space is zero by the Kawamata-Viehweg vanishing theorem ([13], [21]). The third space is also zero by the same vanishing theorem applied on the smooth (but possibly disconnected) variety Z (taking into account of the choice of D"), and hence the middle space is zero. Q.E.D.

<u>Corollary</u> (to lemma 3). <u>For every $i\in\mathbb{Z}$ set</u> $G_i = v_*(F^{(i)})$. <u>Then</u> $R^b q'_*(G_{-i}) = o$ <u>for every $i\geq 1$ and $b = o,1$.</u>

<u>Proof.</u> Consider the Leray spectral sequence

$$(17) \qquad E_2^{a,b} = R^b q'_*(R^a v_*(F^{(-i)}) \Longrightarrow R^{a+b} q''_*(F^{(-i)}).$$

From (17) we get $R^b q'_*(G_{-i}) = R^b q'_*(v_*(F^{(-i)}) \subseteq R^b q''_*(F^{(-i)})$, and by lemma 3 the last sheaf is zero for every $i\geq 1$ and $b = o,1$. Q.E.D.

<u>Remarks</u>. 1) In the (final part of the) proof of theorem 6 we shall use only the above corollary. However, we needed to prove first lemma 3 because we had to apply the Kawamata-Viehweg vanishing theorem on the <u>smooth</u> projective varieties X" and Z.

2) From the definition of the $F^{(i)}$'s one immediately infers that for every open subset $U\subseteq X''$ and $i,j\geq o$ one has natural maps

$$H^\circ(U,F^{(i)})\otimes H^\circ(U,F^{(j)}) \longrightarrow H^\circ(U,F^{(i+j)}),$$

and hence one has a natural structure of a graded k-algebra on

$\bigoplus_{i=o}^{\infty} H^\circ(V,G_i)$, where V is an arbitrary open subset of X'.

3) At this point we want to thank the referee who kindly pointed out to us that in the earlier version of this paper we had incorrectly defined the sheaves $F^{(i)}$ by $F^{(i)} = F^i \otimes O_{X''}(-[iD''/s])$ (instead of $F^{(i)} = F^i \otimes O_{X''}([-iD''/s])$). With this (incorrect) definition statements similar to lemma 3 and its corollary still hold, but remark 2 (which shall be needed below) fails to be true. Fortunately, with the corrected definition of the $F^{(i)}$'s we had to make only minor changes in the proof of lemma 3 and its corollary.

Proof of theorem 6, continued. Having the corollary of lemma 3, we can finally conclude the proof as follows. Recalling (14), we distinguish two cases:

Case s = 1. Replacing E by $E \otimes M^{-1}$, we may assume that $L \cong O_Y(1)$. Then by the corollary of lemma 3, $R^b q'_*(O_{X'}(-Y)) = o$ for b = o,1. Now, the exact sequence

$$o \longrightarrow O_{X'}((i-1)Y) \xrightarrow{t} O_{X'}(iY) \longrightarrow O_Y(i) \longrightarrow o,$$

(where $t \in H^\circ(X', O_{X'}(Y))$ is a global equation of Y in X') yields the exact sequence ($i \geq o$)

$$R^1 q'_*(O_{X'}((i-1)Y)) \longrightarrow R^1 q'_*(O_{X'}(iY)) \longrightarrow R^1 p_*(O_Y(i)).$$

Since by [8], chap. III (2.1.15), $R^1 p_*(O_Y(i)) = o$ for every $i \geq o$, and since we know that $R^1 q'_*(O_{X'}(-Y)) = o$, by induction on i we get that $R^1 q'_*(O_{X'}(iY)) = o$ for every $i \geq o$. In particular, the above exact sequence yields for every $i \geq o$ the exact sequence

$$(18_i) \quad o \longrightarrow q'_*(O_{X'}((i-1)Y)) \xrightarrow{t} q'_*(O_{X'}(iY)) \longrightarrow p_*(O_Y(i)) \longrightarrow o.$$

By [8], chap. III (2.1.15) again, $\bigoplus_{i=o}^{\infty} p_*(O_Y(i)) \cong S(E)$, where S(E) is the symmetric O_B-algebra of E. Denoting by $S = \bigoplus_{i=o}^{\infty} q'_*(O_{X'}(iY))$, from (18_i) we get $S/tS \cong S(E)$. Since S(E) is generated by its homogeneous part of degree one, and since deg(t) = 1, it follows that the graded O_B-algebra S is generated by $S_1 = q'_*(O_{X'}(Y))$. In particular, the natural homomorphism $S(F) \longrightarrow S$ is surjective, where $F = q'_*(O_{X'}(Y))$. On the other hand, since $q'_*(O_{X'}(-Y)) = o$ and $\bigoplus_{i=o}^{\infty} p_*(O_Y(i)) \cong S(E)$, by induction on i in (18_i) we infer that S_i is a locally free O_B- module of rank $\binom{n+i+1}{i}$ for every $i \geq o$. It follows that the surjective maps $S^i(F) \longrightarrow S_i$ are all isomorphisms because $S^i(F)$ and S_i are vector bundles of the same rank. Thus $S \cong S(F)$, and recalling that Y is a q'-ample Cartier divisor on X' (lemma 2), we get that X' is isomorphic to the projective bundle associated to F. The exact sequence (18_1) becomes

$$(19) \qquad o \longrightarrow O_B \longrightarrow F \longrightarrow E \longrightarrow o.$$

a) Suppose first that the exact sequence (19) is not split. Then a result of Gieseker (see [7], theorem 2.2, or also [5], (4.16)) together with the fact that $O_Y(1)$ is ample (which means that the vector bundle E is ample), show that F is also ample, or equivalently, $O_{X'}(Y) = O_{P(F)}(1)$ is ample (and not only q'-ample). Since X' = X" is a desingularization of X whose exceptional locus does not intersect Y, this locus must be zero-dimensional. In other words, $f:X" = X' \longrightarrow X$ has finite fibres, and hence, by the Zariski's main theorem, f is an isomorphism. In particular, X is nonsingular, and this contradicts the hypotheses of theorem 6. Therefore case a) is impossible.

b) Therefore the exact sequence (19) is split. Then $F \cong E \oplus O_B$, and the surjection $E \oplus O_B \longrightarrow O_B$ yields the zero section $B \xrightarrow{i} V(E) = \mathrm{Spec}(S(E)) \hookrightarrow X' \cong P(F)$, where the the second morphism is the natural open immersion whose complement is Y = P(E) (see [8], chap. II, §8). Since E is an ample vector bundle on B, by Grauert's criterion of ampleness for vector bundles (see [1o], (3.5)), the zero section $i(B) \subset X'$ can be blown down to get the projective variety $\mathrm{Proj}(\bigoplus_{i=o}^{\infty} H^o(B, S^i(E))[T])$ (with T an indeterminate of degree 1). Since $S^i(E) \cong p_*(O_Y(i))$ for every $i \geq o$, the latter variety is nothing but the cone C(Y,L). Now, the morphism $f:X" = X' \cong P(F) \longrightarrow X$ has to contract the curve i(B) to a point (since Y is ample on X), and hence one gets a morphism $C(Y,L) \longrightarrow X$. Since Y is ample on both C(Y,L) and X, exactly as in case a) we infer that this morphism is in fact an isomorphism, and hence theorem 6 is proved in case s = 1.

<u>Case $s \geq 2$.</u> Let $i \in \mathbb{Z}$ be an arbitrary integer, and set $i = js + r$, with $o \leq r \leq s-1$. Since $v^*(O_{X'}(jY)) \cong O_{X"}(jY)$, by (16) and the projection's formula we get

$$(2o) \qquad G_i \cong O_{X'}(jY) \otimes G_r \text{ , with } G_o = O_{X'} .$$

Furthermore, by (14), (15) and the definition of the sheaves G_i we have

$$(21) \qquad G_i \otimes O_Y \cong O_Y(i) \otimes p^*(M^{-j} \otimes M_r),$$

where $M_o, \dots, M_{s-1}$ are line bundles on B ($M_o \cong O_B$). Then by (2o) and (21), for every $i \geq o$ we have the exact sequence

$$o \longrightarrow G_{i-s} \xrightarrow{t} G_i \longrightarrow O_Y(i) \otimes p^*(M^{-j} \otimes M_r) \longrightarrow o.$$

Let D = Spec(A) be any affine open subset of B such that E/D ≅

$\cong O_D^{n+1}$, $M/D \cong O_D$ and $M_r/D \cong O_D$ for every $o \leqslant r \leqslant s-1$. Set $X'_D = q'^{-1}(D)$ and $Y_D = p^{-1}(D)$. Then the above exact sequence restricted to X'_D becomes

$$o \longrightarrow G_{i-s} \xrightarrow{\ t\ } G_i \longrightarrow O_{Y_{D_1}}(i) \longrightarrow o.$$

Since by the corollary of lemma 3 we have $H^1(X'_D, G_{i-s}) = R^1 q'_*(G_{i-s})/D = o$ for every $i < s$, exactly as in case $s = 1$ we get $H^1(X'_D, G_i) = o$ for every $i \in \mathbb{Z}$, and hence the exact sequence ($i \geqslant o$)

$$(22_i) \quad o \longrightarrow H^\circ(X'_D, G_{i-s}) \xrightarrow{\ t\ } H^\circ(X'_D, G_i) \longrightarrow H^\circ(Y_D, O_Y(i)) \longrightarrow o.$$

Denote again by S the graded A-algebra $S = \bigoplus_{i=o}^{\infty} H^\circ(X'_D, G_i)$. Then $t \in S_s = H^\circ(X'_D, O_{X'}(Y))$ is an homogeneous element of degree s of S. Since $H^\circ(Y_D, O_Y) = A$ and, by lemma 3, $H^\circ(X'_D, O_{X'}(-Y)) = q''_*(O_{X''}(-Y))/D = q''_*((F^{(s)})^{-1})/D = o$, it follows that $S_o = A$. Then $\bigoplus_{i=o}^{\infty} H^\circ(Y_D, O_Y(i)) \cong A[T_o, \dots, T_n]$, where $T_o, \dots, T_n$ are n+1 indeterminates (all of degree one). By (22_1) there are n+1 elements $t_o, \dots, t_n$ of $S_1 = H^\circ(X'_D, G_1)$ such that $t_i/Y_D = T_i$ for $i = o, \dots, n$. Let S' denote the graded A-subalgebra of S generated by $t_o, \dots, t_n$ and t. Then one has a surjective map of graded A-algebras $A[T, T_o, \dots, T_n] \longrightarrow S'$ mapping T into t and T_i into t_i, $i = o, \dots, n$, where the polynomial A-algebra $A[T, T_o, \dots, T_n]$ in n+2 indeterminates is graded by $\deg(T) = s$ and $\deg(T_i) = 1$ for $i = o, \dots, n$. Then it is easy to see that this map is in fact an isomorphism of graded A-algebras.

On the other hand, by considering the exact sequence (22_{is}) ($i \geqslant o$) and using the fact that $H^\circ(X'_D, O_{X'}(-Y)) = o$, an easy induction on i implies that $S'^{(s)} \cong S^{(s)}$, where according to [8], chap. II, $S^{(s)}$ denotes the graded A-algebra such that $(S^{(s)})_i = S_{is}$ for every $i \geqslant o$. Recalling also that Y is a q'-ample Cartier divisor on X', we infer that

$$X'_D \cong \mathrm{Proj}(S) \cong \mathrm{Proj}(S^{(s)}) \cong \mathrm{Proj}(S'^{(s)}) \cong \mathrm{Proj}(S'),$$

or else, that X'_D is isomorphic to the (n+1)-dimensional weighted projective space $P_A(1, \dots, 1, s)$ over A of weights $(1, \dots, 1, s)$. Furthermore, the restriction $q' : X'_D \longrightarrow D$ coincides to the canonical projection of $P_A(1, \dots, 1, s)$ onto $D = \mathrm{Spec}(A)$. In particular, for every $b \in B$, $X'_b = q'^{-1}(b)$ is isomorphic to the weighted projective space $P(1, \dots, 1, s)$ over k and $Y_b = p^{-1}(b)$ is contained in X'_b as the infinite section (i.e. the subvariety $V_+(T)$ of $P(1, \dots, 1, s)$).

Summing up, we showed that there is a closed subset B' of X' such that q' defines an isomorphism of B' on B, $B' \cap X'_D = V_+(T_o, \dots, T_n) \cong \mathrm{Proj}(A[T])$, and $B' \cap X'_b$ is precisely the vertex x_b of the cone $X'_b \cong P(1, \dots, 1, s)$ ($s \geqslant 2$) for every $b \in B$. Let $y \in Y$ be an arbitrary point, and let L(y) be the generating line of the cone $X'_{p(y)}$ joining the points y and $x_{p(y)}$. Then X'-B' is the disjoint union of all $L_y - x_{p(y)}$ ($y \in Y$), and hence we get a well-defined function $\beta: X'-B' \longrightarrow Y$ by putting $\beta(x) = y$ if $x \in L_y - x_{p(y)}$. The above discussion shows that β is in fact an algebraic morphism defined in a neighbourhood V of Y in X' (or in X) which is a retraction of $Y \subset V$. Then using lemma 3 in [3] (cf. also [6], (3.1)), we infer that $X \cong C(Y,L)$ also if $s \geq 2$.

Theorem 6 is completely proved. Q.E.D.

REFERENCES

1. L. Bădescu, On ample divisors, Nagoya Math. J., 86 (1982), 155-171.
2. L. Bădescu, On ample divisors:II, Algebraic Geometry Proceedings, Bucharest 198o, Teubner-Texte Math. 4o, Leipzig 1981, 12-32.
3. L. Bădescu, Hyperplane sections and deformations, Algebraic Geometry Proceedings, Bucharest 1982, Springer Lect. Notes Math. 1o56, 1-33.
4. L. Bădescu, On a criterion for hyperplane sections, Math. Proc. Camb. Phil. Soc., 1o3 (1988), 59-67.
5. T. Fujita, On the hyperplane section principle of Lefschetz, J. Math. Soc. Japan, 32 (198o), 153-169.
6. T. Fujita, Rational retractions onto ample divisors, Scient. Papers Coll. Art. Sci. Univ. Tokyo, 33 (1983), 33-39.
7. D. Gieseker, P-ample bundles and their Chern classes, Nagoya Math. J., 43 (1971), 91-116.
8. A. Grothendieck & J. Dieudonné, Eléments de Géométrie Algébrique, chap. II, III, Publ. Math. IHES, 8, 11 (1961).
9. R. Hartshorne, Ample subvarieties of algebraic varieties, Springer Lect. Notes Math. 156 (197o).
1o. R. Hartshorne, Ample vector bundles, Publ. Math. IHES, 29 (1966), 63-94.
11. H. Hironaka, Resolution of singularities of an algebraic variety over a field of char. zero, Annals Math., 79 (1964), 1o9-326.
12. P. Ionescu, Generalized adjunction and applications, Math. Proc. Camb. Phil. Soc., 99 (1986), 457-472.
13. Y. Kawamata, A generalization of Kodaira-Ramanujam's vanishing theorem, Math. Ann., 261 (1982), 43-46.
14. S. Kleiman & J. Landolfi, Geometry and deformations of special Schubert varieties, Compos. Math., 23 (1971), 4o7-434.

15. S.M. L'vovskii, Prolongation of projective manifolds and deformations, VINITI Preprint, Moskow University, 1987 (in Russian).
16. D. Mumford, Abelian varieties, TATA Lecture Notes Math., Bombay, 1968.
17. H. Pinkham, Deformations of algebraic varieties with G_m-action, Astêrisque, 2o (1974), Sociêtê Math. France.
18. M. Schlessinger, Infinitesimal deformations of singularities, Thesis, Harvard Univ., 1964.
19. M. Schlessinger, Rigidity of quotient singularities, Invent. Math., 14 (1971), 17-26.
2o. M. Schlessinger, On rigid singularities, Rice Univ. Studies, 51 (1973), 147-162.
21. E. Viehweg, Vanishing theorems, Journ. reine angew. Math., 355 (1982), 1-8.
22. J. Wahl, A cohomological characterization of P^n, Invent. Math., 72 (1983), 315-322.
23. J. Wahl, Equisingular deformations of normal surface singularities: I, Annals Math., 1o4 (1976), 325-356.

INCREST Bucharest, Dept. of Mathematics
B-dul Păcii 22o, 79622 Bucharest,
RUMANIA

On k-spanned projective surfaces

Edoardo Ballico
Dip. di Matematica, Università di Trento, 38050 Povo (TN), Italy

This note can be considered as an appendix to [BS], since here we give an improvement of [BS], th.2.4.

First we recall a few notations. We work over the complex number field. Let $T \subset \mathbf{P}^N$ be a scheme of dimension 0; T is called curvilinear if it is contained in a smooth curve, or equivalently if it has embedding dimension at most 1. Let X be a complete variety embedded in a projective space by a linear subspace W of $H^0(S,L)$, $L \in \mathrm{Pic}(X)$. (X,W) (or X if there is no danger of misunderstanding) is called k-spanned if for all curvilinear subschemes T of X with length(T) = k+1, the restriction map from W to $H^0(T,LT)$ is surjective. L is called k-spanned if $(X,H^0(X,L))$ is k-spanned.

Here we prove the following result.

Theorem *Let (S,W) be a k-spanned smooth surface with $k \geq 3$. Then $\dim(W) \geq k+5$.*

Proof. Set $\mathbf{P} := \mathbf{P}(W)$, hence $S \subset \mathbf{P}$. Assume $w := \dim(W) \leq k+4$. Take a general hyperplane H of $\mathbf{P}$ and set $C := H \cap S$. Then C is a smooth, k-spanned curve in H. It is easy to check that the projection from a point of a smooth m-spanned curve, $m \geq 2$, is a smooth (m-1)-spanned curve in the appropriate projective space. After (k-2) general projections, we find a smooth 2-spanned curve Z in a projective space U, $\dim(U) \leq 4$. If $\dim(U) < 4$, this is a contradiction. Assume $\dim(U) = 4$, hence $w = k+4$. Let d, g be respectively the degree and genus of Z. Note that S has sectional genus g. By Castelnuovo's formula for the number of trisecant lines to a smooth curve in $\mathbf{P}^4$ ([LB],p.182) we get $d \leq 8$, $g \leq 5$, and that Z is linearly normal. But then C and S are linearly normal. In [BS],5.1,5.2,5.3, there is a complete classification of all linearly normal surfaces with $k \geq 2$ and sectional genus $g \leq 5$. For the surfaces listed in [BS],§5, with $k \geq 3$, we have always $h^0(S,L) \geq k+5$, proving the theorem.

This note was born in the warm atmosphere of Max-Planck-Institut (Bonn).

References

[BS] M. Beltrametti, A.J. Sommese: *On k-spannedness for projective surfaces*, preprint MPI/88 - 14.

[LB] P. Le Barz: *Formule multi-secantes pour les courbes gauches quelconques*, in Enumerative geometry and classical algebraic geometry, p. 165-197, Progress in Math. **24**, Birkauser.

ON K-SPANNEDNESS FOR PROJECTIVE SURFACES

Mauro Beltrametti
Dipartimento di Matematica
Via L.B. Alberti 4, I-16132 Genova, Italy.

Andrew J. Sommese
Department of Mathematics, University of Notre Dame
Notre Dame, Indiana 46556, U.S.A.

INTRODUCTION. Let L be a line bundle on a smooth connected projective surface S. In this paper we make a general study of pairs (S,L) where L is k-spanned. K-spannedness of a line bundle L is a natural notion of higher order embedding for the map associated to L that was introduced in [3], e.g. L is 0-spanned (1-spanned) iff L is spanned by global sections (very ample).

In § 0 we recall the definition of k-spannedness and the main result of [3], a Reider type numerical criterion for a line bundle to be k-spanned. We also collect a number of results, we continually use.

In § 1 we study k-spannedness on curves proving a number of inequalities linking invariants of the line bundle, the curve and k. We characterize k-spannedness of L on S in terms of the restriction of L to curves on S.

In § 2 we study lower bounds for $h^0(L)$ in terms of k. For $k = 2$ we get the very strong result that $h^0(L) \geq 6$, while for $k \geq 3$ we only get $h^0(L) \geq k+3$. Our proof is based on jet bundle arguments.

In § 3 we give sufficient conditions for a line bundle L on a $\mathbb{P}^1$ bundle over a curve to be k-spanned. The conditions are necessary for $h^1(\mathcal{O}_S) \leq 1$ and almost necessary in general.

In § 4 we use the results of [13] to study when $kK_S + L$ is spanned by global sections. This gives very strong numerical relations a k-spanned line bundle must satisfy.

In § 5 we use the results obtained to classify the pairs (S,L) with $g(L) \leq 5$ and $k \geq 2$ where L is k-spanned and g(L) is the genus of a smooth $C \in |L|$ (see also [7], [9], [10]).

In § 6 we study the dependence of the inequalities between g(L), $c_1(L)^2$, k for k-spanned L with $k \geq 2$ and the birational geography of S.

We would both like to thank the Max-Planck-Institut für Mathematik for making this joint work possible. The second author would also like to thank the University of Notre Dame and the National Science Foundation (DMS 8420 315) for their support.

§ 0. Notation and background material.

We work over the complex numbers $\mathbb{C}$. Throughout the paper, S always denotes a *smooth connected projective surface*. We denote its structure sheaf by $\mathcal{O}_S$ and the *canonical sheaf* of the holomorphic 2-forms by K_S. For any coherent sheaf F on S we shall denote by $h^i(F)$ the complex dimension of $H^i(S,F)$.

Let L be a line bundle on S. L is said to be *numerically effective, nef* for short, if $L\cdot C \geq 0$ for each irreducible curve C on S, and in this case L is said to be *big* if $c_1(L)^2 > 0$, where $c_1(L)$ is the first Chern class of L. We say that L is *spanned* if it is spanned by the space of its global sections $\Gamma(L)$.

(0.1) We fix some more notation.

$\equiv$ (resp. $\sim$) the numerical (resp. linear) equivalence of divisors;

$\chi(L) = \sum(-1)^i h^i(L)$, the Euler characteristic of a line bundle L;

$|L|$, the complete linear system associated to L;

$q(S) = h^1(\mathcal{O}_S)$, the irregularity of S;

$p_g(S) = h^2(\mathcal{O}_S)$, the geometric genus of S;

$\kappa(S)$, the Kodaira dimension of S;

$e(S)$, the topological Euler characteristic of S.

We denote by $J_t(S,L)$, $J_t(C,L)$ the t-th holomorphic *jet bundles* of a line bundle L on S (resp. on a smooth curve C). Recall that $J_t(S,L)$, $J_t(C,L)$ are vector bundles of rank $(t+1)(t+2)/2$, $t+1$ respectively (for general properties of jet bundles we refer to [8] and [11]).

As usual we don't distinguish between locally free sheaves and vector bundles, nor between line bundles and Cartier divisors. Hence we shall freely switch from the multiplicative to the additive notation and viceversa.

(0.2) **The genus formula**. Let L be a nef and big line bundle on S. Then the *sectional genus*, g(L), of L is defined by the equality $2g(L)-2 = (K_S+L)\cdot L$.

It can be easily seen that g(L) is an integer. Furthermore if there exists an irreducibile reduced curve C in |L|, g(L) is simply the arithmetic genus $p_a(C) = 1-\chi(\mathcal{O}_C)$ of C.

(0.3) Let L be a nef and big line bundle on S. We say that the (generically) polarized pair (S,L) is *geometrically ruled* if S is a $\mathbb{P}^1$ bundle, $p : S \longrightarrow R$, over a nonsingular curve R and the restriction L_f of L to a *fibre* f of p is $\mathcal{O}_f(1)$. We shall denote by E a *fundamental section* of p. We say that (S,L) is a *scroll* (resp. a *conic bundle*) over a nonsingular curve R if there is a surjective morphism with connected fibres $p : S \longrightarrow R$, with the property that L is relatively ample with respect to p and there exists some very ample line bundle

M on R such that $K_S \otimes 2L \sim p^*M$ (resp. $K_S \otimes L \sim p^*M$). We also say that (S,L) is a *geometrically ruled conic bundle* if S is a $\mathbb{P}^1$ bundle $p : S \longrightarrow R$ and the restriction of L to a fibre f of p is $\mathcal{O}_f(2)$.

We denote the rational $\mathbb{P}^1$ bundle $\mathbb{P}(\mathcal{O}_{\mathbb{P}^1} \oplus \mathcal{O}_{\mathbb{P}^1}(n))$, $n \geq 0$, by $\mathbb{F}_n$, the *Hirzebruch surface*. Note that the only $\mathbb{P}^1$ bundle which is not a scroll in the above sense is $\mathbb{F}_0 = \mathbb{P}^1 \times \mathbb{P}^1$ with $L = \mathcal{O}_{\mathbb{F}_0}(1,1)$. We say that S is a *Del Pezzo surface* if $-K_S$ is ample.

(0.4) **Castelnuovo's bound.** Let L be a very ample line bundle on a smooth surface S and let C be a general element in |L|. Assume that |L| embeds S in a projective space $\mathbb{P}^N$ and let $d = L\cdot L$. Then $g(L) = g(C)$ and Castelnuovo's Lemma says that (see e.g. [1])

$$g(C) \leq \left[\frac{d-2}{N-2}\right](d-N+1-(\left[\frac{d-2}{N-2}\right]-1)\ \frac{N-2}{2}) \tag{0.4.1}$$

where [x] means the greatest integer $\leq x$. From (0.4.1), writing $d-2/N-2 = (d-2-\varepsilon)/(N-2)$, $0 \leq \varepsilon \leq N-3$, we find that

$$d \geq N/2 + \sqrt{2(N-2)g(L)+((N-4/2)-\varepsilon)^2}$$

this leading to

$$d \geq \begin{cases} N/2 + \sqrt{2(N-2)g(L)+1/4} & \text{if } N-4 \text{ is odd;} \\ N/2 + \sqrt{2(N-2)g(L)} & \text{if } N-4 \text{ is even.} \end{cases} \tag{0.4.2}$$

(0.5) **k-spannedness.** Let L be a line bundle on S (resp. on a nonsingular curve C). We say that L is k-*spanned* for $k \geq 0$ if for any distinct points $z_1,\dots,z_t$ on S (resp. on C) and any positive integers $k_1,\dots,k_t$ with $\sum_{i=1}^{t} k_i = k+1$, the map $\Gamma(L) \longrightarrow \Gamma(L \otimes \mathcal{O}_Z)$ is onto, where $(Z,\mathcal{O}_Z)$ is a 0-dimensional subscheme defined by the ideal sheaf $\mathcal{I}_Z$ where $\mathcal{I}_Z\mathcal{O}_{S,z}$ is $\mathcal{O}_{S,z}$ (resp. $\mathcal{O}_{C,z}$) for $z \notin \{z_1,\dots,z_t\}$ and $\mathcal{I}_Z\mathcal{O}_{S,z_i}$ is generated by $(x_i,y_i^{k_i})$ at z_i, with (x_i,y_i) local coordinates at z_i on S, $i = 1,\dots,t$ (resp. $\mathcal{I}_Z\mathcal{O}_{C,z_i}$ is generated by $y_i^{k_i}$, y_i local coordinate at z_i on C). We call a 0-cycle Z as above *admissible*.

Note that 0-spanned is equivalent to L being spanned by $\Gamma(L)$ and 1-spanned is equivalent to very ample.

(0.5.1) If L is k-spanned on S, then $L\cdot C \geq k$ for every effective curve C on

S, with equality only if $C \simeq \mathbb{P}^1$. Further either $C \simeq \mathbb{P}^1$ or $p_a(C) = 1$ if $\deg L_C = k+1$.

The fact that $L\cdot C \geq k$ is clear from the definition, as well as $h^0(L_C) \geq k+1$. Now, looking at the embedding of C in $\mathbb{P}^N$ given by $\Gamma(L_C)$ one has $\deg L_C = \deg C \geq N \geq k$, so that $C \simeq \mathbb{P}^1$ whenever $\deg L_C = k$ or $\deg L_C = N = k+1$ and $p_a(C) = 1$ if $\deg L_C = k+1$.

(0.5.2) Let L be a k-spanned line bundle on either S or a smooth curve C. We say that $V \subset \Gamma(L)$ *k-spans* L (or V *is a k-spanning set of* L) if for all admissible 0-cycles (Z, O_Z) with $\text{length}(O_Z) = k+1$, the map $V \longrightarrow \Gamma(L \otimes O_Z)$ is onto.

For a given admissible 0-cycle (Z, O_Z) on S we say that a smooth curve C is *compatible* with (Z, O_Z) if:

- $C \supset Z_{red}$;
- for any point $z \in Z_{red}$, where $I_Z O_{S,z} = (x, y^n)$, x,y local parameters at z, then $f - x \in m_z^n$ where f is the local equation of C at z and m_z is the maximal ideal of $O_{S,z}$.

Thus we get the following characterization of k-spannedness, we need to prove the key-Lemma below.

> $V \subset \Gamma(L)$ k-spans L on S if and only if for all smooth connected compatible curves C on S, $\text{Im}(V \longrightarrow \Gamma(L_C))$ k-spans the restriction L_C.

(0.5.3) LEMMA. *Let* L_i *be* k_i*-spanned line bundles either on* S *or on a smooth curve* C *and let* $V_i \subset \Gamma(L_i)$ k_i*-spans* L_i *for* $i = 1,\dots,m$. *Then the image* V *of* $V_1 \otimes \dots \otimes V_m$ *in* $\Gamma(L_1 \otimes \dots \otimes L_m)$ $(k_1 + \dots + k_m)$ *- spans* $L_1 \otimes \dots \otimes L_m$.

Proof. In view of the characterization of k-spannedness given in (0.5.2) we easily see that one can reduce to the curves case. Further the result is clearly reduced to the case $m = 2$. Thus we have to show that, given a 0-cycle $Z = \sum_{i=1}^{t} n_i p_i$ on C where $n_i > 0$, $\sum_{i=1}^{t} n_i = k+1$, the map $V \longrightarrow \Gamma(L_1 \otimes L_2 \otimes O_Z)$ is onto.

To see this fix an index i. Write $n_i = a_i + b_i$ where $a_i > 0$, $b_i > 0$ and $n_j = a_j + b_j$, $j \neq i$, where $a_j \geq 0$, $b_j \geq 0$. Then $\sum_{r=1}^{t} a_r = k_1+1$, $\sum_{r=1}^{t} b_r = k_2+1$ and let $Z_1 = \sum_{r=1}^{t} a_r p_r$, $Z_2 = \sum_{r=1}^{t} b_r p_r$. By the fact that V_1 k_1-spans L_1 we can choose elements $s_1,\dots,s_{i_{a_i}}$ of V_1 whose images in $\Gamma(L_1 \otimes O_{Z_1})$ vanish at p_j to the a_j-th order for $j \neq i$ and which have prescribed $a_i - 1$ jet at p_i. Similarly we can choose elements $u_1,\dots,u_{i_{b_i}}$ of V_2 whose images in $\Gamma(L_2 \otimes O_{Z_2})$ vanish at p_j to the b_j-th order for $j \neq i$ and which have prescribed b_i-1 jet at p_i. Note that the tensor powers of these sections give a space of sections $\mathcal{W}_i$ of $L_1 \otimes L_2$ which vanish at p_j to the n_j-th order for $j \neq i$ and which have prescribed n_i-1 jet at p_i. Now $\mathcal{W}_1,\ldots,\mathcal{W}_t$ clearly generate $\Gamma(L_1 \otimes L_2 \otimes \mathcal{O}_Z)$, so we are done.

Q.E.D.

Let us recall the following numerical characterization of k-spannedness.

(0.6) THEOREM ([3], (2.1)). *Let* L *be a nef and big line bundle on a surface* S *and let* $L\cdot L \geq 4k+5$. *Then either* K_S+L *is* k-*spanned or there exists an effective divisor* D *such that* L-2D *is* $\mathbb{Q}$-*effective*, D *contains some admissible* 0-*cycle of degree* $t+1 \geq k+1$ *where the* k-*spannedness fails and*

$$L\cdot D-t-1 \leq D\cdot D < L\cdot D/2 < t+1.$$

We need the following consequence of the result above (compare with (5.3.4)).

(0.7) PROPOSITION. *Let* S *be a Del Pezzo surface which is a blowing up of* $\mathbb{F}_0$ *or* $\mathbb{F}_1$. *Let* L *be the spanned line bundle on* S, *obtained by pulling back to* S *the pullback of* $\mathcal{O}_{\mathbb{P}^1}(1)$ *to* $\mathbb{F}_1$ *under the bundle projection* $\mathbb{F}_i \longrightarrow \mathbb{P}^1$, $i = 0,1$. *Further assume* $K_S\cdot K_S = 1$. *Then* $K_S^{-t} \otimes L^q$ *is* k-*spanned if* $q \geq 1$ *and* $t \geq k$.

<u>Proof</u>. Let $M = K_S^{-t-1} \otimes L^q$. Note that M is ample and $M\cdot M = (t+1)^2+4(t+1)q \geq$ $\geq (k+1)^2+4(k+1)q$ since $t \geq k$, $K_S^{-1}\cdot L = 2$. Thus $M\cdot M \geq 4k+5$. If $K_S \otimes M = K_S^{-t} \otimes L^q$ is not k-spanned, by the Theorem above one has

$$M\cdot D-k-1 \leq D\cdot D < M\cdot D/2 < k+1$$

for some effective divisor D on S. Now, since $t \geq k$, $M\cdot D/2 < k+1$ gives $K_S^{-1}\cdot D = 1$ and hence

$$D\cdot D \geq M\cdot D-k-1 \geq qL\cdot C \geq 0.$$

Since $K_S\cdot D = 1$, $D\cdot D \geq 1$ by the genus formula. Then by the Hodge index theorem we get $K_S\cdot K_S = D\cdot D = 1$ and also $K_S^{-1} \equiv D$. This leads to the contradiction $qL\cdot D = 2q \leq D\cdot D = 1$.

Q.E.D.

Note that

$$g(K_S^{-t} \otimes L^q) = t(t-1)/2 + (2t-1)\,q + 1. \tag{0.7.1}$$

To the reader's convenience we recall here the following result from [3], we use several times.

(0.8) PROPOSITION ([3], (2.6)). *Let* S *be a Del Pezzo surface. Then* K_S^{-t} *is* k-*spanned for* $k \geq 0$ *if and only if*:

(0.8.1) $t \geq k/3$ *if* $S = \mathbb{P}^2$;

(0.8.2) $t \geq k/2$ *if* $S = \mathbb{P}^1 \times \mathbb{P}^1$;

(0.8.3) $t \geq k+2$ *if* $K_S \cdot K_S = 1$;

(0.8.4) $t \geq k$ *if* $K_S \cdot K_S \geq 3$ *or* $K_S \cdot K_S = 2$ *and* $k \neq 1$.

Further, if $K_S \cdot K_S = 2$, K_S^{-t} *is very ample iff* $t \geq 2$.

(0.9) **.k-reduction.** Let L be a line bundle on S. A pair (S',L') is said to be a k-*reduction* of (S,L) if there is a morphism $\pi : S \longrightarrow S'$ expressing S as S' with a finite set F blown up and $L \sim \pi^*L' - k\pi^{-1}(F)$. Note that $K_S^k \otimes L \sim \pi^*(K_{S'}^k \otimes L')$.

Apart from some cases where $k \geq 2$ is explicitly needed, we carry out for completeness most results for k-spanned line bundles with $k \geq 1$, even though in the "classical" case $k = 1$ they don't give something new.

In § 4 we use extensively the results of [13] . We refer directly the reader to [13] instead of reporting here the results we need. Through the paper we also use well known results describing polarized pairs (S,L) with L of sectional genus g(L) = 0,1; for this we refer e.g. to [5] and [7] .

§ 1. k-spannedness on curves.

Throughout this section we denote by C a nonsingular irreducible curve of genus g(C) and by K_C the canonical divisor of C. Our aim is to express the k-spannedness on C in terms of some useful numerical conditions.

(1.1) LEMMA. *Let* L *be a line bundle on* C. *Then*:

(1.1.1) L *is* k-*spanned if* $\deg L \geq 2g(C) + k$;

(1.1.2) *if* $\deg L = 2g(C)+k-1$, L *is* k-*spanned if and only if* $h^0(L-K_C) = 0$.

<u>Proof</u>. (1.1.1) follows from the definition. Indeed, let $z_1,\dots,z_r$ be r distinct points on C and let $k_1,\dots,k_r$ r non negative integers such that $\sum_{i=1}^{r} k_i = k+1$.

Then $h^0(K_C-L+\sum_{i=1}^{r} k_i z_i) = 0$, so that we have a surjective map $\Gamma(L) \longrightarrow \Gamma(L \otimes O_Z)$, where Z is the 0-cycle defined as $Z = \sum_{i=1}^{r} k_i z_i$; this means that L is k-spanned.

To prove (1.1.2), note that, since $\deg L = 2g(C)+k-1$, we can write $L = K_C \otimes \mathrm{L}$ for some line bundle L of degree $k+1$. Then $h^1(L) = 0$ and hence L is k-spanned if and only if

$$*) \qquad h^1(L-D) = h^0(D-\mathrm{L}) = 0,$$

for every effective divisor D on C with $\deg D = k+1$. We claim that condition *) is equivalent to $h^0(\mathrm{L}) = h^0(L-K_C) = 0$. In fact, for any divisor D as above, $\deg(L-D) = 2g(C)-2$; hence $h^1(L-D) \neq 0$ implies that $L-D \equiv K_C$ that is $\mathrm{L} \equiv D$, so $h^0(\mathrm{L}) \neq 0$. Viceversa, $h^1(L-\mathrm{L}) = h^1(K_C) = 0$ if $h^0(\mathrm{L}) \neq 0$, a contradiction.

Q.E.D.

The following plays a relevant role in the sequel.

(1.2) THEOREM. *Let* L *be a* k-*spanned line bundle on* C *and let* $h^1(L) \neq 0$. *Then:*

(1.2.1) K_C *is* k-*spanned*;

(1.2.2) $h^0(\mathrm{L}) \leq 1$ *for any line bundle* L *with* $\deg \mathrm{L} \leq k+1$;

(1.2.3) $g(C) \geq 2k+1$.

<u>Proof</u>. First, we can assume $h^1(L) = 1$. Indeed, if $h^1(L) = h^0(K_C-L) \geq 2$, we can write $K_C-L \sim \Delta+M$ where $h^0(\Delta) = 1$ and the moving part M is base points free. Then $L' = L+M$ is k-spanned by (0.5.3) and $h^1(L') = h^0(\Delta)$.

To prove (1.2.1), note that K_C is k-spanned if and only if $h^1(K_C-Z) = h^1(K_C) = 1$, for every length $k+1$ 0-cycle Z on C. This easily follows by looking at the exact sequence

$$0 \longrightarrow K_C \otimes O_C(-Z) \longrightarrow K_C \longrightarrow K_C \otimes O_Z \longrightarrow 0.$$

Now, if $h^1(K_C-Z) \geq 2$, clearly $h^0(K_C-L+Z) \geq 2$ since K_C-L is effective and hence by duality $h^1(L-Z) \geq 2$. Again, the k-spannedness of L can be expressed as $h^1(L-Z) = h^1(L)$, this leading to a contradiction. Thus K_C is k-spanned and $h^1(K_C-Z) = h^0(Z)=1$ for every length $t \leq k+1$ 0-cycle Z on C. This gives (1.2.1) and (1.2.2). From the Existence Theorem (see [1], p. 206) we know that for any integer $t \geq (g(C)+2)/2$ there exists a line bundle L on C of degree t and with $h^0(\mathrm{L}) \geq 2$. Therefore it has to be $k+1 < (g(C)+2)/2$, which gives (1.2.3).

Q.E.D.

(1.3) KEY-LEMMA. *Let* L *be a* k-*spanned line bundle on* C. *Then* $h^0(\mathrm{L}) \geq k+2$ *if*

$g(C) > 0$.

Proof. Since L is k-spanned one sees that the k-th holomorphic jet bundle $J_k(C,L)$ is spanned by the image of $\Gamma(L)$ under the natural map $j_k : L \longrightarrow J_k(C,L)$. Since $J_0(C,L) = L$ and $T_C^* \otimes L$ are ample vector bundles we see from the exact sequence

$$0 \longrightarrow T_C^{*(k)} \otimes L \longrightarrow J_k(C,L) \longrightarrow J_{k-1}(C,L) \longrightarrow 0$$

that $J_k(C,L)$ is an ample vector bundle of rank $k+1$. Then $h^0(L) \geq \mathrm{rk}\, J_k(C,L) + \dim C = k+2$.

(1.4) COROLLARY. *Let* L *be a k-spanned line bundle on* C *with* $g(C) > 0$ *and let* $d = \deg L$. *Then:*

(1.4.1) $d \geq k+2$;

(1.4.2) $d \geq 2k+2$ *if* $d \leq 2g(C)$ *with equality only if either* $d = 2g(C)$ *or* $L \sim K_C$, $k = 1$, $g(C) = 3$.

Proof. If $h^1(L) = 0$, then $d-g(C)+1 = h^0(L) \geq k+2$ gives $d \geq k+g(C)+1 \geq k+2$. If $h^1(L) \neq 0$ Clifford's theorem and (1.3) yield $d/2 +1 \geq h^0(L) \geq k+2$, whence $d \geq 2k+2$ and (1.4.1) is proved.

Note that Clifford's inequality holds true also if $h^1(L) = 0$ whenever $d \leq 2g(C)$. Therefore $d \geq 2k+2$ by (1.4.1). Now, $d = 2k+2$ gives the equality in the Clifford's theorem, so we find that $d = 2g(C)$ if $h^1(L) = 0$, and either $L \sim K_C$ or C is a hyperelliptic curve with L a multiple of the unique g_2^1 on C if $h^1(L) \neq 0$. If $L \sim K_C$, $d = 2k+2 = 2g(C)-2$ and $g(C) \geq 2k+1$ by (1.2.3), this leading to $k = 1$, $g(C) = 3$. In the remaining case $d \leq 2g(C)-2$, $L \sim ng_2^1$ and $K_C \sim mg_2^1$ for some positive integers m,n, $m \geq n$. Then $K_C \sim L+(m-n)g_2^1$ would be very ample, a contradiction to hyperellipticity. This proves (1.4.2).

Q.E.D.

(1.5) REMARK (compare with § 6). Note that if L is a k-spanned line bundle on S with $p_g(S) > 0$, then for a general element $C \in |L|$ the restriction $\mathcal{L} = L_C$ verifies the condition $h^1(\mathcal{L})(= h^0(K_{S|C})) \neq 0$, so $\deg \mathcal{L} = L\cdot L \leq 2g(C)-2$. Hence $g(L) \geq 2k+1$ by (1.2.3) and $L\cdot L \geq 2k+3$ if $k \geq 2$ by (1.4.2).

§ 2. A lower bound for $h^0(L)$.

Let L be a k-spanned line bundle on S. In this section we show that the k-spannedness condition forces S to be embedded by $|L|$ in a projective space of dimension at least 5. First of all, note that from Lemma (1.3), we have

$$h^0(L) \geq k+3, \tag{2.1}$$

so the claim is clear if $k \geq 3$.

(2.2) Let L be a k-spanned line bundle on S with $k \geq 2$. Take a point $x \in S$ and let $V_2 \subset \Gamma(L)$ denote the space of the sections of L that vanish to the 2-nd order at x. We *claim* that after chosing a trivialization of L at x, a basis of V_2 can be written in the form $s_1 = q_1+O(3),\ldots,s_t = q_t+O(3)$ where the q_α's are quadratic functions in the local parameters at x and at least 2 of the q_α's are not (identically) zero and have no common factors, $\alpha = 1,\ldots,t$.

Indeed, set $I = \{i, q_i \neq 0\}$, $J = \{j, q_j = 0\}$ and assume that the q_i's have a linear common factor, say u. The maximal ideal m_x of $O_{S,x}$ can be assumed to be of the form $m_x = (u,v)$ for some linear factor v.

We can also assume that on the open set $U_0 = \{x \in S,\ s_0(x) \neq 0,\ s_0 \in \Gamma(L)\}$ a basis for $\Gamma(S,L)$ on U_0 consists of the $h^0(L)-1$ elements $\{u,v,\ldots,s_i,\ldots,s_j,\ldots\}$. Now L is 2-spanned by the assumption so that the map

$$\rho : \Gamma(L) \longrightarrow \Gamma(L \otimes O_x/(u,v^3))$$

is onto. Since clearly u and the s_i's, s_j's belong to Ker ρ we find dim Ker $\rho \geq$ $\geq h^0(L)-2$, which contradicts $\dim \Gamma(L \otimes O_x/(u,v^3)) = 3$.

Note that the claim we proved here shows that *the kernel of the evaluation map* $j_{2,x} : (S \times \Gamma(L))_x \longrightarrow J_2(S,L)_x$ *at* x *induced by* $j_2 : L \longrightarrow J_2(S,L)$ *has dimension at most* $h^0(L)-5$.

We can now prove the following general result.

(2.3) THEOREM. *If the* 2-th *jets bundles of a* $k \geq 2$ *spanned line bundle* L *on* S *don't span* $J_2(S,L)$ *at at least one point, then:*

(2.3.1) $c_1(S)^2 = 2c_2(S)$ *and the tangent bundle of either* S *or an unramified double cover of* S *splits as a direct sum of line bundles;*

(2.3.2) Cokernel $(j_2 : S \times \Gamma(L) \longrightarrow J_2(S,L)) \approx K_S \otimes L$.

<u>Proof</u>. Consider the commutative diagram

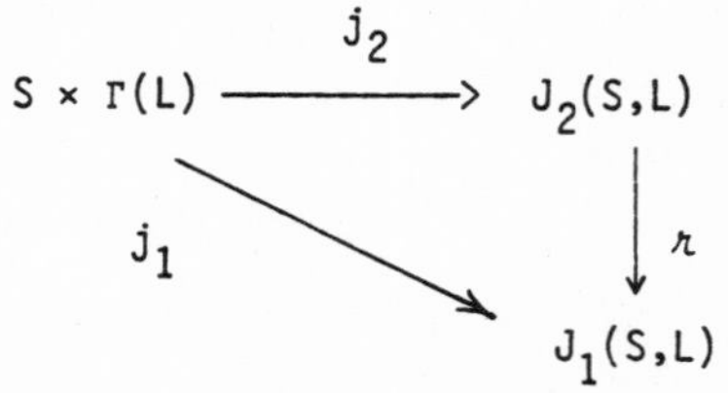

where r denotes the surjective restriction map, whose kernel is $T_S^{*(2)} \otimes L$. Since L is very ample, j_1 is onto, the restriction j of j_2 to the kernel K of j_1 has image contained in $T_S^{*(2)} \otimes L$, so one has a morphism

$$j : K \longrightarrow T_S^{*(2)} \otimes L.$$

Fix a point $x \in S$, local coordinates (z,w) at x and a trivialization of L at x. By the earlier argument (2.2), the image of j in $T_S^{*(2)} \otimes L$ at x is of the form

$$(\mathrm{Im} j)_x = \{\lambda\phi(dz,dw)+\mu\psi(dz,dw);\ \lambda,\mu \in \mathbb{C},\ \phi,\psi \text{ homogeneous quadratic functions without common factors}\}.$$

It is easy to see that any such a special pencil has precisely 2 distinct elements which are squares, e.g. the map $\mathbb{P}^1 \longrightarrow \mathbb{P}^1$ given by (ϕ,ψ) has degree 2 and has precisely two branch points by Hurwitz's theorem. Thus the pencil is given at x by

$$\{\lambda\omega_1^2 + \mu\omega_2^2;\ \lambda,\mu \in \mathbb{C},\ \omega_1,\omega_2 \in T_{S,x}^*\}.$$

Hence two directions on T_S are determined at x. It is easy to check that they vary holomorphically and give a submanifold $A \subset \mathbb{P}(T_S^*) = [T_S - S]/\mathbb{C}^*$, S the 0-section of $T_S \longrightarrow S$, which is a two to one unramified cover of S under π_A, the restriction to A of $\pi : \mathbb{P}(T_S^*) \longrightarrow S$. Now, either A is a union of 2-sections of π or $q_A^{-1}(A) \subset \mathbb{P}(T_A^*)$ is a union of 2 sections of $\mathbb{P}(T_A^*) \longrightarrow A$, where $q_A : \mathbb{P}(T_A^*) \longrightarrow \mathbb{P}(T_S^*)$ is the induced map. In the former case $T_S^* \simeq L_1 \oplus L_2$ for 2 line bundles L_1, L_2 on S; in the latter case $T_A^* \simeq L_1' \oplus L_2'$ for 2 line bundles L_1', L_2' on A. Note that by a well known result of Bott [4], $c_1(L_i)^2 = c_1(L_i')^2 = 0$. Thus $c_1(S)^2 = 2c_2(S)$ in the former case and $c_1(A)^2 = 2c_2(A)$ in the latter case. Since π_A is an unramified 2-sheeted cover in the latter case, it follows that $c_1(A)^2 = 2c_1(S)^2$, $c_2(A) = 2c_2(S)$. Thus in any case $c_1^2(S) = 2c_2(S)$, which proves (2.3.1).

Note that the image $j(K)$ in $T_S^{*(2)} \otimes L$ in the former case is $(L_1^2 \otimes L) \oplus (L_2^2 \otimes L)$ and hence the cokernel is $L_1 \otimes L_2 \otimes L = K_S \otimes L$. In the latter case L_1, L_2 are locally still well defined as in the former case. Nonetheless chosing any open set U such that L_1, L_2 are well defined, $T_U^* = L_1 \oplus L_2$ and $L_1 \otimes L_2$ is canonically identified with K_S. Thus the cokernel of j is always $K_S \otimes L$. So we are done by noting that coker j = coker j_2.

Q.E.D.

As a consequence, we get the result claimed at the beginning of this section.

(2.4) THEOREM. *Let* L *be a* k-*spanned line bundle on* S, $k \geq 2$. *Then* $h^0(L) \geq 6$.

Proof. From (2.1) we know that $h^0(L) \geq 5$ and $h^0(L) \geq 6$ if $k \geq 3$. So let $k = 2$ and assume $h^0(L) = 5$. Then previous argument (2.2) shows that the evaluating map $j_2 : S \times \Gamma(L) \longrightarrow J_2(S,L)$ induced by $j_2 : L \longrightarrow J_2(S,L)$ is injective. Hence we have an exact sequence of vector bundles

$$0 \longrightarrow S \times \Gamma(L) \longrightarrow J_2(S,L) \longrightarrow K_S \otimes L \longrightarrow 0$$

by the above Theorem. Thus $\det(J_2(S,L)) \sim K_S \otimes L$. Now a direct computation, by looking at the exact sequences, $t = 1,2$,

$$(2.4.1) \qquad 0 \longrightarrow T_S^{*(t)} \otimes L \longrightarrow J_t(S,L) \longrightarrow J_{t-1}(S,L) \longrightarrow 0$$

shows that $\det(J_2(S,L)) = K_S^4 \otimes L^6$. Therefore $K_S^3 \otimes L^5 \sim \mathcal{O}_S$ and hence there exists a line bundle M on S such that $M^{-5} \sim K_S$, $M^3 \sim L$. Since $M^3 \sim L$, M is ample so $p(t) = \chi(M^t)$ is a non degenerate degree 2 polynomial. But $M^{-5} \sim K_S$ implies, by Kodaira's vanishing theorem, $p(t) = 0$ for $t = -1,-2,-3,-4$, a contradiction. This proves that $h^0(L) \geq 6$.

Q.E.D.

Next we show that if $h^0(L) = 6$, then $J_2(S,L)$ is generically spanned.

(2.5) PROPOSITION. *With the notation as in* (2.4), *there exists at least one point* $x \in S$ *such that* $j_{2,x} : (S \times \Gamma(L))_x \longrightarrow J_2(S,L)_x$ *is onto.*

Proof. Note that if $(S,L) \approx (\mathbb{P}^2, \mathcal{O}_{\mathbb{P}^2}(2))$ it is well known that $j_2 : S \times \Gamma(L) \longrightarrow J_2(S,L)$ is onto (see e.g. [8] or [11]). Thus we can assume $(S,L) \not\approx (\mathbb{P}^2, \mathcal{O}_{\mathbb{P}^2}(2))$ and let us suppose $j_{2,x}$ to be not onto for any $x \in S$. Then by (2.3.1) there is an exact sequence of vector bundles

$$0 \longrightarrow \mathrm{Ker}\, j_2 \longrightarrow S \times \Gamma(L) \longrightarrow J_2(S,L) \longrightarrow K_S \otimes L \longrightarrow 0$$

and hence Ker j_2 has rank 1, since $\mathrm{rk}(S \times \Gamma(L)) = \mathrm{rk}\, J_2(S,L) = 6$. The total Chern classes verify the relation, where $K = \mathrm{Ker}\, j_2$,

$$(2.5.1) \qquad (1+K)\cdot c(J_2(S,L)) = 1+K_S+L.$$

We know that $c_1(J_2(S,L)) = \det(J_2(S,L)) = 4K_S+6L$ while a long but standard computation, by using sequences (2.4.1), gives us

(2.5.2) $$c_2(J_2(S,L)) = 5c_2(S)+5K_S\cdot K_S+20K_S\cdot L+15L\cdot L.$$

Furthermore from (2.5.1) we obtain

$$K\cdot c_1(J_2(S,L)) + c_2(J_2(S,L)) = 0$$

and hence

(2.5.3) $$c_2(J_2(S,L)) = (3K_S+5L)\cdot(4K_S+6L) = 12K_S\cdot K_S+38K_S\cdot L+30L\cdot L.$$

By combining (2.5.2) with (2.5.3), and noting that $c_1(S)^2 = 2c_2(S)$ by (2.3.1), we find

(2.5.4) $$L\cdot L = 3c_2(S)+12(g(L)-1).$$

Note also that K_S+L is nef. Otherwise (S,L) would be either $(\mathbb{P}^2,\mathcal{O}(2))$, $(\mathbb{P}^2,\mathcal{O}(1))$, $(\mathbb{P}^1\times\mathbb{P}^1,\mathcal{O}(1,1))$ or a scroll, contradicting $(S,L) \neq (\mathbb{P}^2,\mathcal{O}(2))$ or the fact that L is at least 2-spanned. Therefore $K_S\cdot K_S+4(g(L)-1) \geq L\cdot L$; hence (2.5.2) and $K_S\cdot K_S = 2c_2(S)$ lead to

(2.5.5) $$c_2(S)+8(g(L)-1) \leq 0.$$

Clearly $g(L) \neq 0$ since $k \geq 2$ and $(S,L) \neq (\mathbb{P}^2,\mathcal{O}(2))$. Similarly $g(L) \neq 1$: otherwise (S,L) would be either a scroll, contradicting again $k \geq 2$, or a Del Pezzo surface, contradicting $c_2(S) \leq 0$. Thus $g(L) \geq 2$, so $2c_2(S) = K_S\cdot K_S < 0$ and therefore $\chi(\mathcal{O}_S) < 0$. This implies that S is birationally ruled, so $K_S\cdot K_S \leq 8(1-q(S))$, and the Riemann-Roch theorem yields

(2.5.6) $$c_2(S) \geq 4-4q(S).$$

Hence from (2.5.5), (2.5.6) we infer that $g(L) \leq (q(S)+1)/2$. Now, since (S,L) is neither $(\mathbb{P}^2,\mathcal{O}(1))$, $(\mathbb{P}^2,\mathcal{O}(2))$, $(\mathbb{P}^1\times\mathbb{P}^1,\mathcal{O}(1,1))$ nor a scroll, it has to be $g(L) > q(S)$ (see e.g. [12]). So we get $q(S) = 0$, contradicting $\chi(\mathcal{O}_S) < 0$. This proves the Proposition.

Q.E.D.

Now, certain arguments that we have not been able to make rigorous, together with the fact that $(S,L) \simeq (\mathbb{P}^2,\mathcal{O}_{\mathbb{P}^2}(2))$ whenever j_2 is an isomorphism by a result due to Sommese [11], suggest the following

(2.6) **Conjecture.** Let L be a k-spanned line bundle on S, $k \geq 2$. Then $h^0(L) = 6$ if and only if $(S,L) \simeq (\mathbb{P}^2,\mathcal{O}_{\mathbb{P}^2}(2))$.

§ 3. k-spannedness on geometrically ruled surfaces.

Throughout this section, S is assumed to be a geometrically ruled surface over a nonsingular curve R of genus $g(R)$. As usual, E, f denote a section of minimal self-intersection $E^2 = -e$ and a fibre of the ruling. Here we find some sufficient numerical conditions for a line bundle L on S to be k-spanned. In some case, such conditions come out to be also necessary.

First we consider the case $g(R) = 0$.

(3.1) PROPOSITION. *Let* $S = \mathbb{F}_r$ *be a Hirzebruch surface of invariant* $r \geq 1$ *and let* $L \equiv aE+bf$ *be a line bundle on* S. *Then* L *is* k-*spanned if and only if* $a \geq k$ *and* $b \geq ar+k$.

<u>Proof</u>. If L is k-spanned, then $L \cdot f = a \geq k$ and $L \cdot E = -ar+b \geq k$. To show the converse, write

$$L \sim k(E+(r+1)f)+(a-k)(E+rf)+(b-(ar+k))f$$

and note that $E+(r+1)f$ is very ample and $E+rf$, f are spanned (see e.g. [6], p. 379, 382). Then we are done by (0.5.3).

Q.E.D.

(3.2) REMARK. On a quadric $\mathbb{F}_0 = \mathbb{P}^1 \times \mathbb{P}^1$ is clear that a line bundle L of type (a,b) is k-spanned if and only if $a \geq k$, $b \geq k$. Indeed, $\mathcal{O}_{\mathbb{P}^1 \times \mathbb{P}^1}(a,b)$ is $k = \min(a,b)$-spanned.

Thus we can assume $g(R) > 0$. Recall that $K_S \equiv -2E+(2g(R)-2-e)f$ where $e = -E \cdot E$ is the invariant of S.

(3.3) PROPOSITION. *Let* S *be a geometrically ruled surface with invariant* $e \geq 0$ *and* $q(S) > 0$. *Let* $L \equiv aE+bf$ *be a line bundle on* S. *Then* L *is k-spanned if*

$$a \geq k; \; b \geq ae+2q(S)-2+\max(k+2,e).$$

<u>Proof</u>. First note that $E+ef$ is nef; indeed we see that $(E+ef) \cdot B \geq 0$ for every irreducibile curve B on S, recalling that for such a curve B, $B \neq E, f$, $B \equiv \alpha E+\beta f$ with $\alpha > 0$, $\beta \geq \alpha e$ ([6], p. 382). Now let

$$M = L-K_S \equiv (a+2)E+(b+e-2(q(S)-1))f.$$

Then $M \cdot M = (2b-4(q(S)-1)-ae)(a+2) \geq 2(k+2)^2$ and hence $M \cdot M \geq 4k+5$ for $k \geq 1$. Further M is nef; indeed

$$M \equiv (a+2)(E+ef)+(b-ae-e-2(q(S)-1))f$$

and both $E+ef$, f are nef. Thus if L is not k-spanned, Theorem (0.6) applies to say that there exists an effective divisor D such that

$$M\cdot D-k-1 \leq D\cdot D < M\cdot D/2 < k+1.$$

We can write $D \equiv xE+yf$ where $x=D\cdot f \geq 0$, $y=D\cdot(E+ef) \geq 0$. Now $M\cdot D=x(b-ae-e-2(q(S)-1))+y(a+2)$, then from $M\cdot D/2 < k+1$ and the assumptions made on a and b we get $y(k+2) < 2(k+1)$ which leads to $y = 0,1$. If $y = 0$, $D\cdot D = -ex^2+2xy \geq M\cdot D-k-1$ yields

$$-ex+x(k+2)-k-1 \leq -ex^2$$

and $x \geq 1$ since $y = 0$. Hence $ex(x-1)+2x \leq 0$, a contradiction. If $y = 1$, $D\cdot D \geq M\cdot D-k-1$ gives

$$-ex+1 \leq -ex^2$$

that is $xe(x-1)+1 \leq 0$, again a contradiction.

Q.E.D.

(3.4) PROPOSITION. *Let* S *be a geometrically ruled surface of invariant* $e < 0$ *and* $q(S) > 0$. *Let* $L \equiv aE+bf$ *be a line bundle on* S. *Then* L *is* k-*spanned if*

$$a \geq k;\ b \geq ae/2 + 2q(S)+k.$$

Proof. First, note that $E+(e/2)f$ is nef. Indeed one sees that $(E+(e/2)f)\cdot B \geq 0$ for every irreducible curve B on S, recalling that for such a curve B, $B \neq E,f$, $B \equiv \alpha E+\beta f$ with either $\alpha = 1$, $\beta \geq 0$ or $\alpha \geq 2$, $\beta \geq \alpha e/2$ ([6], p. 382). Let

$$M = L-K_S \equiv (a+2)E+(b-(2q(S)-2)+e)f.$$

Then $M\cdot M = (2b-4(q(S)-1)+2e-(a+2)e)(a+2) \geq 2(k+2)^2$ and hence $M\cdot M \geq 4k+5$ for $k \geq 1$. Further, by writing

$$M \equiv (a+2)(E+\frac{e}{2}f) + (\frac{2b-4(q(S)-1)-ae}{2})f$$

we see that M is nef. Thus if L is not k-spanned, there exists by (0.6) an effective divisor D such that

$$M\cdot D-k-1 \leq D\cdot D < M\cdot D/2 < k+1.$$

We can write $D \equiv x(E+(e/2)f)+yf$ where $x \in \mathbb{Z}$, $2y \in \mathbb{Z}$ and further $x = D\cdot f \geq 0$, $D\cdot(E+(e/2)f) = y \geq 0$. Here

$$M\cdot D = x(\frac{-ae}{2} + b-2(q(S)-1))+y(a+2),$$

then from $M\cdot D/2 < k+1$ and the assumptions made on a and b we find $(k+2)(x+y)/2 < k+1$, which gives $x+y < 1$. Therefore, since $y \in \mathbb{Z}$ if $x = 0$, the only possible cases are $(x,y) = (1,1/2), (1,0), (0,1)$ and an easy check shows that they contradict $D\cdot D = 2xy \geq M\cdot D-k-1$.

Q.E.D.

In the special case when $q(S) = 1$ something more can be said.

(3.5) REMARK ($q(S)=1$). Assume S is a geometrically ruled surface over a curve of genus $g(R) = 1$. Then a standard but rather long computation, following the lines of the previous proofs, gives us necessary and sufficient conditions for a line bundle $L \equiv aE+bf$ on S to be k-spanned. We state here the results, omitting the proof for shortness.

(3.5.1) if $e = -1$, L is k-spanned if and only if $a \geq k$, $2b+a \geq 2(k+2)$;

(3.5.2) if $e \geq 0$, L is k-spanned if and only if $a \geq k$, $b \geq k+2+ae$.

(3.6) REMARK (conic bundle case). It is worth to point out that if (S,L) is a conic bundle on a curve R of genus $q(S)$ and L is a k-spanned line bundle on S, then $k \leq 2$. Further, if $q(S) \geq 1$, $k = 2$ and S is geometrically ruled then

$$g(L) \geq q(S)+3.$$

Indeed $L \equiv 2E+bf$, so $L\cdot f = 2$ and hence $k \leq 2$. If $k = 2$, $\deg L_E = L\cdot E = b-2e \geq 4$ by (1.4.1) while the genus formula gives us $g(L) = b+2q(S)-1-e$. Therefore $g(L) \geq 2q(S)+e+3$, so we are done since $e \geq -q(S)$.

§ 4. The k-th adjunction mapping.

Let L be a k-spanned line bundle on S. It is useful to use [13] to find for which positive integers t, $t \leq k$, the line bundle tK_S+L is very ample or spanned. The results of this section are essentially corollaries of the analogous results for very ample line bundles contained in [13], we used over and over (especially Theorems (0.1), (0.2) and (2.1)).

For $t = 1$, we have the following

(4.1) THEOREM. *Let* L *be a k-spanned line bundle on* S *with* $k \geq 2$. *Then* (S,L) *contains no lines and* K_S+L *is very ample unless either:*

(4.1.1) $k = 2$ *and* (S,L) *is a geometrically ruled conic bundle;*

(4.1.2) $k = 2$ *and either* $(S,L) \simeq (\mathbb{P}^2, \mathcal{O}_{\mathbb{P}^2}(2))$, $g(L)=0$, *or* $(S,L) \simeq$
$\simeq (\mathbb{P}^1\times\mathbb{P}^1, \mathcal{O}_{\mathbb{P}^1\times\mathbb{P}^1}(2,2))$, $g(L)=1$;
(4.1.3) $k = 2$ *and* S *is a Del Pezzo surface with* $K_S\cdot K_S = 2$, $L \sim K_S^{-2}$, $g(L)=3$;
(4.1.4) $k = 3$ *and* $(S,L) \simeq (\mathbb{P}^2, \mathcal{O}_{\mathbb{P}^2}(3))$, $g(L)=1$.

Proof. First note that L is in fact 2-spanned or 3-spanned in all cases listed above. This is clear if (S,L) is either as in (4.1.1), (4.1.2) or (4.1.4), while (0.8) shows that L is 2-spanned in case (4.1.3).

Now if L is k-spanned, clearly (S,L) contains no lines since $L\cdot C \geq k$ for any curve C on S. Then (S,L) cannot be either $(\mathbb{P}^2, \mathcal{O}_{\mathbb{P}^2}(1))$, a scroll, or not relatively minimal. Thus by looking over the lists in [13] we see that K_S+L is very ample unless either (S,L) is in one of the cases listed above or

i) S is a $\mathbb{P}^1$ bundle over an elliptic curve with invariant $e = -1$ and $L \simeq \xi^3$, ξ the tautological line bundle;

ii) S is a Del Pezzo surface with $K_S^{-3} \sim L$, $K_S\cdot K_S = 1$;

iii) S is a Del Pezzo surface with $K_S^{-1} \sim L$.

In case i), ξ is an effective elliptic curve and $L\cdot\xi = 3$, which contradicts (1.4.1).

In case ii), the general element $E \in |K_S^{-1}|$ is an elliptic curve and $L\cdot E = \deg L_E = 3$ contradicts again (1.4.1).

In case iii), since (S,L) is relatively minimal it has to be either $(S,L) \simeq (\mathbb{P}^1\times\mathbb{P}^1, \mathcal{O}(2,2))$ or $(S,L) \simeq (\mathbb{P}^2, \mathcal{O}_{\mathbb{P}^2}(3))$ as in class (4.1.2) or (4.1.4) respectively. This completes the proof.

Q.E.D.

(4.2) REMARK. Let L be a k-spanned line bundle on S with $k \geq 2$ and assume K_S+L very ample. Then $(K_S+L)\cdot L \geq k$ so by the genus formula

$$g(L) \geq 1 + k/2.$$

Note also that if (S, K_S+L) is not relatively minimal and ℓ is a line with $\ell^2=-1$, then $(K_S+L)\cdot\ell = 1$ gives $L\cdot\ell = 2$, hence $k = 2$.

(4.3) **Special classes.** To go on it is convenient to analyze first the very ampleness and the spannedness of tK_S+L, L k-spanned line bundle and t positive integer, in three particular cases.

(4.3.1) Let $S = \mathbb{P}^2$. Then $L \sim \mathcal{O}_S(k)$; hence $tK_S+L \sim \mathcal{O}_S(-3t+k)$ is very ample iff $t < k/3$ and spanned iff $t \leq k/3$.

(4.3.2) Let $S = \mathbb{P}^1\times\mathbb{P}^1$. Then $L \sim \mathcal{O}_S(a,b)$ with $k = \min(a,b)$. Therefore $tK_S+L \sim \mathcal{O}_S(a-2t, b-2t)$ is very ample if and only if $t < k/2$ and spanned if and only if $t \leq k/2$.

(4.3.3) Let S be a $\mathbb{P}^1$ bundle and let f be a fibre. Here we can assume by induction on t that $(t-1)K_S+L$ is very ample. Then by [13] (see in particular (1.4)) we know that:

i) tK_S+L is very ample unless either $((t-1)K_S+L)\cdot f = 1,2$ or S is a $\mathbb{P}^1$ bundle over an elliptic curve with invariant $e = -1$, $(t-1)K_S+L \sim 3\xi$, ξ the tautological line bundle and $((t-1)K_S+L)\cdot f = 3$;

ii) tK_S+L is ample and spanned unless $((t-1)K_S+L)\cdot f = 1$ or 2 and is spanned unless $((t-1)K_S+L)\cdot f = 1$.

Thus, recalling that $L\cdot f \geq k$, an easy check gives us the following

(4.3.3.1) for a positive integer t, tK_S+L is:

- very ample if $t < k/2$, unless S is a $\mathbb{P}^1$ bundle over an elliptic curve of invariant $e = -1$ and L is described as in i) above; further in this case tK_S+L is very ample if $t \leq (k-2)/2$;
- ample and spanned if $t < k/2$;
- spanned if $t \leq k/2$.

Thus we can assume now that S is neither $\mathbb{P}^2$, $\mathbb{P}^1\times\mathbb{P}^1$ nor a $\mathbb{P}^1$ bundle.

(4.4) THEOREM. *Let* L *be a* k-*spanned line bundle on* S *with* $k \geq 2$ *and let* S *be neither* $\mathbb{P}^2$, $\mathbb{P}^1\times\mathbb{P}^1$ *nor a* $\mathbb{P}^1$ *bundle. Then, for a positive integer* t, $t \leq k-1$, *we have:*

(4.4.1) tK_S+L *is very ample unless* $t = k-1$, S *is a Del Pezzo surface with* $K_S\cdot K_S = 2$ *and* $L \sim K_S^{-k}$;

(4.4.2) $(t+1)K_S+L$ *is spanned unless* $t = k-1$ *and* (S,L) *is as in* (4.4.1). *Further if* $(t+1)K_S+L$ *is spanned, the morphism associated to* $\Gamma((t+1)K_S+L)$ *has a* 2-*dimensional image unless* $t = k-1$ *and either* S *is a Del Pezzo surface with* $L \sim K_S^{-k}$, $K_S\cdot K_S \neq 1$ *or* $(S,(k-1)K_S+L)$ *is a conic bundle,* $L\cdot f = 2k$.

Proof. By induction on t, we can assume $(t-1)K_S+L$ to be very ample. Note that by [13], (0.1) we can also assume tK_S+L to be spanned: otherwise S is as in one of the above examples. Note also that S does not contain lines ℓ such that $\ell^2 = -1$ and $((t-1)K_S+L)\cdot\ell = 1$ since $t \leq k-1$ and $L\cdot f \geq k$. If tK_S+L is not very ample we see by [13], (0.2) and (2.1) that the only possibilities are:

i) $K_S^{-1} \sim (t-1)K_S+L$;

ii) $(S,(t-1)K_S+L)$ is a conic bundle;

iii) $K_S^{-2} \sim (t-1)K_S+L$ and $K_S\cdot K_S = 2$.

iv) $K_S^{-3} \sim (t-1)K_S+L$ and $K_S\cdot K_S = 1$.

In case i), either S contains a line $\ell, \ell^2 = K_S\cdot\ell = -1$, and hence $L\cdot\ell = tK_S^{-1}\cdot\ell = t \geq k$, $S = \mathbb{P}^2$ or $S = \mathbb{P}^1\times\mathbb{P}^1$, a contradiction.

In case ii) we can assume $L\cdot f \geq 2k$, f a fibre of the ruling; otherwise each fibre would be irreducible and hence S would be a $\mathbb{P}^1$ bundle. Then $((t-1)K_S+L)\cdot f = 2$ contradicts $t \le k-1$.

In case iv), $L \sim K_S^{-(t+2)}$ is k-spanned if and only if $t+2 \ge k+2$ by (0.8), this contradicting once again $t \le k-1$.

In case iii), $L \sim K_S^{-(t+1)}$ is k-spanned if and only if $t+1 \ge k$, again by (0.8). Hence $t = k-1$ and $L \sim K_S^{-k}$. This proves (4.4.1).

Now, by [13], (0.1) we see that $(t+1)K_S+L$ is spanned whenever tK_S+L is very ample under the assumptions made on S.

Finally by [13], (0.2) we see that, if $(t+1)K_S+L$ is spanned, the morphism associated to $\Gamma((t+1)K_S+L)$ has a 2-dimensional image unless either $K_S^{-1} \sim tK_S+L$ or (S,tK_S+L) is a conic bundle. In both cases we find $t = k-1$ and we are done. Note that if $K_S^{-1} \sim tK_S+L$, there exists some line ℓ with $\ell^2=K_S\cdot\ell = -1$ since S is neither $\mathbb{P}^2$ nor $\mathbb{P}^1\times\mathbb{P}^1$.

Q.E.D.

The above result gives us rather strong numerical conditions for k-spannedness.

(4.5) COROLLARY. *Let* L *be a* k*-spanned line bundle on* S, *write* $d = L\cdot L$ *and let* $g(L)$ *be the sectional genus of* L. *Further assume* S *be neither* $\mathbb{P}^2$, $\mathbb{P}^1\times\mathbb{P}^1$, *a* $\mathbb{P}^1$ *bundle, nor a Del Pezzo surface with* $K_S\cdot K_S = 2$, $L \sim K_S^{-k}$ *as in* (4.4.1). *Then we have:*

(4.5.1) $d \le (k^2K_S\cdot K_S+4k(g(L)-1))/(2k-1)$;

(4.5.2) $d \le 2k(g(L)-1)/(k-1)$;

(4.5.3) $d \le kK_S\cdot K_S+2(g(L)-1)$ *if* $\kappa(S) \ge 0$.

<u>Proof</u>. By (4.4.2), kK_S+L is nef. Then $(kK_S+L)^2 \ge 0$, $(kK_S+L)\cdot L \ge 0$ and $K_S\cdot(kK_S+L) \ge 0$, together genus formula (0.2), give (4.5.1), (4.5.2) and (4.5.3) respectively.

Q.E.D.

§ 5. A classification of (S,L) for small values of g(L).

As an application of the previous results we classify in this section the polarized pairs (S,L) where L is a k-spanned line bundle on S with $k \ge 2$ and sectional genus $g(L) \le 5$. Note that in [9], [10] a complete classification is carried out for $g(L) \le 6$ if $k = 1$ (see also [7]).

The cases $g(L) \le 3$ are easy consequences of Theorem (4.1).

(5.1) PROPOSITION. *Let* L *be a* k*-spanned line bundle on* S *with* $k \ge 2$ *and sectional* *genus* $g(L) \leq 3$. *Then we have:*

(5.1.1) $g(L) = 0$, $k = 2$, $(S,L) \approx (\mathbb{P}^2, \mathcal{O}_{\mathbb{P}^2}(2))$;

(5.1.2) $g(L) = 1$; *either* $k = 2$ *and* $(S,L) \approx (\mathbb{P}^1\times\mathbb{P}^1, \mathcal{O}_{\mathbb{P}^1\times\mathbb{P}^1}(2,2))$ *or* $k = 3$ *and* $(S,L) \approx (\mathbb{P}^2, \mathcal{O}_{\mathbb{P}^2}(3))$;

(5.1.3) $g(L) = 2$, $k = 2$ *and either* $(S,L) \approx (\mathbb{F}_0, \mathcal{O}_{\mathbb{F}_0}(a,b))$ *with* $(a,b) = (2,3),(3,2)$, *or* $(S,L) \approx (\mathbb{F}_1, 2E+4f)$;

(5.1.4) $g(L) = 3$; *either* $k = 4$ *and* $(S,L) \approx (\mathbb{P}^2, \mathcal{O}_{\mathbb{P}^2}(4))$ *or* $k = 2$ *and either* (S,L) *is isomorphic to* $(\mathbb{F}_0, \mathcal{O}_{\mathbb{F}_0}(a,b))$, *with* $(a,b) = (2,4),(4,2)$, $(\mathbb{F}_1, 2E+5f)$, $(\mathbb{F}_2, 2E+6f)$, *or* S *is a Del Pezzo surface with* $K_S^{-2} \sim L$, $K_S \cdot K_S = 2$.

Proof. Let $g(L) \leq 1$. Then K_S+L is not very ample in view of (4.2), so that (S,L) is as in (5.1.1) or (5.1.2) by Theorem (4.1).

Let $g(L) = 2$. Note that $L\cdot L \geq 4$ by (1.4.1) and hence $p_g(S) = 0$ by the genus formula. Therefore

$$h^0(K_S+L) = \chi(K_S+L) = 2 - q(S),$$

so that K_S+L is not very ample. Thus by combining (4.1) and (3.6) we see that (S,L) is a geometrically ruled conic bundle over $\mathbb{P}^1$ and $k = 2$. We have $K_S \equiv 2E-(2+r)f$; if $L \equiv 2E + bf$, $b \geq 2r+2$ by (3.1) and the genus formula $2 = (K_S+L)\cdot L = 2(b-2-r)$ gives $b = r+3$. Then either $r = 0$, $b = 3$ or $r = 1$, $b = 4$.

Let $g(L) = 3$. We know that $h^0(L) \geq 6$ by (2.4), so $d = L\cdot L \geq 7$ by Castelnuovo's bound (0.4.2). The same argument as above shows that $p_g(S) = 0$ and $h^0(K_S+L) = 3$, hence if K_S+L is very ample we have $(S, K_S+L) \approx (\mathbb{P}^2, \mathcal{O}_{\mathbb{P}^2}(1))$ that is $(S,L) \approx (\mathbb{P}^2, \mathcal{O}_{\mathbb{P}^2}(4))$ and $k = 4$. If K_S+L is not very ample again by (4.1) we see that $k=2$ and either S is a Del Pezzo surface with $K_S\cdot K_S = 2$, $L \sim K_S^{-2}$ or (S,L) is a geometrically ruled conic bundle. In this case $S = \mathbb{F}_r$ by (3.6). If $L \equiv 2E+bf$, $b \geq 2r+2$ by (3.1) and the genus formula $4 = (K_S+L)\cdot L = 2(b-2-r)$ gives $b = 4+r$. Then $r \leq 2$ and we are done.

Q.E.D.

(5.2) PROPOSITION. *Let* L *be a* k-*spanned line bundle on* S *with* $k \geq 2$ *and sectional genus* $g(L) = 4$. *Then either:*

(5.2.1) $k = 3$, $S = \mathbb{P}^1\times\mathbb{P}^1$ *and* $L \sim \mathcal{O}_S(3,3)$;

(5.2.2) $k = 2$, S *is a cubic surface in* $\mathbb{P}^3$ *and* $L \sim \mathcal{O}_S(2)$; *or,*

(5.2.3) $k = 2$, *either* $S = \mathbb{F}_r$ *with* $r \leq 3$, $L \equiv 2E+(5+r)f$ *or* S *is a* $\mathbb{P}^1$ *bundle over an elliptic curve of invariant* $e = -1$ *and* $L \equiv 2E+2f$.

Proof. One has $h^0(L) \geq 6$ by (2.4) then $d = L\cdot L \geq 8$ by Castelnuovo's bound (0.4.2). Therefore the genus formula and the Riemann-Roch theorem give us

$$h^0(K_S+L) = \chi(K_S+L) = 4-q(S).$$

Then if K_S+L is very ample, it has to be $q(S) = 0$ and $|K_S+L|$ embeds S as a surface of degree $d' = (K_S+L)^2$ in $\mathbb{P}^3$. Hence $K_S \sim \mathcal{O}_S(d'-4)$ and $L \sim \mathcal{O}_S(5-d')$. Now since $p_g(S) = 0$ and L is at least 2-spanned the only possible cases are $d' = 2,3$. If $d'=2$ we get class (5.2.1). If $d' = 3$, S is a cubic in $\mathbb{P}^3$ and $L \sim \mathcal{O}_S(2)$. Note that L is 2-spanned since $L\cdot\ell = 2$ for a line ℓ on S, so we find class (5.2.2).

If K_S+L is not very ample (S,L) is a geometrically ruled conic bundle by (4.1) and $q(S) = 0,1$ by (3.6). Let $L \equiv 2E+bf$.

If $q(S) = 0$, $S = \mathbb{F}_r$ and $b \geq 2r+2$ by (3.1). The genus formula $6 = (K_S+L)\cdot L = 2(b-2-r)$ gives $b = 5+r$. Then $r \leq 3$.

If $q(S) = 1$, $K_S \equiv -2E-ef$, $e = -E^2$, $b - 2e \geq 4$ by (1.4.1) and the equality $6 = (K_S+L)\cdot L = 2(b-e)$ yields $b = 3+e$, hence $e = -1$. An easy check by using (0.6) shows that $L \equiv 2E+2f$ is 2-spanned (see also (3.5.1)).

Q.E.D.

In the remaining case $g(L) = 5$, Theorem (2.4) plays a relevant role.

(5.3) PROPOSITION. *Let L be a k-spanned line bundle on S with $k \geq 2$ and sectional genus $g(L) = 5$. Then either:*

(5.3.1) $k = 2$ *and* $|L|$ *embeds* S *in* $\mathbb{P}^5$ *as a K3 surface of degree 8, a complete intersection of three quadrics;*

(5.3.2) $k = 2$, $(S,L) \simeq (\mathbb{F}_1, 3E+5f)$;

(5.3.3) $k = 2$, *S is a Del Pezzo surface*, $L \sim -2K_S$, $K_S\cdot K_S = 4$;

(5.3.4) $k = 2$, S *is the blowing up* $\sigma : S \longrightarrow \mathbb{F}_r$ *of* $\mathbb{F}_r$, $r = 0,1$, *along* 7 *distinct points* p_i , $L \sim \sigma^*(4E+(2r+5)f)-2\sum_{i=1}^{7} P_i$, $P_i = \sigma^{-1}(p_i)$;

(5.3.5) $k = 2$, $(S,L) \simeq (\mathbb{F}_r, 2E+(6+r)f)$, $r \leq 4$; *or*,

(5.3.6) $k = 2$, *S is a* $\mathbb{P}^1$ *bundle over an elliptic curve*, $L \equiv 2E+(e+4)f$, $e = 0, -1$.

<u>Proof</u>. Since $h^0(L) \geq 6$ by (2.4), Castelnuovo's inequality (0.4.2) gives now $d = L\cdot L \geq 8$.

First, let us assume K_S+L very ample. We distinguish two cases, according to the value of $p_g(S)$.

If $p_g(S) > 0$ it has to be $d = 8$ by the genus formula and hence $K_S\cdot L = 0$ so that $K_S \equiv 0$. From [12], § 3 we know that $5 = g(L) \geq h^0(L)+q(S)-1$ and hence $h^0(L)=6$, $q(S) = 0$. Thus $|L|$ embeds S as a degree 8 K3 surface in $\mathbb{P}^5$. Further $k = 2$ in view of (1.2.3). Note that S is a complete intersection of three quadrics. Indeed, if not, it is known that a general element $C \in |L|$ contains a g^1_3 (see e.g. [2], p. 142). Now $K_C \sim L_C$ is 2-spanned and $h^0(D) \leq 1$ for any divisor D on C with $\deg D \leq 3$ by (1.2), this contradicting C to be trigonal.

If $p_g(S) = 0$, the Riemann-Roch theorem yields $h^0(K_S+L) = \chi(K_S+L) = 5-q(S)$, which gives $q(S) = 0$, $h^0(K_S+L) = 5$. Then $|K_S+L|$ embeds S in $\mathbb{P}^4$ as a surface of degree $d' = (K_S+L)^2$ and one has (see [6], p. 434)

(5.3.7) $$d'^2-5d'-10(g(L)-1)+12\chi(\mathcal{O}_S) = 2K_S\cdot K_S.$$

Now the usual Hodge index theorem yields $dd' \leq [L\cdot(K_S+L)]^2 = 64$, so that $d' \leq 6$. From (5.3.7) and the equalities

$$g(K_S+L) = d' - g(L) + 2;$$
$$d' = K_S\cdot K_S+2(2g(L)-2)-d = K_S\cdot K_S+16-d$$

a purely numerical check gives us for d, d', $K_S\cdot K_S$, $g(K_S+L)$ the values as in the table below

cases	$K_S\cdot K_S$	d'	$g(K_S+L)$	d
i)	8	3	0	21
ii)	4	4	1	16
iii)	1	5	2	12
iv)	-1	6	3	9

In case i), (S,K_S+L) is a $\mathbb{P}^1$ bundle $\mathbb{F}_r$ over $\mathbb{P}^1$ with $L \equiv 3E+bf$ and $d = 21$ leads to $2b-3r = 7$, hence $r \neq 0$. Since $b \geq 3r+k$ by (3.1) we find $3r \leq 7-2k$ which gives $k = 2$, $r = 1$; so we obtain class (5.3.2).

In case ii), (S,K_S+L) is either a Del Pezzo surface or a scroll over an elliptic curve, this contradicting $q(S) = 0$. Then $L \sim K_S^{-2}$ and we know from (0.8) that K_S^{-2} is 2-spanned. So we get class (5.3.3).

In case iii), let C be a general element of $|K_S+L|$. Then from the exact sequence

$$0 \longrightarrow K_S \longrightarrow K_S^2 \otimes L \longrightarrow K_C \longrightarrow 0$$

we find $h^0(2K_S+L) = h^0(K_C) = g(K_S+L) = 2$. By [13], (0.1), (0.2) we know that $2K_S+L$ is spanned and, since $(2K_S+L)^2 = 0$, (S,K_S+L) is a conic bundle over $\mathbb{P}^1$. Further, since $K_S\cdot K_S=1=K_{\mathbb{F}_r}\cdot K_{\mathbb{F}_r}-7$, we see that S is obtained as the blowing up $\sigma : S \longrightarrow \mathbb{F}_r$ of $\mathbb{F}_r$ at 7 distinct points p_i's. Then $K_S+L \sim \sigma^*M - \sum_{i=1}^{7} P_i$, $P_i = \sigma^{-1}(p_i)$, for some ample line bundle $M \equiv 2E+bf$ on $\mathbb{F}_r$. Therefore $b \geq 2r+1$ (see [6], p. 380) and $g(M) = g(K_S+L) = 2$, so the genus formula for M gives $b = r+3$. Thus $r \leq 2$ and $L \sim \sigma^*(4E + (2r+5)f) - 2\sum_{i=1}^{7} P_i$. Now, case $r = 2$ is clearly excluded since

$4E + 9f$ is not 2-spanned by (3.1) and, if E' denotes the proper transform under σ, $L\cdot E' \leq (4E + 9f)\cdot E = 1$. Thus Proposition (0.7) applies to say that L is 2-spanned. Indeed L is of the form $K_S^{-t} \otimes \mathcal{L}^q$ with $t = 2$, $q = 1$ and $\mathcal{L}$ the inverse image under σ of the pullback of $\mathcal{O}_{\mathbb{P}^1}(1)$ to $\mathbb{F}_r$ under a bundle projection $\mathbb{F}_r \longrightarrow \mathbb{P}^1$. This gives class (5.3.4).

In case iv), again from [13], (0.1), (0.2) we know that $2K_S+L$ is spanned and, since $(2K_S+L)^2 = 1$, $|2K_S+L|$ gives a birational morphism $\sigma : S \longrightarrow \mathbb{P}^2$. Further since $K_S\cdot K_S = -1$ we see that σ is the blowing up of $\mathbb{P}^2$ along 10 distinct points p_i's and $L \sim \sigma^*\mathcal{O}_{\mathbb{P}^2}(7) - 2\sum_{i=1}^{10} P_i$, P_i's the exceptional divisors. Let γ be a cubic plane curve passing through 9 of the points p_i's. Note that γ does not contain the remaining point. Otherwise the proper transform γ' of γ under σ belongs to $|-K_S|$ and hence $h^0(-K_S) \geq 1$. Since $k \geq 2$ this contradicts the fact that $-L\cdot K_S = 1$. Then $\gamma'\cdot L = 3$, so γ' is irreducible. Fix now four points of the p_i's and take six plane cubic curves C_j as above, passing through the four fixed points and with $C_j'\cdot L = 3$, $j=1,\ldots,6$. Since $h^0(L) = 6$, we can choose an element $A \in |L|$ whose image $\sigma(A)$ passes through the four fixed points, so that $A\cdot C_j' \geq 4$, $j=1,\ldots,6$. It thus follows that $\sigma(A)$ contains the cubics C_j's and this clearly contradicts $L\cdot A = L\cdot L = 9$.

Thus we can assume K_S+L not very ample. Then (S,L) is a geometrically ruled conic bundle by (4.1) with irregularity $q(S) = 0,1$ or 2 in view of (3.6).

Note that the case $q(S) = 2$ does not occur. Indeed the equalities $(K_S+L)^2 = 0$, $(K_S+L)\cdot L = 8$, $K_S\cdot K_S = 8(1-q(S))$ give $d = 8$ if $q(S) = 2$, a contradiction.

If $q(S) = 0$, the genus formula $8 = (K_S+L)\cdot L = 2(b-2-r)$ yields $b = 6+r$ where $L \equiv 2E+bf$, $r = -E^2$. Then, since $b \geq 2r+2$ by (3.1), we find $r \leq 4$ and we are in class (5.3.5).

If $q(S) = 1$, by using again the genus formula one has $b = 4+e$, where $L \equiv 2E+bf$, $e = -E^2$ and $\deg L_E = b-2e \geq 4$ by (1.4.1). Thus we find either $e = 0$, $b = 4$ or $e = -1$, $b = 3$. Note that in both cases L is 2-spanned in view of (3,3), (3,4) and we are in class (5.3.6).

Q.E.D.

(5.5) REMARK. If the conjecture (2.6) is true, then (5.3.1) does not occur. We attempted without success to show that the restriction L of $\mathcal{O}_{\mathbb{P}^5}(1)$ to S, S equal to the complete intersection of three quadrics in $\mathbb{P}^5$, is only 1-spanned. It should be noted that there exist such S which contain a line, ℓ, of $\mathbb{P}^5$ and for these, since $L\cdot\ell = 1 < 2$, it follows that L is not 2-spanned. In general though there are no lines on such an intersection of quadrics.

(5.6) REMARK (compare with § 6). Let L be a k-spanned line bundle on S with $k \geq 2$

and assume $p_g(S) \geq 2$. Then $g(L) \geq 2k+1$ by (1.2.3). In the extremal case $g(L) = 2k+1$ the inequality

$$p_g(S) \leq k - 3$$

holds true, hence in particular $\chi(\mathcal{O}_S) \leq k-2$. To see this, recall that $L \cdot L \geq 2k+3$ by (1.5), so the genus formula reads $K_S \cdot L \leq 2k-3$. Thus we are done after showing that $K_S \cdot L \geq p_g(S) + k$. Indeed, $h^0(K_S - L) = 0$ since $(K_S - L)\cdot L < 0$ so that $h^0(K_{S|C}) \geq p_g(S)$. Now if the p_i's are $p_g(S) - 2$ different points, on S, we have $h^0(K_{S|C} - \sum_i p_i) \geq 2$. Therefore $\deg K_{S|C} - p_g(S) + 2 = K_S \cdot L - p_g(S) + 2 \geq k+2$ by (1.2.2).

§ 6. Geography of surfaces and k-spannedness.

In this section we study the relation between k-spannedness of a line bundle L on S and the birational geometry of S. We aim for a broad picture. The arguments we use clearly give much sharper bounds in particular cases. Since the case of very ample line bundles is well studied we make the blanket assumption that $k \geq 2$. Through this section we shall use repeatedly almost all the results we stated in § 1 as well as the genus formula (0.2) and property (0.5.1). We also use a number of well known results on the birational classification of surfaces for which we refer to [2]. We shall write d instead of $L \cdot L$.

(6.1) Let $\pi : S \longrightarrow S'$ be a morphism of S to a minimal model S'. Let $L' = (\pi_* L)^{**} = [\pi(C)]$ where C is a smooth element of $|L|$. Note $K_S \sim \pi^* K_{S'} + \sum_{i=1}^{r} n_i p_i$ where the p_i's are the irreducible components of the positive dimensional fibres of π, $n_i \geq 1$, $r = e(S) - e(S')$. Further $n_i = 1$ for all i if and only if $\pi : S \longrightarrow S'$ is a simple blowing up of a finite set of r points. From this we easily obtain the following simple lemma.

(6.1.1) LEMMA. *One has* $L \cdot K_S \geq k(e(S) - e(S')) + L \cdot \pi^* K_{S'}$, *with equality if and only if* (S', L') *is a k-reduction of* (S, L).

(6.1.2) COROLLARY. *If* $\kappa(S) \geq 0$, *then* $L \cdot K_S \geq k(e(S) - e(S'))$. *If further* $\kappa(S) \geq 1$ *and* $h^0(K_S^t) > 0$ *for some* $t > 0$ *then* $L \cdot K_S \geq k(e(S) - e(S')) + (k+1)/t$.

Proof. It follows from (6.1.1) by noting that $K_{S'}$ is nef and the general element A of $|\pi^* t K_{S'}|$ has positive arithmetic genus, so that $L \cdot A \geq k+1$.

Q.E.D.

(6.2) THEOREM. *Assume* $\kappa(S) \geq 0$. *Then* $d \geq 2k + 3$. *Further* $g(L) \geq 2k + 1$ *unless possibly if* S *is minimal,* $p_g(S) = 0$ *and* $q(S) = 0$ *or* 1. *If* $g(L) \leq 2k$ *and* $\kappa(S) = 2$ *then* $q(S) = 0$, $1 \leq K_S \cdot K_S \leq 9$, $d \geq (5k + 10)/2$ *and* $g(L) \geq (3k + 8)/2$.

Proof. Let C be a general element of $|L|$. Since $\kappa(S) \geq 0$, $d \leq 2g(L) - 2$ so it follows that $d \geq 2k + 3$. If $h^1(L_C) \neq 0$ then $g(L) \geq 2k + 1$. Thus we can assume that $h^1(L_C) = 0$ and therefore $p_g(S) = 0$. Since $\chi(\mathcal{O}_S) \geq 0$ we conclude that $q(S) = 0$ or 1. Further, by the Riemann-Roch theorem

$$d = h^0(L_C) + g(L) - 1 = 2h^0(L_C) + K_S \cdot L \tag{6.2.1}$$

whence

$$g(L) - 1 = h^0(L_C) + K_S \cdot L. \tag{6.2.1)'$$

If S were non minimal, $K_S \cdot L \geq k$ by (6.1.1). Hence $g(L) \geq k+2+k$ by (6.2.1)'. Therefore we can assume further that S is minimal. Now, let $g(L) \leq 2k$. If $\kappa(S)=2$, then $K_S \cdot K_S \geq 1$ and $\chi(\mathcal{O}_S) > 0$, while $p_g(S) = 0$ implies $q(S) = 0$ and hence $\chi(\mathcal{O}_S)=1$. Thus $K_S \cdot K_S \leq 9$ by the Miyaoka-Yau inequality. The Riemann-Roch theorem gives $h^0(K_S^2) \geq 2$. It thus follows that $K_S \cdot L \geq (k + 1)/2$ by (6.1.1). Actually $K_S \cdot L \geq$ $\geq (k + 2)/2$ since otherwise we would have a pencil of rational or elliptic curves on S. Then by (6.2.1), (6.2.1)' we find $d \geq (5k + 10)/2$ and $g(L) \geq (3k + 8)/2$.

Q.E.D.

(6.3) THEOREM. *Let* S *be a* $\mathbf{P}^1$ *bundle* $p : S \longrightarrow R$ *over a curve* R *of genus* $q(S)$. *Then*

(6.3.1) $d \geq 2k^2$ *and* $g(L) \geq (k - 1)^2$ *if* $q(S) = 0$;

(6.3.2) $d \geq k(k + 2)$ *and* $g(L) \geq k(k + 1)/2$ *if* $q(S) = 1$;

(6.3.3) $d \geq 2k + 4$ *and* $g(L) \geq 2k + 1$ *if* $q(S) \geq 2$.

Proof. Let E be a section of p of minimal self-intersection and f a fibre of p, so $L \equiv aE + bf$.

If $q(S) = 0$, $E^2 = -r$ and $b \geq ar + k$, $a \geq k$ by (3.1). Hence $d = L \cdot L = a(2b - ar) \geq$ $\geq k(b + k) \geq 2k^2$. Similarly $g(L) \geq 2(k - 1)^2$.

Let $q(S) = 1$, $E^2 = -e$. Here $a \geq k$ and either $b \geq ae + k + 2$ if $e \geq 0$ or $2b - ae \geq k + 2$ if $e = -1$ (see (3.5)). So $d = L \cdot L = a(2b-ae) \geq 2(k + 2)k$ if $e \geq 0$ and $d \geq k(k + 2)$ if $e = -1$. Further $2g(L) - 2 = (a - 1)(2b - ae) \geq 2(k - 1)(k + 2)$ if $e \geq 0$ and $2g(L) - 2 \geq (k - 1)(k + 2)$ if $e = -1$. In either case $d \geq k(k + 2)$ and $g(L) \geq k(k + 1)/2$.

Let $q(S) \geq 2$. We know from (4.3.3) that $kK_S + 2L$ is nef. Hence $k^2K_S \cdot K_S + 4kK_S \cdot L + 4L \cdot L \geq 0$. Now $K_S \cdot K_S = 8 - 8q(S)$, then

$$4k(2g(L) - 2) \geq (4k - 4)d + (8q(S) - 8)k^2$$

and also

$$g(L) - 1 \geq (k - 1)d/2k + (q(S) - 1)k.$$

If $h^1(L_C) \neq 0$ we are done. Hence we can assume $h^1(L_C) = 0$, so that $d = h^0(L_C) + g(L) - 1 \geq g(L) + k + 1$. Thus

$$g(L) - 1 \geq (k - 1)(g(L) + k + 1)/2k + (q(S) - 1)k$$

which gives

$$(k + 1)(g(L) - 1)/2k \geq (k - 1)(k + 2)/2k + (q(S) - 1)k$$

or

$$g(L) \geq (k-1)(k+2)/(k+1) + 2k^2/(k+1) + 1 = 3k - 1 \geq 2k + 1.$$

Finally $d \geq g(L) + k + 1$ yields $d \geq 4k > 2k + 3$.

Q.E.D.

(6.4) THEOREM. *If* $K_S \cdot K_S \leq -x < 0$ *and* S *is not a* $\mathbf{P}^1$ *bundle then*

$$d \geq 2k + 3;\ g(L) \geq k(1 + x/4) + 3/4k.$$

<u>Proof</u>. Now $kK_S + L$ is nef by (4.4), so we find

$$-k^2x + 2k(2g(L) - 2) - (2k - 1)d \geq 0$$

and also

(6.4.1) $$g(L) - 1 \geq kx/4 + (2k - 1)d/4k.$$

If $d > 2g(L) - 2$ we get

$$g(L) - 1 \geq kx/4 + (2k - 1)(2g(L) - 2)/4k + (2k - 1)/4k$$

or

$$(g(L) - 1)/2k \geq kx/4 + (2k - 1)/4k$$

and also

$$g(L) \geq k^2x/2 + k + 1/2$$

which gives $d \geq k^2x + 4k - 1 \geq 2k + 3$. If $d < 2g(L) - 2$, then $d \geq 2k + 3$ and, by (6.4.1), $g(L) \geq k(1 + x/4) + 3/4k$. Note that $k^2x/2 + k + 1/2 \geq k(1 + x/4) + 3/4k$.

Q.E.D.

(6.5) THEOREM. *If* $\chi(\mathcal{O}_S) < 0$ *and* S *is not a* $\mathbb{P}^1$ *bundle then*

$$d \geq 2k + 3; \quad g(L) > k(2q(S) - 1) + k/4.$$

Proof. Since $\chi(\mathcal{O}_S) < 0$, S is a ruled surface with $q(S) > 1$ and $K_S \cdot K_S < 8(1-q(S)) < 0$. Use Theorem (6.4) with $x = 8q(S)-7$.

Q.E.D.

It mainly remains to consider rational and elliptic surfaces.

(6.6) LEMMA. *If* S *is rational and* $K_S \cdot K_S \geq 0$ *then* $h^0(K_S^{-1}) > 0$.

Proof. S is either $\mathbb{P}^2$ or a blowing up of $\mathbb{F}_r$. An easy calculation shows that $h^0(K_{\mathbb{F}_r}^{-1}) \geq 9$. Each time a point is blown up on a surface, the number of sections of the anticanonical line bundle decreases by at most 1, so that $h^0(K_S^{-1}) \geq h^0(K_{\mathbb{F}_r}^{-1}) - \#$ where # denotes the number of blowing ups. Thus since $\# = K_{\mathbb{F}_r} \cdot K_{\mathbb{F}_r} - K_S \cdot K_S \leq 8$ the Lemma is proven.

Q.E.D.

(6.7) PROPOSITION. *Assume* S *is not a* $\mathbb{P}^1$ *bundle. Then:*

(6.7.1) $d \geq k^2$ *and* $g(L) \geq k(k - 1)/2 + 1$ *if* $K_S \cdot K_S \geq 0$ *and* S *is rational;*

(6.7.2) $d \geq 2k + 3$ *and* $g(L) \geq 2k + 1$ *if* $K_S \cdot K_S \leq -4$;

(6.7.3) $d \geq 4k$ *and* $g(L) \geq 2k-1$ *if* $K_S \cdot L \leq -4$.

Proof. Since $kK_S + L$ is nef by (4.4) one has

(6.7.4) $$d \geq -kK_S \cdot L.$$

If S is rational and $K_S \cdot K_S \geq 0$, Lemma (6.6) gives $h^0(K_S^{-1}) > 0$, hence $-K_S \cdot L \geq k$ so $d \geq k^2$. Further

(6.7.5) $$2g(L) - 2 \geq -(k - 1)K_S \cdot L \geq k(k - 1)$$

whence $g(L) \geq k(k - 1)/2 + 1$. This proves (6.7.1).

Now (6.7.2) follows from (6.4) while (6.7.4) and (6.7.5) yield (6.7.3).

Q.E.D.

(6.8) REMARK. Note that if $kK_S + L$ is nef, by writing $(kK_S + L)^2 = k^2K_S \cdot K_S + (2k - 1)K_S \cdot L + 2g(L) - 2 \geq 0$ we find

$$2g(L) - 2 \geq -k^2K_S \cdot K_S - (2k - 1)K_S \cdot L.$$

Therefore if $K_S \cdot L \leq 0$ and $K_S \cdot K_S < 0$ one has

$$g(L) \geq k^2/2 + 1 \quad \text{and} \quad d(\geq -k^2K_S \cdot K_S - 2kK_S \cdot L) \geq k^2.$$

(6.9) THEOREM. *If* S *is rational,* $d \geq k^2$. *Further* $g(L) \geq k(k-1)/2 + \min(1,k-2)$

if $K_S \cdot L \leq 0$ *and* $g(L) > 5k/4$ *if* $K_S \cdot L > 0$.

Proof. If S is a $\mathbb{P}^1$ bundle use (6.3). If S is not a $\mathbb{P}^1$ bundle, use (6.7.1) if $K_S \cdot K_S \geq 0$; (6.8) if $K_S \cdot K_S < 0$ and $K_S \cdot L \leq 0$; (6.4) with $x = 1$ if $K_S \cdot K_S < 0$ and $K_S \cdot L > 0$.

Q.E.D.

(6.10) THEOREM. *Let* S *be an elliptic ruled surface but not a* $\mathbb{P}^1$ *bundle. Then* $d \geq k^2$, $g(L) \geq (k^2+2)/2$ *unless* $K_S \cdot L > 0$. *If* $K_S \cdot L > 0$ *then* $d \geq 2k + 3$ *and* $g(L) > 5k/4$.

Proof. We know that $kK_S + L$ is nef by (4.4). Then

$$2g(L) - 2 \geq -k^2K_S \cdot K_S - (2k - 1)K_S \cdot L$$

as in (6.8) with $K_S \cdot K_S < 0$. So if $K_S \cdot L \leq 0$ we find $g(L) \geq (k^2+2)/2$ and also $d \geq 2g(L) - 2 \geq k^2$. If $K_S \cdot L > 0$ we use (6.4).

Q.E.D.

REFERENCES

[1] E. Arbarello, M. Cornalba, P.A. Griffiths, J. Harris, *Geometry of Algebraic Curves Volume* I, Grundlehren, 267 Springer-Verlag (1985).

[2] A. Beauville, *Surfaces algébriques complexes*, Astérisque 54 (1978).

[3] M. Beltrametti, P. Francia, A.J. Sommese, *On Reider's method and higher order embeddings*, Duke Math. Journal, April 1989.

[4] R. Bott, *On a topological obstruction to integrability*, Proc. Symp. Pure Math. XVI, (1970), 127-131.

[5] T. Fujita, *On polarized manifolds whose adjoint bundles are not semipositive*, Advanced Studies in Pure Math. 10 (1987), Algebraic Geometry, Sendai, 1985, 167-178.

[6] R. Hartshorne, *Algebraic Geometry*, G.T.M. 52, Springer-Verlag (1977).

[7] P. Ionescu, *Embedded projective varieties of small invariants*, Proceedings of the Week of Algebraic Geometry, Bucharest 1982, Lectures Notes in Math., Springer-Verlag, 1056 (1984).

[8] A. Kumpera, D. Spencer, *Lie equations, Vol.* I: *General theory*, Princeton Univ. Press, New Jersey, 1972.

[9] E.L. Livorni, *Classification of algebraic nonruled surfaces with sectional genus less than or equal to six*, Nagoya Math. J., 100 (1985), 1-9.

[10] E.L. Livorni, *Classification of algebraic surfaces with sectional genus less than or equal to six III: Ruled surfaces with* $\dim \phi_{K_X \otimes L}(X)=2$, Math. Scand., 59 (1986), 9-29.

[11] A.J. Sommese, *Compact Complex Manifolds Possessing a Line Bundle with a Trivial Jet Bundle*, Abh. Math. Sem. Univ. Hamburg, 55(1985), 151-170.

[12] A.J. Sommese, *Ample divisors on Gorenstein varieties*, Proceedings of Complex Geometry Conference, Nancy 1985, Revue de l'Institut E. Cartan, 10 (1986).

[13] A.J. Sommese, A. Van de Ven, *On the adjunction mapping*, Math. Ann., 278 (1987), 593-603.

[14] E. Ballico, *A characterization of the Veronese surface*, to appear in Proc. A.M.S.

[15] E. Ballico, *On k-spanned projective surfaces*, in this volume.

Note. Very recently some improvements have been obtained by E. Ballico. In [14] a slight modification of Conjecture (2.6) is proved. Furthermore, let L be a k-spanned line bundle on a smooth surface with $k \geq 3$. Then in [15] the better lower bound $h^0(L) \geq k+5$ is given (compare with (2.1)). In both the papers [14] and [15] our results of Section 5 are also used.

In our new paper "Zero cycles and k-th order embeddings of smooth projective surfaces" we define a k-very ample line bundle on a smooth projective surface, S, as a line bundle, L, such that given any length k+1 zero dimensional subscheme $(Z, \mathcal{O}_Z)$ on S the restriction map

$$\Gamma(L) \to \Gamma(L \otimes \mathcal{O}_Z)$$

is onto. This definition is stronger than that of k-spannedness, but in the paper mentioned above we show that the key criterion for k-spannedness, Theorem (0.6) of this paper, holds for k-very ampleness. This means that all the results in this paper hold for k-very ampleness. We are currently preparing a sequel to this paper where we give other new and stronger consequences of k-very ampleness.

ON THE HYPERPLANE SECTIONS OF RULED SURFACES

Aldo Biancofiore
Dipartimento di Matematica, Università degli Studi de L'Aquila
Via Roma Pal.Del Tosto, 67100 L'Aquila, Italia

Introduction

Let L be a line bundle on a connected, smooth, algebraic, projective surface X. In this paper we have studied the following questions:

1) Under which conditions is L spanned by global sections? I.e. if $\phi_L:X\to\mathbf{P}^N$ denotes the map associated to the space $\Gamma(L)$ of the sections of L, when is ϕ_L a morphism?

2) Under which conditions is L very ample? I.e. when does ϕ_L give an embedding?

This problem arise naturally in the study, and in particular in the classification, of algebraic surfaces (see [3],[5],[6],[8],[9],[10]). In this paper we have restricted our attention to the case in which X is gotten by blowing up s distinct points $y_1,\dots,y_s\in Y$, where Y is a geometrically ruled surface. If we denote by $P_1,\dots,P_s$ the corresponding exceptional curves then a line bundle L on X is of the form $L\equiv\pi^*(L^\wedge) - \sum_{j=1,\dots,s} t_jP_j$ where $\pi:X\to Y$ is the blowing up morphism with center $y_1,\dots,y_s$, and $L^\wedge$ is a line bundle on Y. Partial answers to the questions (1) and (2) in the case in which X is a Hirzebruch surface are in [1] when $t_1=\dots=t_s=1$. In [4] it was studied the very ampleness of $L^\wedge$.

In §0 we explain our notation and collect background material. In §1 we give sufficient conditions under which L is spanned or very ample. In §2 we find some special properties of rational ruled surfaces. In §3 we refine the results found in §1 for rational ruled surfaces under the hypothesis of general position of the points $y_1,\dots,y_s$.

We would like to thank A.J.Sommese for very useful discussions.

§0 Background Material.

(0.0) Let L be a line bundle on a smooth connected projective surface X. Let $M=L-K_X$, where K_X is the canonical line bundle on X.

(0.1) In order to semplify our notations we give the following definitions: Let X and L be as in (0.0).

1. We say that L is "0-very ample" if L is spanned by global sections.

2. We say that L is "1-very ample" if L is very ample.

(0.2) Definition: For every $m\in N$, denote by D_M the set of all divisors $E\subseteq X$, such that $E\neq 0$ and mE is effective. Moreover we set $D=\bigcup_{m\in N} D_m$ and $D_m=\{E\in D_1 \mid M-2E\in D\}$.

(0.3) Theorem (Reider): Let X,L and M be as in (0.0). Assume that:
1) $M\in D$; 2) $M^2\geq 5+4i$; 3) $(M-E)\cdot E\geq 2+i$ for any $E\in D_1$ and i=0,1. Then L is i-very ample.

Proof: See [2].

(0.4) Throughout this section we will always assume that X,L and M are as in (0.0). The following results have been proved in [2,§1]. Let $E\in D_1$. Then $E=E_1+...+E_k$ where E_j, j=1,...,k are all the irreducible and reduced components of E. Denote by $\mathcal{E}_i$, i=0,1, the set of all $E\in D_1$ such that either k=1 or if $k\geq 2$ then the following inequalities must be satisfied

(0.4.1) $$\sum_{j=1,\dots,k} E_j\cdot(E-E_j)\geq(k-1)(2+i)+1$$

and

(0.4.2) $$E'\cdot E''\geq 2 \qquad \text{if } E=E'+E'' \quad \text{and} \quad E',E''\in D_1.$$

(0.4.3) If any $E\in \mathcal{E}_i\cap D_M$, i=0,1, verify the inequality

(0.4.4) $$(M-E)\cdot E\geq 2+i$$

then (0.4.4) holds also for any $E\in D_M$.

(0.4.5) Lemma: Let $E\in\mathcal{E}_i$, i=0,1. Then $g(E)\geq 0$, where $g(E)=1+(E+K_X).E/2$.

(0.4.6) Remark: Let $E\in D_1$. Then

1) $(M-E)\cdot E=L.E-2g(E)+2$; 2) If g(E)=0 then $E\in\mathcal{E}_i$ if and only if E is smooth. Moreover if L is i-very ample then $L\cdot E\geq i$.

(0.4.7) Lemma: Let $E\in D_M$, g(E)=1 and L be very ample. Then $L\cdot E\geq 3$.

(0.4.8) Let $E\in D_M$. Since

(0.4.9) $$M^2=4E\cdot(M-E)+(M-2E)^2,$$

then $E.(M-E)\geq 2+i$ if and only if $M^2\geq 5+4i+(M-2E)^2$. Moreover from (0.4.9) assuming

(0.4.10) $$\begin{cases} M^2\geq 5+4i \\ (M-E)\cdot E\leq 1+i \end{cases}$$

it follows that

(0.4.11) $$(M-2E)^2\geq 1.$$

(0.4.12) Lemma: Let $E \in \mathcal{D}_M$, i=0,1. Assume that $E^2 \geq 0$, $(M-2E)\cdot E \geq 0$ and that (0.4.10) holds. Then one of the following is satisfied: 1) i=0, $E^2=0$, $M\cdot E=1$; 2) i=1, $E^2=0$, $M\cdot E=1,2$; 3) i=1, $E^2=1$, $M \equiv 3E$.

(0.4.13) Lemma: Let $M^2 \geq 5+4i$ and $E^2 \geq -1$ for any $E \in \mathcal{E}_i \cap \mathcal{D}_M$ such that g(E)=0. If there is $E \in \mathcal{E}_i \cap \mathcal{D}_M$ such that g(E)=1, $E^2=0$ and $1 \leq M\cdot E < 1+i$, then L is not i-very ample.

§*1. Ruled Surfaces.*

(1.0) Let $y_1,\dots,y_s$ be distinct points on a geometrically ruled surface Y. Let $\pi: X \to Y$ express X as Y with $y_1,\dots,y_s \in Y$ blown up. Denote by P_j, j=1,...,s the corresponding exceptional curves. If $t_1,\dots,t_s \in \mathbf{N}$ we set $L \equiv aC_0+bf-\sum_{j=1,\dots,s} t_jP_j$ where $a \geq 0$. W.l.o.g. we can assume $t_1 \geq \dots \geq t_s \geq 1$. Since $K_X \equiv -2C_0+(2q-2-e)f+\sum_{j=1,\dots,s} P_j$ it follows that $M=L-K_X \equiv (a+2)C_0+(b-2q+2+e)f-\sum_{j=1,\dots,s}(t_j+1)P_j$ where $q=g(C)=h^{1,0}(C)=h^{1,0}(X)$ denotes the irregularity of X. Any divisor E on X is such that $E \equiv xC_0+yf-\sum_{j=1,\dots,s} \alpha_jP_j$. For all the notations about ruled surfaces see [7].

Throughout this section X,L and M are supposed to be as in (1.0).

(1.0.1) Lemma: Let $M^2>0$ and $a \geq 0$. Then $M \in \mathcal{D}$.

Proof: From $h^2(\alpha M)=h^0(K_X-\alpha M)=0$ and from the Riemann-Roch theorem it follows that $h^0(\alpha M) \geq \chi(\mathcal{O}_X)+(1/2)(\alpha^2M^2-\alpha M.K_X)>0$, for $\alpha >> 0$. /.

(1.1) Let $\Lambda=\{(x,e) \in \mathbf{Z} x \mathbf{Z} \mid x \geq 0 \text{ and } e \geq -q\}$, $\Lambda_1=\{(x,e) \in \Lambda \mid x \geq 2 \text{ and } -q \leq e<0\}$ and $\Lambda_0=\Lambda-\Lambda_1$.
Let $\mathcal{D}'$ denote the set of all divisors E on X such that
1) $x \geq 0$

$$2)\ y \geq \begin{cases} 0 & \text{if } (x,e) \in \Lambda_0 \\ (1/2)xe & \text{if } (x,e)\ \Lambda_1 \end{cases}$$

$$3)\ \alpha_j \leq \begin{cases} x+y & \text{if } (x,e) \in \Lambda_0 \\ x+y-(1/2)xe & \text{if } (x,e) \in \Lambda_1. \end{cases}$$

Then

(1.1.1) $$\mathcal{D}'\supseteq\mathcal{D}.$$

Moreover if $\mathcal{D}'_M=\{E\in\mathcal{D}_1 \mid M-2E\in\mathcal{D}'\}$ then $\mathcal{D}'_M\supseteq\mathcal{D}_M$. Let $M-2E\equiv zC_0+wf-\sum_{j=1,\dots,s}\lambda_jP_j$ and $\Gamma_k=\{(x,e)\in\Lambda_k \mid x\leq(a+2)/2\}$ for k=0,1. Then, for any $E\in\mathcal{E}_i\cap\mathcal{D}_M$, we have

1) $$0\leq x\leq(a+2)/2$$

2) $$y\geq\begin{cases}0 & \text{if } (x,e)\in\Gamma_0\\ xe/2 & \text{if } (x,e)\in\Gamma_1\end{cases}$$

and

$$y\leq\begin{cases}(b+2+e-2q)/2 & \text{if } (z,e)\in\Gamma_0\\ (b+2+e-2q-ze/2)/2 & \text{if } (z,e)\in\Gamma_1\end{cases}$$

3)

$$\alpha_j\leq\begin{cases}x+y & \text{if } (x,e)\in\Gamma_0\\ x+y-xe/2 & \text{if } (x,e)\in\Gamma_1\end{cases}$$

and

$$\alpha_j\geq\begin{cases}(t_j+1-z-w)/2 & \text{if } (z,e)\in\Gamma_0\\ (t_j+1-z-w+ze/2)/2 & \text{if } (z,e)\in\Gamma_1.\end{cases}$$

Note that $z=a+2-2x$ and $w=b-2q+2+e-2y$.

(1.1.3) Lemma: Assume that $M^2\geq5+4i$, i=0,1. Let $E\in\mathcal{E}_i\cap\mathcal{D}_M$ such that $E^2\geq0$. If $E\cdot(M-E)\leq1+i$ then either $E^2=0$ and $1\leq M\cdot E\leq1+i$ or i=1, $E^2=1$ and $M\equiv3E$.

Proof: By (0.4.12) it is sufficient to show that $E\cdot(M-2E)\geq0$. Assume that $E\cdot(M-2E)=x(w-ze/2)+z(y-xe/2)-\sum_{j=1,\dots,s}\alpha_j\lambda_j<0$. From (0.4.11) it follows that $(x(w-ze/2)+z(y-xe/2))^2<(\sum_{j=1,\dots,s}\alpha_j\lambda_j)^2$ $\leq(\sum_{j=1,\dots,s}\alpha_j^2)(\sum_{j=1,\dots,s}\lambda_j^2)\leq(2x(y-xe/2)-E^2)(2z(w-ze/2)-1)$ which implies $0\leq(x(w-ze/2)+z(y-xe/2))^2$ $<-2z(w-ze/2)E^2-\sum_{j=1,\dots,s}\alpha_j^2<0$. Thus we get a contradiction. /.

(1.1.4) Lemma: Assume $M^2\geq5+4i$, i=0,1. If either z=0 or z>0 and $w\leq ze/2$ then

(1.1.5) $$(M-E)\cdot E\geq2+i.$$

Proof: $(M\text{-}2E)^2=z(2w\text{-}ze)\text{-}\Sigma_{j=1,\dots,s}\lambda_j^2$. If either $z=0$ or $z>0$ and $w\leq ze/2$ then $(M\text{-}2E)^2\leq 0$. Thus (1.1.5) follows from (0.4.9). /.

(1.2) Lemma: Let $M^2\geq 5+4i$, $i=0,1$ be such that $z\geq 1$ and $w>ze/2$. If $E\equiv xC_0+yf\text{-}\Sigma_{j=1,\dots,s}\alpha_jP_j\in\mathcal{E}_i$ then: 1) If $x=0$ then $0\leq y\leq 1$; 2) If $x=y=0$ then $\Sigma_{j=1,\dots,s}\alpha_j=-1$ and $\alpha_j\leq 0$, $j=1,\dots,s$; 3) If $x\geq 1$ then $0\leq\alpha_j\leq x$, $j=1,\dots,s$; 4) If $E^\wedge\equiv xC_0+yf\text{-}\Sigma_{j=1,\dots,s}\beta_jP_j$ where $\beta_j=\mathrm{Min}\{\alpha_j,(t_j+1)/2\}$ then $E^\wedge\in\mathcal{E}_i$ and

(1.2.1) $$E^\wedge\cdot(M\text{-}E^\wedge)\leq E\cdot(M\text{-}E).$$

Moreover if $M\text{-}2E\in D'$ then also $M\text{-}2E^\wedge\in D'$.

Proof: 1) If $x=0$ and $y\geq 2$ then $g(E)<0$ and by (0.4.5) it follows that $E\notin\mathcal{E}_i$. 2) Since E is effective and $E\neq 0$ then $\alpha_j\leq 0$. Moreover if $\Sigma_{j=1,\dots,s}\alpha_j\leq -2$ then $g(E)<0$ and using (0.4.5) we see again that $E\notin\mathcal{E}_i$. 3) Assume $\alpha_j\geq x+1$ for some $j\in\{1,\dots,s\}$. Then there are effective divisors $E_1\equiv f\text{-}P_j$ and $E_2=E\text{-}E_1$ such that $E_1\cdot E_2\leq 0$. Hence $E\notin\mathcal{E}_i$. If $\alpha_j<0$, for some $j\in\{1,\dots,s\}$ then $E_1=P_j$ and $E_2=E\text{-}E_1$ are effective divisors such that $E_1.E_2\leq 0$. Thus also in this case $E\notin\mathcal{E}_i$. 4) If $\alpha_j=1$ for any $j\in\{1,\dots,s\}$ then there is nothing to prove. Assume $\alpha_t\geq 2$ for some $t\in\{1,\dots,s\}$, then we claim that $E+P_t\in\mathcal{E}_i$.

To prove the claim we have to prove that $E+P_t$ satisfies (0.4.1) and (0.4.2). Let $E_{k+1}=P_t$ and $E=E_1+\dots+E_k$. Then $\Sigma_{j=1,\dots,k+1}E_j\cdot(E+P_t\text{-}E_j)=\Sigma_{j=1,\dots,k}E_j\cdot(E\text{-}E_j)+2P_t\cdot E\geq(k\text{-}1)(2+i)+1+2\alpha_t\geq k(2+i)+1$. Therefore (0.4.1) is satisfied. Let E' and E" be effective divisors on X such that E=E'+E". In order to prove the claim it is enough to prove

(1.2.2) $$(E'+P_t)\cdot E''\geq 2.$$

If $E''\cdot P_t\geq 0$ then (1.2.2) is verified since $E'\cdot E\geq 2$. So we can assume $E''\cdot P_t<0$. Let $F'=E'+P_t$ and $F''=E''\text{-}P_t$. Then F' and F" are effective divisors such that F'+F"=E. Therefore $F'\cdot F''\geq 2$. Since $(E'+P_t)\cdot E''=F'\cdot(F''+P_t)=F'\cdot F''+F'\cdot P_t$ and $E'\cdot P_t=\alpha_t\text{-}E''\cdot P_t\text{-}1\geq 2$, we obtain (1.2.2). So the claim is proved. Now by induction on $n=\Sigma_{j=1,\dots,s}(\alpha_j\text{-}\beta_j)$ and by the above claim, it follows that $E^\wedge\in\mathcal{E}_i$. Moreover, since $(M\text{-}2E^\wedge)\cdot P_j=\rho_j=t_j+1\text{-}2\beta_j$, then

$$\rho_j=\begin{cases}\lambda_j & \text{if } ((t_j+1)/2)\geq\alpha_j\\ 1 & \text{if } ((t_j+1)/2)\geq\alpha_j \text{ and } t_j \text{ is even}\\ 0 & \text{if } ((t_j+1)/2)\geq\alpha_j \text{ and } t_j \text{ is odd.}\end{cases}$$

Now it is easy to check that

$$\rho_j \le \begin{cases} z+w & \text{if } (z,e)\in\Gamma_0 \\ z+w-ze/2 & \text{if } (z,e)\in\Gamma_1. \end{cases}$$ /.

(1.2.3) Definition: We say that $L\equiv aC_0+bf-\sum_{j=1,\dots,s} t_jP_j$ satisfies the property $(\mathcal{P}_i)$, i=0,1, if for any $E\equiv f-\sum_{j=1,\dots,s}\alpha_jP_j\in\mathcal{D}_1$ with $0\le\alpha_j\le 1$, then $L.E=a-\sum_{j=1,\dots,s}\alpha_jt_j\ge i$, i.e. if for any fiber D on Y we have $\sum_{j\in\Lambda_D}t_j\le a-i$, where $\Lambda_D=\{j\in[1,\dots,s]/y_j\in D\}$.

(1.2.4) Remark: a) Let $E\equiv xC_0+yf-\sum_{j=1,\dots,s}\alpha_jP_j\in\mathcal{E}_i\cap\mathcal{D}_M$. If x=y=0 then from (1.2) it follows that $(M-E).E\ge 3$. b) If L does not satisfy $(\mathcal{P}_i)$ then L is not i-very ample. Moreover if L satisfies $(\mathcal{P}_i)$ then $(M-E).E\ge 3$, for any $E\equiv f-\sum_{j=1,\dots,s}\alpha_jP_j\in\mathcal{D}_1$.

Denote by T_i the set of all $E\equiv xC_0+yf-\sum_{j=1,\dots,s}\alpha_jP_j\in\mathcal{E}_i\cap\mathcal{D}_M$ such that

1) $1\le x<(a+1)/2$

2) $(b-2q+2+e-(ze+1)/2)/2\ge y\ge \begin{cases} 0 & \text{if } (x,e)\in\Gamma_0 \\ xe/2 & \text{if } (x,e)\in\Gamma_1 \end{cases}$

3) $\mathrm{Min}\{x,(t_j+1)/2\}\ge\alpha_j\ge \begin{cases} \mathrm{Max}\{0,(t_j+1-z-w)/2\} & \text{if } (z,e)\in\Gamma_0 \\ \mathrm{Max}\{0,(t_j+1-z-w+(ze/2)\}/2 & \text{if } (z,e)\in\Gamma_1 \end{cases}$

(1.2.5) Theorem: Let i=0,1. Assume that $M^2\ge 5+4i$, L satisfies property $(\mathcal{P}_i)$ and that $(M-E)\cdot E\ge 2+i$ for any $E\in T_i$ such that $E^2<0$. Then L is i-very ample unless there is $E\in T_i$ such that either $E^2=0$ and $i\le M\cdot E\le 1+i$ or i=1, $E^2=1$ and $M\equiv 3E$.

Proof: It follows from (0.3), (0.4.3), (1.1.3), (1.1.4), (1.2) and (1.2.4). /.

(1.3) Theorem: Let $e\ge 0$. Assume that L satisfies $(\mathcal{P}_i)$, i=0,1. If

(1.3.1) $$b\ge ae+2q+i+\sum_{j=1,\dots,s}t_j$$

then L is i-very ample.

Proof: In order to show that L is i-very ample, we need to check that 1) $M^2\ge 5+4i$ and 2) $(M-E)\cdot E\ge 2+i$ for any $E\in\mathcal{E}_i\cap\mathcal{D}_M$.

1) $M^2=2(a+2)(b+2-2q-ae/2)-\sum_{j=1,\dots,s}(t_j+1)^2\ge(a+2)ae+2(a+2)(2+i)+\sum_{j=1,\dots,s}(2t_j(a+2)-(t_j+1)^2)$.

If a=i then s=0 and $M^2 \geq 2(2+i)^2=8+10i$. Assume $a\geq 1+i$. Then $1\leq t_j\leq a-i$, $j=1,\dots,s$ and $M^2\geq 12(1+i)+\sum_{j=1,\dots,s}((t_j+1)^2+2it_j-2)\geq 2(6+s)(1+i)\geq 5+4i$.

2) $(M-E)\cdot E=(a+2-2x)y-(a+1-x)xe+(b+2-2q)x-\sum_{j=1,\dots,s}\alpha_j(t_j+1-\alpha_j)\geq(a+2-2x)y+(x-1)xe+(2+i)x+\sum_{j=1,\dots,s}(xt_j-\alpha_j(t_j+1-\alpha_j))$. If x=0 then from (1.2) it follows that $0\leq y\leq 1$. If x=y=0 then $\sum_{j=1,\dots,s}\alpha_j=-1$ and therefore $(M-E)\cdot E\geq 3$. If x=0 and y=1 by (P_i) we get that $(M-E)\cdot E\geq 2+i$. Now assume $x\geq 1$. Due to (1.2) we can consider $0\leq\alpha_j\leq \mathrm{Min}\{x,(t_j+1)/2\}$. Since $y\geq 0$ then $(M-E)\cdot E\geq 2+i+\sum_{j=1,\dots,s}(xt_j-\alpha_j(t_j+1-\alpha_j))$.

Thus we need to estimate $xt_j-\alpha_j(t_j+1-\alpha_j)$, $j=1,\dots,s$. To do this we have to consider two cases: (A) $((t_j+1)/2\ )<x$, (B) $x\leq(t_j+1)/2$.

(A) Since $\alpha_j\leq(t_j+1)/2$ then $xt_j-\alpha_j(t_j+1-\alpha_j)\geq xt_j-((t_j+1)/2)^2\geq 0$.

(B) Since $\alpha_j\leq x\leq(t_j+1)/2$ then $xt_j-\alpha_j(t_j+1-\alpha_j)\geq x^2-x\geq 0$.

Hence in both cases $xt_j-\alpha_j(t_j+1-\alpha_j)\geq 0$. Therefore $(M-E)\cdot E\geq 2+i$ when $x\geq 1$. /.

(1.3.2) Remark: The bound (1.3.1) is sharp. It can be improved if not all y_j, j=1,..,s lie on $D\equiv C_0$.

(1.3.3) Corollary: Assume $e\geq 0$ and s=0. If

(1.3.4) $a\geq i$ and $b\geq ae+2q+i$

then L is i-very ample. Moreover if $q\leq 2i$ then L is i-very ample if and only if (1.3.4) holds.

(1.3.5) For i=0,1, we let

$$\delta_i=\delta_i(L)=\begin{cases} 0 & \text{if } a=0 \\ \mathrm{Max}\{ae/2,-7/6-(s+i)/3\} & \text{if } a=1 \\ \mathrm{Max}\{ae/2,-(11+4i+4s+3s_2)/8\} & \text{if } a=2 \\ \mathrm{Max}\{ae/2,\ -1-i/2-s+s_1/2\} & \text{if } a\geq 3 \end{cases}$$

where s_k is the number of $j\in\{1,\dots,s\}$ such that $t_j=k$. We note that: 1) if a=0 then i=0 and s=0; 2) if a=1 and i=1 then s=0; 3) if a=2 and i=1 then $s_2=0$.

(1.4) Theorem: Let e<0. Assume that L satisfies (P_i),i=0,1. If

(1.4.1) $b\geq ae/2+2q+i+\sum_{j=1,\dots,s}t_j+\delta_i$

then L is i-very ample.

Proof: We have to check that: 1) $M^2\geq 5+4i$; 2) $(M-E)\cdot E\geq 2+i$ for any $E\in\mathcal{E}_i\cap\mathcal{D}_M$.

1) We have $M^2=2(a+2)(b+2-2q-ae/2)-\Sigma_{j=1,\ldots,s}(t_j+1)^2\geq 2(a+2)(2+i+\delta_i+\Sigma_{j=1,\ldots,s}t_j-(1/2)\Sigma_{j=1,\ldots,s}((t_j+1)^2/(a+2)))$. If $a=i$ then $s=0$ and $M^2\geq 2(2+i)(2+i+\delta_i)\geq 5+4i$. If $a=1+i$ then $t_j=1$ for $j=1,\ldots,s$ and $M^2\geq 2(3+i)(2+i+\delta_i+s(1+i)/(3+i))\geq 5+4i$ since

$$\delta_i\geq\begin{cases}-1-1/6-s/3 & \text{if } i=0\\ -2+1/8-s/2 & \text{if } i=1.\end{cases}$$

If $a=2+i$ then $1\leq t_j\leq 2$ for $j=1,\ldots,s$ and $s=s_1+s_2$. Thus, in this case, $M^2\geq 2(4+i)(2+i+\delta_i+s_1(2+i)/(4+i)+s_2(7+4i)/(8+2i))\geq 5+4i$. Since

$$\delta_i\geq\begin{cases}-11/8-s/2-3s_2/8 & \text{if } i=0\\ -3/2-s+s_1/2 & \text{if } i=1.\end{cases}$$

Assume now $a\geq 3+i$. Since $(1/(a+2))\leq 1/5$ and $((t_j+1)/(a+2))\leq 1$ then $M^2\geq 2(5+i)(2+i+\delta_i+\Sigma_{j=1,\ldots,s}t_j-(1/2)\Sigma_{j=1,\ldots,s}((t_j+1)/(a+2))(t_j+1))\geq 2(5+i)(1+i/2-s+s_1/2+\Sigma_{k\geq 1}ks_k-2s_1/(5+i)-9s_2/(10+2i)-(1/2)\Sigma_{k\geq 3}(k+1)s_k\geq 2(5+i)(1+i/2+(1+i)s_1/(10+2i)+(1+2i)s_2/(10+2i)+\Sigma_{k\geq 3}(k-3)s_k/2\geq 5+4i$.

2) The case $x=0$ is trivial. If $x\geq 1$ we have $(M-E)\cdot E=(a+2-2x)y-(a+1-x)xe+(b+2-2q)x-\Sigma_{j=1,\ldots,s}\alpha_j(t_j+1-\alpha_j)\geq(a+2-2x)(y-xe/2)+x(2+i+\delta_i+\Sigma_{j=1,\ldots,s}(t_j-\alpha_j(t_j+1-\alpha_j)/x))$. Let $x=1$. Then $y\geq 0$ and $0\leq\alpha_j\leq 1$. Hence $(M-E)\cdot E\geq 2+i+\delta_i-ae/2\geq 2+i$ since $\delta_i\geq ae/2$. Assume now $x\geq 2$. Then $a\geq 2x-1\geq 3$, $y\geq xe/2$, $\alpha_j\leq\mathrm{Max}\{x,(t_j+1)/2\}$ and $(M-E)\cdot E\geq x(2+i+\delta_i+\Sigma_{j=1,\ldots,s}(t_j-(t_j+1-\alpha_j)\alpha_j/x))$. Thus $(M-E)\cdot E\geq 2+i$. /.

(1.4.2) Corollary: Let $s=0$ and $e<0$. If $a\geq i$ and

(1.4.3) $$b\geq ae/2+2q+i+\mathrm{Max}\{ae/2,-1-i/2\}$$

then L is i-very ample. On the other hand if L is i-very ample and either $q=1$, $a\geq 1$ and $e=-1$ or $q=2$, $a=1$ and $e<0$, then (1.4.3) holds.

Proof: The fact that (1.4.3) implies that L is i-very ample is a particular case of (1.4). Now assume that L is i-very ample. Let $q=1$ then there is an effective divisor $E\equiv 2C_0-f$ with $g(E)=1$ (see [12], [4]). Hence we must have $L\cdot E=-ae+2b\geq 3$ and $L\cdot C_0=-ae+b\geq 3$ i.e. $b\geq -a+3$ and $b\geq -a/2+3/2$. Thus (1.4.3) holds. If $q=2$ and $a=1$, then $L\cdot C_0=-e+b\geq 5$ which implies (1.4.3). /.

§2 *Rational Ruled Surfaces.*

(2.0) Throughout this section we will always assume that X,Y,L and M are as in (1.0) and we will always let $q=0$ and $M^2 \geq 5+4i$, $i=0,1$.

(2.0.1) Lemma: Let $E \equiv xC_0+yf-\Sigma_{j=1,\dots,s}\alpha_j P_j \in \mathcal{E}_i \cap \mathcal{D}'_M$. Assume $x \geq 1$. Then either $x=1$ and $y=0$ or

$$(2.0.2) \qquad y \geq \begin{cases} xe & \text{if either } x=1 \text{ and } e \geq 0 \text{ or } x \geq 2 \text{ and } e \leq 2 \\ xe+2-e & \text{if } x \geq 2 \text{ and } e \geq 2. \end{cases}$$

Proof: Let $E^\wedge \equiv xC_0+yf$ be an effective divisor on Y. Let $\mu_j(E^\wedge)$ denote the multiplicity of $E^\wedge$ at y_j. Then $\mu_j(E^\wedge) > \alpha_j$ and $E \equiv \pi^*(E^\wedge)-\Sigma_{j=1,\dots,s}\alpha_j P_j$. Assume now that $(x,y) \neq (1,0)$. By [7, Prop.V.2.20 p.382] we have that $E^\wedge$ is not irreducible when $y \leq xe-1$. Let $x=1$. If $y \leq e-1$ then $E^\wedge$ is not irreducible and there are $F_1 \equiv f$ and $F_2 = E^\wedge - F_1$ effective divisors such that $F_1.F_2 \leq 1$. Setting $E_k = \pi^*(F_k)$ we have that E_k, $k=1,2$ are effective divisors where $E_1+E_2=E$ and $E_1.E_2 \leq 1$. Hence $E \notin \mathcal{E}_i$. Therefore (2.0.2) holds. Assume now that $x \geq 2$ and $y < \mathrm{Min}\{xe, xe+2-e\}$ Then by the long cohomology sequence associated to the short exact sequence $0 \to L(E^\wedge - C_0) \to L(E^\wedge) \to L(E^\wedge | C_0) \to 0$ we have that $F_1 = C_0$ and $F_2 = E^\wedge - C_0$ are effective divisors and $F_1.F_2 = y-(x-1)e \leq 1$. Setting $E_j = \pi^*(F_j)$ we have that E_j, $j=1,2$, are effective, where $E=E_1+E_2$ and $E_1.E_2 \leq 1$. Hence $E \notin \mathcal{E}_i$. So (2.0.2) holds also in this last case. /.

(2.0.3) Definition: We say that $L \equiv aC_0+bf-\Sigma_{j=1,\dots,s}t_j P_j$ satisfies the property $(\mathcal{Q}_i)$, $i=0,1$, if for any $E \equiv C_0-\Sigma_{j=1,\dots,s}\alpha_j P_j \in \mathcal{D}_1$ with $0 \leq \alpha_i \leq 1$, we have $L \cdot E = b-ae-\Sigma_{j=1,\dots,s}\alpha_j P_j \geq i$.

(2.0.4) Remark: If L does not satisfy $(\mathcal{Q}_i)$ then L is not i-very ample. If L satisfies $(\mathcal{Q}_i)$ then $(M-E) \cdot E \geq 3$, for any $E \equiv C_0-\Sigma_{j=1,\dots,s}\alpha_j P_j \in \mathcal{D}$.

(2.0.5) Denote by S_i the set of all $E \equiv xC_0+yf-\Sigma_{j=1,\dots,s}\alpha_j P_j \in \mathcal{E}_i \cap \mathcal{D}'_M$ such that

1) $1 \leq x \leq (a+1)/2$

$$2) \; (b+2+e-(ze+1)/2)2 \geq y \geq \begin{cases} \mathrm{Max}\{1, xe\} \text{ if either } x=1 \text{ and } e \geq 0 \text{ or } x \geq 2 \text{ and } e \leq 2 \\ xe+2-e \text{ if } x \geq 2 \text{ and } e \geq 2 \end{cases}$$

3) $\mathrm{Min}\{x,(t_j+1)/2\}\geq\alpha_j\geq\mathrm{Max}\{0,(t_j+1-z-w)/2\}$; where $z=a+2-2x$ and $w=b+2+e-2y$.

(2.0.6) Theorem: Let $a\geq0$, $M^2\geq5+4i$. Assume that L satisfies (P_i) and (Q_i) and that $(M-E)\cdot E\geq2+i$ for any $E\in S_i$ such that $E^2<0$. Then L is i-very ample unless there is $E\in S_i$ such that either $E^2=0$ and $i\leq M\cdot E\leq1+i$ or $i=1$, $E^2=1$ and $M\equiv3E$.

Proof: It follows from (1.2.5) and (2.0.4). /.

(2.1) Let $D\equiv xC_0+yf$ be an effective divisor on Y. If either $x<0$ or $y<0$ then $h^0(D)=0$. Therefore we can assume w.l.o.g. that $x\geq0$ and $y\geq0$.

(2.1.1) Proposition: Let $D\equiv xC_0+yf$ be as in (2.1). Then

$$\text{(2.1.2)}\quad h^0(D)=\begin{cases}(k+1)(y+1)-k(k+1)e/2 & \text{if } ke-1\leq y\leq(k+1)e-2 \text{ and } 0\leq k\leq x-1\\ \\ (x+1)(y+1)-x(x+1)e/2 & \text{if } y\geq xe-1.\end{cases}$$

Proof: Let $D'=D-C_0$. We can consider the long cohomology sequence

$$\text{(2.1.3)}\qquad 0\to H^0(D')\to H^0(D)\to H^0(D|C_0)\to H^1(D')\to\ldots$$

associated to the short exact sequence

$$0\to[D']\to[D]\to[D|C_0]\to0.$$

Claim: We have

$$\text{(2.1.4)}\qquad h^0(D)=h^0(D')+h^0(D|C_0)$$

Proof of the claim: If $h^0(D|C_0)=0$ then (2.1.4) is trivial. So we can assume $h^0(D|C_0)\neq0$. Let $x=0$. Then from $H^0(D)\cong H^0(D|C_0)$ and $h^1(D)=h^1(D|C_0)=0$ it follows that $h^1(D')=0$. Thus (2.1.3) implies (2.1.4). Let $x\geq1$. Since $\deg(D|C_0)=y-xe$ then $[D|C_0]\equiv\mathcal{O}_{\mathbf{P}^1}(y-xe)$. Thus $y\geq xe$. Moreover from ([4,Theorem(1.1)]) it follows that if $y\geq(x-1)e-1$ then $h^1(D')=0$. Hence (2.1.3) implies (2.1.4) also in this case. Thus the claim is proved. /.

From (2.1.4) it follows that

$$h^0(D)=h^0(yf)+\sum_{k=0,\ldots,x-1}h^0((D-kC_0)|C_0)=h^0(\mathcal{O}_{\mathbf{P}^1}(y))+\sum_{k=1,\ldots,x}h^0(\mathcal{O}_{\mathbf{P}^1}(y-ke)).$$

which together with

$$h^0(\mathcal{O}_{\mathbf{P}^1}(\delta))=\begin{cases}\delta+1 & \text{if } \delta\geq0\\ \\ 0 & \text{if } \delta<0.\end{cases}$$

implies (2.1.2). /.

(2.2) For any $E \equiv xC_0+yf-\sum_{j=1,\dots,s}\alpha_j P_j \in S_i$ let α and τ be defined by

(2.2.1) $$\alpha=\alpha(E)=h^0(E+\sum_{j=1,\dots,s}\alpha_j P_j)-1$$

and

(2.2.2) $$\tau=\tau(E)=xe-y-1.$$

Then

(2.2.3) $$\tau(E)\leq\begin{cases}-1 & \text{if either } x=1 \text{ and } e\geq 0 \text{ or } x\geq 2 \text{ and } 0\leq e\leq 2\\ e-3 & \text{if } x\geq 2 \text{ and } e\geq 2.\end{cases}$$

Moreover from Riemann-Roch Theorem it follows that

(2.2.4) $$\alpha(E)-\sum_{j=1,\dots,s}\alpha_j(\alpha_j+1)/2=\begin{cases}E\cdot(E-K_X)/2 & \text{if } \tau\leq -1\\ E\cdot(E-K_X)/2+\tau(E) & \text{if } 0\leq\tau\leq e-3.\end{cases}$$

§3. *Rational Ruled Surfaces: General Position.*

(3.0) Throughout this section we will always assume that X,Y,L and M are as in (1.0) and that $q=0$. For any divisor D on Y, such that $D\not\equiv 0$, we let $\mu_j=\mu_j(D)$ denote the multiplicity of D at y_j.

(3.0.1) Definition: We say that $y_1,\dots,y_s$ are "in general position" w.r.t. L if for any divisor $F\equiv xC_0+yf$ of Y such that: 1) F is irreducible and reduced; 2) $0\leq x\leq(a+1)/2$; 3) $0\leq y\leq(b-ae/2+(x+1)e+2)/2$; 4) $\mu_j(F)\leq(t_j+1)/2$, $j=1,\dots,s$; then

(3.0.2) $$(1/2)\sum_{j=1,\dots,s}\mu_j(F)(\mu_j(F)+1)\leq h^0(F)-1.$$

(3.0.3) Remark: If $2\geq t_1\geq\dots\geq t_s$ then $\mu_j(F)\leq 1$ and (3.0.2) becomes

(3.0.4) $$\sum_{j=1,\dots,s}\mu_j(F)\leq h^0(F)-1.$$

(3.1) Lemma: Let $y_1,\dots,y_s$ be in general position w.r.t. L. Let $E\equiv xC_0+yr-\sum_{j=1,\dots,s}\alpha_j P_j\in\mathcal{E}_i$, $i=0,1$, be such that $x\leq(t_j+1)/2$, $j=1,\dots,s$. Then

(3.1.1) $$(1/2)\sum_{j=1,\dots,s}\alpha_j(\alpha_j+1)\leq\alpha(E).$$

Proof: If x=y=0 then $\sum_{j=1,...,s}\alpha_j=-1$ and $\alpha_j\leq 0$, j=1,...,s. Hence (3.1.1) is satisfied since $\alpha(E)=0$. Let $E=E_1+...+E_k$, where E_t, t=1,...,k are all the irreducible and reduced components of E. We can assume $k\geq 2$ since for $k\leq 1$ (3.1.1) is satisfied by definition. Let $E_t=x_tC_0+y_tf-\sum_{j=1,...,s}\alpha_{t,j}P_j$. We can consider the following two cases: 1) either e=0 or $e\geq 1$ and $(x_t,y_t)\neq(1,0)$ for t=1,...,k; 2) $e\geq 1$ and there is at least one $t\in\{1,...,k\}$ such that $(x_t,y_t)=(1,0)$. In case (1) $y_t\geq x_te$, t=1,...,s, since E_t is irreducible and reduced. Therefore

$$y=\sum_{t=1,...,s}y_t\geq\sum_{t=1,...,s}x_te=xe.$$

Then by (2.2.4) it follows that (3.1.1) is equivalent to

(3.1.2) $$E\cdot(E-K_X)\geq 0.$$

If $E\cdot(E-K_X)<0$ then $0>E\cdot(E-K_X)=\sum_{t=1,...,k}E_t\cdot(E_t-K_X)+\sum_{t=1,...,k}E_t\cdot(E-K_t)\geq(k-1)(2+i)+1\geq 0$ which gives a contradiction. Therefore (3.1.1) holds. In case (2) we note that $E_t\equiv C_0-\sum_{j=1,...,s}\alpha_{t,j}P_j$ satisfies (3.1.1). Hence $\alpha_{t,j}=0$, j=1,...,s. Therefore we can write $E\equiv mC_0+E^\wedge$ where $E^\wedge\equiv(x-m)C_0+yf-\sum_{j=1,...,s}\alpha_jP_j$ satisfies case (1). Thus $E^\wedge\cdot(E^\wedge-K_X)\geq 0$. Now it is easy to see that $E\cdot(E-K_X)\geq E^\wedge\cdot(E^\wedge-K_X)$. /.

(3.1.3) Lemma: Let $y_1,...,y_s$ be in general position w.r.t. L. Let $E\equiv xC_0+yf-\sum_{j=1,...,s}\alpha_jP_j\in\mathcal{E}_i\cap\mathcal{D}'_M$ be such that $(x,y)\neq(1,0)$. Then

(3.1.4) $$E\cdot K_X\leq g(E)-1\leq E^2 \qquad \text{if } \tau\leq 0$$

and

(3.1.5) $$E\cdot K_X-\tau(E)\leq g(E)-1\leq E^2+\tau(E) \qquad \text{if } 0\leq\tau\leq e-3.$$

Proof: It follows from (2.2.4) since $g(E)=1+E\cdot(E+K_X)/2$ and $y_1,...,y_s$ are in general position w.r.t. L. /.

(3.2) Proposition: Assume that $M^2\geq 5+4i$ and that $y_1,...,y_s$ are in general position w.r.t. L. Consider $E\equiv xC_0+yf-\sum_{j=1,...,s}\alpha_jP_j\in\mathcal{E}_i\,\mathcal{D}'_M$ such that

$$g(E)\geq\begin{cases}1 & \text{if } \tau\leq 0\\ 1+\tau(E) & \text{if } 0\leq\tau\leq e-3.\end{cases}$$

If $(M-E)\cdot E\leq 1+i$ then either

$$E^2=0,\ 1\le M\cdot E\le 1+i \text{ and } g(E)=\begin{cases}1 & \text{if } \tau\le 0\\ 1+\tau(E) & \text{if } 0\le\tau\le e\end{cases}$$

or

$$E^2=1,\ i=1,\ M\equiv 3E \text{ and } g(E)\le\begin{cases}2 & \text{if } \tau\le 0\\ 2+\tau(E) & \text{if } 0\le\tau\le e-3.\end{cases}$$

Proof: By (3.1.4) and (3.1.5) we have $E^2\ge 0$. Now the statement follows from (1.1.3). /.

(3.2.1) Lemma: Assume that $M^2\ge 5+4i$, $e\le 3$ and that $y_1,\ldots,y_s$ are in general position w.r.t. L. Consider $E\in\mathcal{E}_i\cap\mathcal{D}_M$ such that $g(E)=1$, $E^2=0$ and $1\le M\cdot E\le 1+i$, then M is not i-very ample.

Proof: We have $M\cdot E=(M-E)\cdot E=L\cdot E-2g(E)+2=L\cdot E$. If i=1 the statement follows from (0.4.7). Assume i=0. Then $M\cdot E=L\cdot E=1$. Moreover if $F\in\mathcal{E}_i\cap\mathcal{D}_M$ with $g(F)=0$ then $F^2\ge -1$ since $y_1,\ldots,y_s$ are in general position. Now the statement follows from (0.4.13). /.

(3.3) Theorem: Let $M^2\ge 5+4i, i=0,1$, and let $y_1,\ldots,y_s$ be in general position w.r.t. L. Suppose that if i=1 then for any $E\in\mathcal{E}_1\cap\mathcal{D}_M$ such that

$$g(E)=\begin{cases}2 & \text{if } \tau\le 0\\ 2+\tau(E) & \text{if } 0\le\tau\le e-3\end{cases}$$

we have either $E^2\ne 1$ or $M\not\equiv 3E$. If $0\le e\le 3$, then L is i-very ample if and only if for any $E\in\mathcal{E}_i\cap\mathcal{D}_M$ such that $0\le g(E)\le 1$ then $L\cdot E\ge 2g(E)+i$. If $e\ge 4$, then L is i-very ample if for any $E\in\mathcal{E}_i\cap\mathcal{D}_M$ such that

$$0\le g(E)\le\begin{cases}1 & \text{if } \tau\le 0\\ 1+\tau(E) & \text{if } 0\le\tau\le e-3\end{cases}$$

then $L\cdot E\ge 2g(E)+i$.

Proof: It is a consequence of (0.4.7), (3.2) and (3.2.1). /.

(3.4) Theorem: Let $M^2\ge 5+4i$ and let $y_1,\ldots,y_s$ be in general position w.r.t. L. Suppose that $a\ge 2t_1$ and that if i=1 then for any $E\in\mathcal{E}_i\cap\mathcal{D}_M$ such that

$$g(E)=\begin{cases}2 & \text{if } \tau\le 0\\ 2+\tau(E) & \text{if } 0\le\tau\le e-3\end{cases}$$

we have either $E^2\neq 1$ or $M\equiv 3E$. If $e=0$, then L is i-very ample if

(3.4.1) $$b\geq 2t_1+1 \text{ and } a\geq 2t_1+1.$$

If $e\geq 1$, then L is i-very ample if

(3.4.2) $$b\geq ae+i$$

(3.4.3) $$b\geq (e+2)t_1+1+i$$

(3.4.4) $$b\geq (e-2)(a+1)/2+(3e+7)t_1/4+(1+i)/2$$

(3.4.5) $$b\geq (e-2)(a+2)/3+(3e+4)t_1/3+(i-2)/3$$

are all satisfied.

(3.4.6) Remark: If $e=1,2$ then (3.4.3) implies (3.4.4) and (3.4.5); if $e=3$ then (3.4.4) implies (3.4.5) and if $e\geq 3$ then (3.4.4) implies (3.4.3).

Proof of theorem (3.4): Assume that there is $E\equiv xC_0+yf-\sum_{j=1,\dots,s}\alpha_jP_j\in\mathcal{E}_i\cap\mathcal{D}_M$ $i=0,1$, such that $L\cdot E\leq 2g(E)-1+i$ and

(3.4.7) $$0\leq g(E)\leq \begin{cases} 1 & \text{if } \tau\leq 0 \\ 1+\tau(E) & \text{if } 0\leq\tau\leq e-3. \end{cases}$$

Then (3.1.3) implies

(3.4.8) $$(L+t_1K_X)\cdot E\leq \begin{cases} (2+t_1)(g(E)-1)+1+i & \text{if } \tau\leq 0 \\ (2+t_1)(g(E)-1+\tau)-2\tau+1+i & \text{if } 0\leq\tau\leq e-3. \end{cases}$$

Moreover

(3.4.9) $$(L+t_1K_X)\cdot E\geq x(b-2t_1-ae/2)+(a-2t_1)(y-xe/2)\geq x(b-(2+e)t_1)-(1+\tau)(a-2t_1).$$

At first we consider the case $e=0$. Then $\tau<0$. W.l.o.g. we may suppose $b\geq a$. By (3.4.8) and (3.4.9) we get

$$a\leq 2t_1+(2+t_1)/(x+y)(g(E)-1)+(1+i)/(x+y).$$

Since $x+y\geq 1$ if $g(E)=0$ and $x+y\geq 4$ if $g(E)=1$ then $a\leq 2t_1$ which contradicts (3.4.1). Consider now the case $e\geq 1$. If $x=0$ then $y=1$ and $L\cdot E\geq 1$ since $a\geq 2t_1$. If $x=1$ and $y=0$ then $\alpha_1=\dots=\alpha_s=0$. Hence $L\cdot E\leq 2g(E)-1+i$ which contradicts (3.4.2). From now on we assume that $x\geq 1$ and that (2.0.2) holds. We have

(3.4.10) $$g(E)\leq (1-x)(2+\tau-xe/2)$$

which combined with (3.4.7) implies that:

1) If $e\geq 1$, $g(E)=0$ and $\tau\leq -1$ then $x\geq 1$; 2) If $e\geq 1$ and either $g(E)=0$ and $0\leq\tau\leq e-3$ or $g(E)=1$ then $x\geq 2$; 3) If $e\geq 3$ and $1\leq g(E)\leq (e-1)/2$ then $x\geq 2$; 4) If $e\geq 3$ and $e/2\leq g(E)\leq e-2$ then $x\geq 3$. Assume that $\tau(E)\leq -1$. From (3.4.8) and (3.4.9) it follows that

(3.4.11) $$b\leq (2+e)t_1+A/x$$

where $A=(2+t_1)(g(E)-1)+1+i$. If $g(E)=0$ then $x\geq 1$ and $A<0$; if $g(E)=1$ then $x\geq 1$ and $A=1+i$. In both cases (3.4.11) contradicts (3.4.3). Assume now that $\tau\geq 0$. Then we have $e\geq 3$ and $x\geq 2$. From (3.4.8) and (3.4.9) it follows that

(3.4.12) $$b\leq(2+e)t_1+B/x.$$

where $B=\tau(a-2t_1)+t_1(g(E)-1+\tau)+2(g(E)-1)+a-2t_1+1+i$.

Assume $g(E)=0$. Then $B=\tau(a-t_1)+a-3t_1-1+i$. If $B<0$ then (3.4.12) contradicts (3.4.3). If $B\geq 0$ then (3.4.12) becomes $b\leq(e-2)a/2+(e+4)t_1/2+(i-1)/2$ which contradicts (3.4.4). If $1\leq g(E)\leq(e-1)/2$ then $B>0$ and (3.4.12) becomes $b\leq(e-2)(a+1)/2+t_1(3e+7)/4+i/2$ which contradicts (3.4.4). If $e/2\leq g(E)\leq e-2$ then $e\geq 4$, $x\geq 3$ and $B>0$. Therefore (3.4.12) implies $b\leq(e-2)(a+2)/2+t_1(3e+4)/3-1+i/3$ which contradicts (3.4.5). /.

References

[1] Bese E., On the Spannedness and Very Ampleness of Certain Line Bundles on the Blow-ups of $\mathbf{P}^2$ and F_r, Math.Ann. **262**, 225-238(1983).

[2] Biancofiore A., On the hyperplane sections of blow ups of complex projective plane, To appear on Can.J. Math.

[3] Biancofiore A. and Livorni E.L.,On the iteration of the adjunction process in the study of rational surfaces, Ind.Univ.Math.J. Vol.**36**,No.1, 167-188 (1987).

[4] Biancofiore A. and Livorni E.L.,On the genus of a hyperplane section of a geometrically ruled surface, Annali di Mat. Pura ed Appl. (IV), Vol.**CXLVII**, 173-185 (1987).

[5] Biancofiore A. and Livorni E.L.,On the iteration of the adjunction process for surfaces of negative Kodaira dimension, To appear on Manuscripta Mathematica.

[6] Biancofiore A. and Livorni E.L., Algebraic ruled surfaces with low sectional genus, Ricerche di Matematica Vol.XXXVI,facs.1°,17-32 (1987)

[7] Hartshorne R., Algebraic Geometry, Springer Verlag, New York (1977).

[8] Livorni E.L., Classification of algebraic surfaces with sectional genus less than or equal to six. II Ruled surfaces with dim $\Phi_{K_X\otimes L}(X)=1$, Can.J. Math. Vol.**XXXVII**,No.4,1110-1121(1986).

[9] Livorni E.L., Classification of algebraic surfaces with sectional genus less than or equal to six. II Ruled surfaces with dim $\Phi_{K_X\otimes L}(X)=2$, Math. Scand. **59**, 9-29(1986).

[10] Livorni E.L., On the existence of some surfaces, Preprint.

[11] Reider I., Vector bundles of rank 2 and linear systems on algebraic surfaces, Ann. of Math. **127**, 309-316(1988).

[12] Sommese A.J. and Van de Ven A., On the adjunction mapping, Math. Ann. **278**, 593-603(1987).

FOOTNOTES TO A THEOREM OF I. REIDER

Fabrizio Catanese *
Dipartimento di Matematica dell' Università di Pisa
via Buonarroti 2, 56100 PISA

Introduction.

_ After the suggestion of one of the editors of these Proceedings, we publish this article which essentially reproduces a letter I wrote to Igor Reider on november 1986, after giving a seminar at the Institute Mittag-Leffler on Reider's results which have appeared in [Rei].

§1 is devoted to giving a more general and precise version of a result stated in an article by Griffiths and Harris ([G-H] , [Rei]), applying a construction due to Serre to construct vector bundles on algebraic surfaces starting from 0-dimensional subschemes failing to impose independent conditions to certain linear systems ; this version has not been superseded by the results appearing in Tjurin's new article ([Tju]) .

§2 supplies instead details for the proof (proposition 3 of [Rei]) that m-canonical systems , if m is at least 3 , give (in characteristic 0 , and with a couple of exceptions) embeddings of the canonical models of surfaces of general type ; for these details Reider defers the reader to the above quoted letter , and this result is due to cooperation with Torsten Ekedahl who in fact devised the final trick to solve the combinatorial problem to which the proof had been reduced ([Ek]).

It is a pleasure to acknowledge the warm hospitality and stimulating atmosphere I found at the Mittag-Leffler Institute in september '86 , and to thank the organizers of the Conference for their kind invitation .

§1 ZERO CYCLES ON SURFACES AND RANK 2 BUNDLES.

In this section X shall be a projective normal Gorenstein surface over an

---

*) Partly supported by 40% MPI , and a member of GNSAGA of CNR

algebraically closed field k , i.e. X is normal and Cohen-Macaulay and ω_X, the dualizing sheaf of X , is an invertible sheaf ; we shall denote by K a Cartier divisor associated to ω_X .

We let Z be a purely 0-dimensional subscheme of X ; Z shall also be called a 0-cycle.

If we assume further that Z is a local complete intersection (l.c.i. , for short) , then , denoting , for $p \in \text{supp}(Z)$, by R_p the local ring $\mathcal{O}_{Z,p}$ of Z at p , and by R(Z) the direct sum of the R_p 's , we have that

(1.1) $R= R_p$ is a 0-dimensional Gorenstein ring , in other terms , there is a non-degenerate pairing $R \times R \longrightarrow k$ given by local duality .

(1.2) $R=R_p$ has a natural decreasing filtration , given by the powers of the maximal ideal of p , and the last non zero term of this filtration is called the <u>socle</u> of R , and shall be denoted by $S = S_p$. The condition that R be a Gorenstein ring implies that S is a 1-dimensional k-vector space.

(1.3) We recall moreover that the pairing (1.1) is compatible with the algebra structure on R , i.e. , for $f,g \in R$, $< f, g > = < 1 , fg >$, and therefore the socle S is just the annihilator of the maximal ideal $\mathcal{M}_p$ of $R = R_p$.

In the sequel , given a k-vector space V , we shall denote by $V^\vee$ its dual.

<u>Theorem 1.4</u>

Let X be a Gorenstein surface and Z a 0-cycle on X ; let L be a Cartier divisor on X and [L] the invertible sheaf associated to the Cartier divisor .

If J_Z denotes the ideal sheaf of Z , we may consider the exact sequence

(*) $H^0([K+L]) \xrightarrow{r} H^0([K+L]|_Z) \longrightarrow H^1(J_Z[K+L])$,

and consider an isomorphism of the middle term with R(Z) (given by some local trivialization of [K + L]) .

Then there is an isomorphism between

i) the group of extensions $0 \longrightarrow \mathcal{O}_X \longrightarrow E \longrightarrow J_Z[L] \longrightarrow 0$,
modulo the subgroup of extensions $0 \longrightarrow \mathcal{O}_X \longrightarrow E' \longrightarrow [L] \longrightarrow 0$
(giving E as the subsheaf of E' defined as the preimage of $J_Z[L]$) .

ii) the group of linear forms $\alpha \in R(Z)^{v}$ vanishing on the image of $H^0([K+L])$.

Moreover, in the above isomorphism, E is locally free if and only if Z is a l.c.i. and ,writing α_p for the restriction of α to R_p, α_p does not vanish on the socle S_p of R_p.

<u>Proof</u>. Dualizing the exact sequence (*), we obtain that the group of linear forms α as in ii) is isomorphic to the space $H^1(J_Z[K+L])^{v}$ modulo the subspace $H^1([K+L])^{v}$, and we conclude for the first assertion since these two vector spaces are naturally isomorphic to $\mathrm{Ext}^1(J_Z[L], \mathcal{O}_X)$, resp. to $\mathrm{Ext}^1([L], \mathcal{O}_X)$.

We denote by α^* an extension in $\mathrm{Ext}^1(J_Z[L], \mathcal{O}_X)$ inducing α.

We have to see when does the extension α^* give a locally free sheaf E. First of all, since E has rank 2, if E is locally free, then Z is locally defined by two equations, so Z must be a l.c.i. . Moreover, the local to global spectral sequence for Ext provides a natural map :

$$\mathrm{Ext}^1(J_Z[L], \mathcal{O}_X) \longrightarrow H^0(\mathcal{E}xt^1(J_Z[L], \mathcal{O}_X)) \cong H^0(\mathcal{E}xt^2(\mathcal{O}_Z[L], \mathcal{O}_X)) \cong \mathrm{Ext}^2(\mathcal{O}_Z[K+L], [K]) \cong H^0([K+L]|_Z)^{v} \cong R(Z)^{v}$$

(the last two isomorphisms being respectively given by Serre duality on X and by the chosen trivialization of $[K+L]$ around Z).

The given extension α^* thus naturally maps to α, with α_p giving a local extension $0 \longrightarrow \mathcal{O}_{X,p} \longrightarrow E_p \longrightarrow J_{Z,p} \longrightarrow 0$ as follows .

Using local duality we can identify R_p^{v} with R_p, hence we can pick a function g around p whose class in R_p represents α_p.

Moreover, since Z is a l.c.i. , the ideal J_Z is locally generated by two functions h_1, h_2, and then E_p is given as the cokernel of the homomorphism of free sheaves associated to the transpose of the row

(g, h_1, h_2) so that we have an exact sequence

$$0 \longrightarrow \mathcal{O}_{X,p} \longrightarrow \mathcal{O}_{X,p}^3 \longrightarrow E_p \longrightarrow 0$$

and the embedding of $\mathcal{O}_{X,p}$ in E_p is induced by the isomorphism of $\mathcal{O}_{X,p}$ with the first factor of $\mathcal{O}_{X,p}^3$ (hence the quotient of E_p by $\mathcal{O}_{X,p}$, if h is the column with coefficients h_1, h_2, is isomorphic to

$\mathcal{O}_{X,p}{}^2 / h\mathcal{O}_{X,p}$, and thus to $J_{Z,p}$ as desired) .

It is now clear that E_p is locally free if and only if g does not vanish at p , i.e. its class does not annihilate the socle S_p of R_p .

Q.E.D.

Remark 1.5. If H^1 ([K + L]) = 0 , then for each $\alpha \varepsilon$ R(Z $)^v$ there is a unique extension α^* inducing α .

Example 1.6 If Z is a cycle of length 2 supported at a smooth point p of X , then there do exist local coordinates (x, y) such that J_Z is generated by (x^2 , y) . The socle S coincides with the maximal ideal of R , and such a locally free extension exists if and only if S is not contained in the image of the restriction map r from H^0 ([K + L]) . I.e. , either p is a base point and Im (r) = 0 , or p is not a base point and r is not onto .

Example 1.7. If Z consists of m distinct smooth points , p_1 , .. p_m , then E is locally free iff α_p is non zero for each p = p_1 , .. p_m .
In this case we have a non trivial extension (by which we mean , not obtained from an extension $0 \longrightarrow \mathcal{O}_X \longrightarrow E' \longrightarrow [L] \longrightarrow 0$) if and only if the points p_i are projectively dependent via the rational map associated to the linear system |K + L| , or ,more precisely , if the linear functionals e_i , for i= 1, ..m , given by evaluation at p_i (and in fact only defined up to a scalar multiple) are linearly dependent ; this is in fact the condition that r be not surjective .

We obtain a locally free sheaf if no p_i is a base point of | K + L | and if , q_i being the image point of p_i , there does exist among the q_i's a relation of linear dependence with all the coefficients different from zero.

To understand what this geometrical condition means, we may assume that q_1 , .. q_h is a maximal set of linearly independent elements among the q_i's : then , since the given field k is infinite , such a relation of linear dependence exists if and only if $h < m$ and the remaining q_j's do not all lie in one of the coordinate hyperplanes of the projective space of dimension (h-1) spanned by the points q_1, ..q_h .

Remark 1.8. The following observation came out in a conversation I had with Mauro Beltrametti . Assume that X is smooth and that Z is a 0-cycle for which the restriction map r is not onto , whereas for each subscheme Z' of Z the restriction map r' is onto . Then the image of r is a hyperplane in R(Z) , hence there is a unique nonzero linear form α vanishing on Im (r), and a corresponding extension E is locally free (implying that Z must be a l.c.i.). In fact , otherwise E is contained in its double dual E' which is locally free , and gives an extension $0 \longrightarrow \mathcal{O}_X \longrightarrow E' \longrightarrow J_{Z'}[L] \longrightarrow 0$ where now Z' is a proper subscheme of Z . By assumption this sequence is split locally at Z , hence also the extension giving E is locally split, a contradiction.

The following lemma is essential in order to be able to prove that the adjoint linear systems | K + L | give embeddings of X .

Lemma 1.9. If p is a smooth point of X and $H^0([L])$ surjects onto $\mathcal{O}_Z$ for each l.c.i. 0-cycle Z of length 2 supported at p , then | L | gives an embedding at p .

Proof. Let M_p be the maximal ideal of the local ring $\mathcal{O}_{X,p}$: if $H^0([L])$ does not surject onto $\mathcal{O}_{X,p} / \mathcal{M}_p^2$, by our assumption , the image is 2-dimensional and intersects $\mathcal{M}_p/\mathcal{M}_p^2$ in a 1 -dimensional subspace W. Thus we obtain a contradiction by considering the length 2 cycle Z defined by $\mathcal{M}_p^2$ and by W .

Q.E.D.

Remark 1.10 The lemma does not hold already for a A_1 singularity . In fact , if $H^0(\mathcal{M}_p[L])$ does not surject onto $\mathcal{M}_p/\mathcal{M}_p^2$, then the image is contained in a 2- plane W in $\mathcal{M}_p/\mathcal{M}_p^2$ .Unfortunately W and $\mathcal{M}_p^2$ generate a length 2 , but not a l.c.i. cycle , because if the line W^v in the Zariski tangent space is tangent to X , then J_Z is not locally generated by two elements .

§ 2 PLURICANONICAL EMBEDDINGS OF SURFACES OF GENERAL TYPE

In this section k is an algebraically closed field of characteristic 0 and X is the canonical model of a surface of general type : thus X is a normal Gorenstein projective surface with ω_X ample , and if S is a minimal resolution of singularities of X , S is a minimal surface of general type.

To a singular point p of X there corresponds a divisor E on S , called a fundamental cycle , and consisting, with suitable multiplicities , of all the curves mapping down to p (hence these are all curves which have 0 intersection number with K) . The main property we want to mention here (cf. [Ar] for more details) is that there is a natural isomorphism (given by pull -back) between $\mathcal{O}_{X,p} / \mathcal{M}_p^2$ and $H^0 (\mathcal{O}_{2E}) \cong H^0 (\mathcal{O}_{2E} (mK))$, and therefore a pluricanonical system $| \omega_X^m |$ gives an embedding at p if and only if the sequence

(2.1.) $0 \longrightarrow H^0 ([mK -2E]) \longrightarrow H^0 ([mK]) \longrightarrow H^0 (\mathcal{O}_{2E} (mK)) \longrightarrow 0$

is exact .

Assume that $m > 1$: then $H^1 ([mK]) = 0$ (cf . [Bom]) , and the exactness of (2.1.) amounts to the vanishing $H^1 ([mK -2E]) = 0$.

Lemma 2.2. If E is a fundamental cycle on a minimal surface of general type S , then $H^1 ([mK -2E]) = 0$, provided $m > 3$, or $m=3$, $K^2 > 2$.

Proof. At page 188 of [Bom] (proof of theorem 3 , where E is though denoted Z) , it is shown that the desired vanishing holds if $H^0 ([(m -1) K -2E])$ is not zero , and one has moreover $m^2 K^2 > 9$, $m + K^2 > 4$.

We can therefore assume that $H^0 ([(m -1) K -2E]) = 0$.
Since also $H^2([(m -1) K -2E]) = 0$ (in fact the dual space is $H^0 ([(2-m)K + 2E])$, which is zero for $m > 2$, otherwise we would have an effective divisor with negative intersection number with K), the conclusion is that , by the Riemann-Roch formula , $1/2 (m-1)(m-2) K^2 -4 + \chi$ is non-positive.
Since $K^2 > 2$, $m > 2$, $\chi > 0$, the only possibility is that $m = K^2 = 3$, $\chi = 1$.
If $H^1 ([mK -2E])$ is non zero , recalling that $m = 3$, we have a non-trivial extension

(@) $0 \longrightarrow \mathcal{O}_S \longrightarrow E \longrightarrow \mathcal{O}_S (2K - 2E) \longrightarrow 0.$

We obtain immediately that $H^0 (E)$ has dimension 1 , whereas $c_1^2 (E) = 4$ and $c_2 (E) = 0$, hence E is numerically unstable (cf. [Bog] , [Rei]) and we have a Bogomolov destabilizing extension

(#) $0 \longrightarrow \mathcal{O}_S (M) \longrightarrow E \longrightarrow J_Z (D) \longrightarrow 0$,

where Z is a 0-cycle , and the divisor M - D is in the positive cone .

Recall also that M + D is linearly equivalent to 2K - 2E .
Therefore $K (M - D) > 0$,and $KM + KD = 2 K^2 = 6$, hence $K M > 3$,

while $K E < 3$.
As a consequence we get $H^0([-M]) = 0$: tensoring both exact sequences (@) and (#) by $\mathcal{O}_S(-M)$, we obtain that $H^0(E(-M))$ is at least 1-dimensional and is a subspace of $H^0([D])$, so that we may assume D is an effective divisor.

Recall though that by our assumption $H^0([M+D]) = 0$,hence $H^0([M])$ $= 0$ too. We noticed that $3 = K^2 < K M$, hence $H^0([K-M]) = 0$, and dually $H^2([M]) = 0$, so that the Riemann-Roch formula gives us that $1/2(M^2 - MK) + 1$ is a nonpositive number, i.e. $M^2 < KM - 1$.
We have $(M+D)^2 = 4(K-E)^2 = 4(K^2 + E^2) = 4 = D^2 + 2MD + M^2 < D^2 + 2MD + KM - 1$; since $KM + KD = 6$, we obtain

$$(2.3) \quad KD < 2MD + D^2 + 1.$$

<u>Claim 2.4.</u> $D = 0$.
<u>Proof of the claim.</u> By the Index theorem $3D^2 = K^2D^2$ is less than or equal to $(KD)^2$ which is in turn at most 4, by a previously obtained inequality, thus D^2 is at most 1.
Observe now that $c_2(E) = 0$, thus $\deg(Z) + MD = 0$, and MD is non positive. Looking at (2.3), since KD is non negative, we immediately obtain that $MD = 0 = \deg(Z)$. Again, (2.3) gives that KD is at most 1, hence $3D^2$ is bounded by $(KD)^2$ which is at most 1, hence D^2 is nonpositive. Once again (2.3) gives $KD = D^2 = 0$, and we conclude that $D = 0$ since the selfintersection form is strictly negative definite for the effective divisors orthogonal to K.

<u>Q.E.D.for the claim</u>

By the claim (we noticed also that $\deg Z = 0$), (#) reduces to an extension of the following form

$$0 \longrightarrow \mathcal{O}_S(2K-2E) \longrightarrow E \longrightarrow \mathcal{O}_S \longrightarrow 0.$$

Since the first term of the above sequence has no global sections, by our assumption, the above sequence is easily seen to give a splitting of (@), a contradiction.

<u>Q.E.D. for Lemma2.2</u>

<u>Corollary 2.5.</u> If X is the canonical model of a surface of general type, then the mth pluricanonical system $|\omega_X^{m}|$ gives an embedding of X whenever $m > 4$, or m=4, $K^2 > 1$, m=3 and $K^2 > 2$.

Proof. The proof follows theorem 1 of [Rei] , lemma 1.9. , and lemma 2.2..

Q.E.D.

REFERENCES.

[Ar] - Artin,M. : On isolated rational singularities of surfaces , *Amer. Jour. Math.* , 88, (1966) , 129 - 136.

[Bog] - Bogomolov, F.A. : Holomorphic tensors and vector bundles on projective varieties, (translated in) *Math.U.S.S.R. Izvestija*,13, (1979),499-555.

[Bom] - Bombieri,E.: Canonical models of surfaces of general type ,*Publ. Math.I.H.E.S.* ,42,(1973) , 171-219.

[Cat] - Catanese,F. : Canonical rings and "special" surfaces of general type, *Proc.of Symp. in Pure Math.*,46, (1987) , 175-194.

[Ek] - Ekedahl,T. : Letter to the author, october 1986 .

[G-H] - Griffiths,P.-Harris,J. : Residues and zero-cycles on algebraic varieties, *Ann. of Math.(2)* ,108 ,(1978) , 461-505.

[Rei] - Reider,I. : Vector bundles of rank 2 and linear systems on algebraic surfaces, *Ann. of Math.* , 127 , (1988), 309-316.

[Tju] - Tjurin,A.: Cycles,curves and vector bundles on an algebraic surface, *Duke Math. Jour.* , 54 ,(1987), 1-26.

AN OBSTRUCTION TO MOVING MULTIPLES OF SUBVARIETIES

Herbert Clemens
Mathematics Department., Univsity of Utah
Salt Lake City, Utah 84112, USA

§0. Introduction

In this paper, we consider the following situation. Let X be a reduced, irreducible complex analytic variety and let

$$Y \subseteq X$$

be a compact subvariety which is reduced, equidimensional of dimension q and connected, and suppose that X is smooth at each generic point of Y. When we say that *a multiple of* Y *moves in* X, we mean that we have a commutative diagram

$$\begin{array}{ccccc} Z & \xrightarrow{\subseteq} & W & \xrightarrow{\pi} & \Delta \\ {\scriptstyle i}\downarrow & & \downarrow{\scriptstyle f} & & \\ Y & \xrightarrow{\subseteq} & X & & \end{array} \tag{0.1}$$

where π is proper and flat onto the unit disc Δ with $Z = \pi^*(\{0\})$, W is reduced, irreducible and normal, and every component of Z dominates Y.

We wish to contruct obstructions to the existence of a diagram (0.1) in terms of the dominant map i and the higher order neighborhoods of Y in X. The idea is to get some obstructions which can be filtered so that the graded pieces do not depend on higher-order neighborhoods of Z in W.

We begin in §1 by geometrically considering "normal differential operators to Y" at "good points." In §2, we translate all this into the formalism of local cohomology and define the higher-order obstructions formally. Intuitively, if we have a deformation (0.1), then it gives r-th order "normal differential operators to Y" defined along Z for each positive integer r. By the formula for higher derivatives (see(1.2)), the "symbol" of this operator must be the r-th power of the symbol of the operator for r = 1. The obstructions measure whether this relation on symbols is possible. In §3, we compute the first two obstructions in the case in which X is a Cartier divisor on some variety P and a deformation (0.1) with X replaced by P is given. We see that these obstructions reduce in this case exactly to the condition that the image of Z under f remain inside X to appropriate order as Z moves in W.

The author would like to thank J. Kollár for many helpful discussions.

§1. "Fibering" a neighborhood

To frame the problem in geometric terms, we begin by supposing that, in addition to the assumptions made in §0, we assume that X is a projective variety. We let X' denote a small regular neighborhood of Y in X, and we wish to "fiber" X' over Y. To do this, we take a very ample linear system on projective space and let $|D|$ be the restriction of a generic q-dimensional linear sub-system to X. For each $y \in Y$, let B_y be the base locus of the hyperplane in $|D|$ consisting in divisors which pass through y. Then B_y meets Y transversely at y except for y lying in a divisor $F \subseteq Y$. By changing the q-plane $|D|$ in the original very ample system, F varies in a algebraic system whose only fixpoints are the singular points of Y and the singular points of X which lie on Y. If U is a sufficiently small open set in X' containing

$$Y' = Y - F,$$

then there is an analytic fibration

$$\rho: U \longrightarrow Y'$$

whose fibre at y is the unique *component* D_y of the unique $B_y \cap U$ containing y:

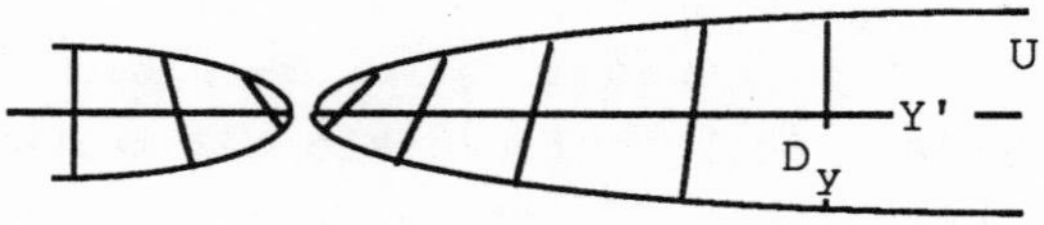

Suppose that the q-cycle m'Y moves in an analytic family Y_s in X. Then for fixed $y \in Y'$ and small s, the scheme $Y_s \cap D_y$ is finite of degree m', and, by the implicit function theorem, there is an analytic map x(t), $s = t^{m'}$, such that $x(t) \in Y_s \cap D_y$ for all t.

(1.1) **Lemma**: Let

$$f(t) = (x_1(t), \ldots, x_c(t))$$

be a non-trivial complex-analytic mapping from a disc Δ to the polydisc Δ^c which takes 0 to $(0, \ldots, 0)$. Let $\mathfrak{D}_1$ be the geometric fibre at $t = 0$ of the sheaf of differential operators on Δ, and let $\mathfrak{D}_c$ be the geometric fibre at $(x_1, \ldots, x_c) = (0, \ldots, 0)$ of the sheaf of differential operators on Δ^c, and let

$$f_*: \mathfrak{D}_1 \longrightarrow \mathfrak{D}_c$$

be the map induced by f. Then there is a non-trivial homogeneous first-order operator

$$a_1 \partial/\partial x_1 + \ldots + a_c \partial/\partial x_c$$

in the image of f_* all of whose powers lie in Gr(image f_*).

Proof: Let $D = \partial/\partial t$. Recall that $f_*(\partial^m/\partial t^m) =$

$$(1.2) \qquad \Sigma_r \sum_{I=(i_1,\ldots,i_r)} c_I \, (D^{i_1} x_{k_1}) \cdots (D^{i_r} x_{k_r}) \frac{\partial^r}{\partial x_{k_1} \cdots \partial x_{k_r}}$$

where the usual summation convention is used in the k's and I runs over all partitions of m into r positive integers. For notational simplicity, we do the rest of the proof in the case in which c = 2. We write

$$x(t) = a' t^m + \text{higher powers}$$
$$y(t) = b' t^n + \text{higher powers}$$

where, by linear change of coordinates, we can assume that $m < n$. Then $f_*(\partial^m/\partial t^m) = (\partial^m x/\partial t^m) \cdot \partial/\partial x$. Also, in the expression for $f_*(\partial^{rm}/\partial t^{rm})$, the coefficients of operators $\partial^s/\partial(x,y)^s$ for $s > r$ are products of s terms of the form $\partial^{m_i} x/\partial t^{m_i}$ or $\partial^{m_i} y/\partial t^{m_i}$ with

$\Sigma_i m_i = s$ and each $m_i > 0$. So some $m_i < m$, and the coefficient must be zero. Similarly, the coefficients of all operators $\partial^r/\partial(x,y)^r$ must be zero unless the coefficient has the form $(\partial^m x/\partial t^m)^r$. But this coefficient only occurs in front of the operator $\partial^r/\partial x^r$.

Note that if we move the q-plane $|D|$ in our original linear system, then, at least at smooth points of Y and X, the corresponding operators $a_1\partial/\partial x_1 + \ldots + a_c\partial/\partial x_c$ paste according to the rule for the normal bundle of Y in X.

§2. Moving a multiple of the submanifold

We begin by restating our hypotheses on $Y \subseteq X$. We assume only that X is a reduced and irreducible analytic variety, and that Y is a connected (reduced and equidimensional) compact subvariety of X. We assume that X is smooth at the generic point of each component of Y. We assume that "a multiple of Y moves in X," that is, we have the following commutative diagram:

(2.1)
$$\begin{array}{ccccc} Z & \xrightarrow{\subseteq} & W & \xrightarrow{\pi} & \Delta \\ i\downarrow & & \downarrow f & & \\ Y & \xrightarrow{\subseteq} & X & & \end{array}$$

where

1) W is normal, irreducible, of dimension q+1, and π is proper and flat with reduced fibres over a disc $\Delta = \{t \in \mathbf{C}: |t| < 1\}$.

2) the reduced analytic variety $Z = \pi^*(0)$ ideal-theoretically,

3) f is a generically finite morphism,

4) $i: Z \longrightarrow Y$ is generically finite on each component of Z.

Let $W^\wedge$ denote the formal completion of W along Z. Any global section of $\mathcal{O}_W/\mathcal{I}_Z{}^s$ induces an $\mathcal{O}_W$-module endomorphism of $\mathcal{O}_W/\mathcal{I}_Z{}^s$, and so the group

$$\mathrm{Ext}^1(\mathcal{O}_W/\mathcal{I}_Z{}^s,\ \mathcal{O}_W)$$

has the structure of an $H^0(\mathcal{O}_{W^\wedge})$-module. Let

$$\phi\colon \mathrm{Ext}^1(\mathcal{O}_W/\mathcal{I}_Z{}^s,\ \mathcal{O}_W) \longrightarrow \mathrm{Ext}^1(\mathcal{O}_W/\mathcal{I}_Z,\ \mathcal{O}_W)$$

be such that the natural morphism

$$\mathrm{Ext}^1(\mathcal{O}_W/\mathcal{I}_Z,\ \mathcal{O}_W) \longrightarrow \mathrm{Ext}^1(\mathcal{O}_W/\mathcal{I}_Z{}^s,\ \mathcal{O}_W)$$

composes with ϕ to give the identity on

$$\mathrm{Ext}^1(\mathcal{O}_W/\mathcal{I}_Z,\ \mathcal{O}_W) \approx H^0(N_{Z/W}{}^*) \approx \mathbf{C}.$$

We "differentiate" $f \in H^0(\mathcal{O}_{W^\wedge})$ by defining the action of $\xi \in \mathrm{Ext}^1(\mathcal{O}_W/\mathcal{I}_Z{}^s,\ \mathcal{O}_W)$ on f to be given by the formula

$$\phi(f\cdot\xi).$$

In the case in which Z is a smooth point on a curve W, this differentiation can be identified (non-canonically) with the action of

$$\Sigma_{r<s}\, a_r\, \partial^r/\partial t^r$$

on (formal) functions discussed in §1.

Let $m = \max\{m'\colon f^*\mathcal{I}_Y \to \mathcal{I}_Z{}^{m'}\}$ where the arrow is induced by the natural map $f^*\mathcal{O}_Y \to \mathcal{O}_Z$. Then, for $r < s$, we have morphisms

$$f^*(\mathcal{I}_Y{}^r/\mathcal{I}_Y{}^s) \longrightarrow \mathcal{I}_Z{}^{rm}/\mathcal{I}_Z{}^{sm}.$$

Let Z(k) denote a the k-th order neighborhood of Z, that is, the scheme with functions given by $\mathcal{O}_W/\mathcal{I}_Z{}^{k+1}$. Recall that that there is a double complexes of sheaves whose two filtrations have E_2-terms

$$\mathcal{E}xt_W{}^p(\mathcal{F},\ \mathcal{E}xt_W{}^q(\mathcal{O}_{Z(k)},\ \mathcal{O}_W))$$

and

$$\mathcal{E}xt_W{}^q(\mathcal{T}or^B{}_p(\mathcal{F},\mathcal{O}_{Z(k)}),\ \mathcal{O}_W)$$

respectively, and that, if $\mathcal{F} = \mathcal{I}_Z{}^{k'}/\mathcal{I}_Z{}^{k''}$, the only non-zero terms occurring at E_2 in either complex occur when $q = 1$. So,

$$\mathcal{E}xt_W^1(\mathcal{T}or^B_p(\mathcal{I}_Z^{k'}/\mathcal{I}_Z^{k''}, \mathcal{O}_{Z(k)}), \mathcal{O}_W) \approx$$

$$\mathcal{E}xt_W^p(\mathcal{I}_Z^{k'}/\mathcal{I}_Z^{k''}, \mathcal{E}xt_W^q(\mathcal{O}_{Z(k)}, \mathcal{O}_W)).$$

Let A be an open set in X and $Y' = A \cap Y$, and let $B = f^{-1}(A)$ and $Z' = B \cap Z$. We let $Y'(k)$ denote the k-th order neighborhood of Y' in A, and similarly for Z'. Using the above "$\mathcal{E}xt$-$\mathcal{T}or$" isomorphism and the natural maps

$$\mathrm{Ext}_B{}^i(f^*\mathcal{F}, \mathcal{G}) \longrightarrow \mathrm{Ext}_A{}^i(\mathcal{F}, f_*\mathcal{G}),$$

which is an isomorphism for $i = 0$, we construct the following diagram for $k \geq sm-rm$:

(2.2)

$$\begin{array}{ccc}
\mathrm{Ext}^1_\Delta(\mathcal{O}_\Delta/\mathcal{I}^{sm+1}_{\{0\}}, \mathcal{O}_\Delta) & = & \mathrm{Ext}^1_W(\mathcal{O}_W/\mathcal{I}^{sm+1}_Z, \mathcal{O}_W) \\
 & & \downarrow \\
 & & \mathrm{Ext}^1_B(\mathcal{O}_B/\mathcal{I}^{sm+1}_{Z'}, \mathcal{O}_B) \\
 & & \downarrow \\
 & & \mathrm{Ext}^1_B(\mathcal{I}^{rm}_{Z'}/\mathcal{I}^{sm+1}_{Z'}, \mathcal{O}_B) \\
 & & \updownarrow = \\
 & & \mathrm{Hom}_B(\mathcal{I}^{rm}_{Z'}/\mathcal{I}^{sm+1}_{Z'}, \mathcal{O}_{Z'(k)}) \\
 & & \updownarrow = \\
 & \swarrow \text{ (diagonal arrow)} & \mathrm{Hom}_B(\mathcal{I}^{rm}_{Z'}/\mathcal{I}^{sm+1}_{Z'}, \mathcal{I}^{k+rm-sm}_{Z'}/\mathcal{I}^{k+1}_{Z'}) \\
 & & \downarrow \\
\mathrm{Hom}_A(\mathcal{I}^r_{Y'}/\mathcal{I}^{s+1}_{Y'}, f_*\mathcal{I}^{k+rm-sm}_{Z'}/\mathcal{I}^{k+1}_{Z'}) & = & \mathrm{Hom}_B(f^*(\mathcal{I}^r_{Y'}/\mathcal{I}^{s+1}_{Y'}), \mathcal{I}^{k+rm-sm}_{Z'}/\mathcal{I}^{k+1}_{Z'}) \\
\downarrow & & \downarrow \\
\mathrm{Hom}_A(\mathcal{I}^r_{Y'}/\mathcal{I}^{s+1}_{Y'}, f_*\mathcal{O}_{Z'(k)}) & = & \mathrm{Hom}_B(f^*(\mathcal{I}^r_{Y'}/\mathcal{I}^{s+1}_{Y'}), \mathcal{O}_{Z'(k)})
\end{array}$$

If $r = s$ and $k = 0$, the map

$$\mathrm{Ext}_W{}^1(\mathcal{O}_W/\mathcal{I}_Z{}^{rm+1}, \mathcal{O}_W) \longrightarrow \mathrm{Hom}_Y(\mathcal{I}_Y{}^r/\mathcal{I}_Y{}^{r+1}, i_*\mathcal{O}_Z)$$

induced by the diagonal arrow above, will be called the *symbol map*, and we will denote it by σ. Notice that, if Y is a local complete intersection in X,

$$\mathrm{Hom}_Y(\mathcal{I}_Y{}^r/\mathcal{I}_Y{}^{r+1}, i_*\mathcal{O}_Z) = H^0(Z;\ i^*S^rN_{Y/X}).$$

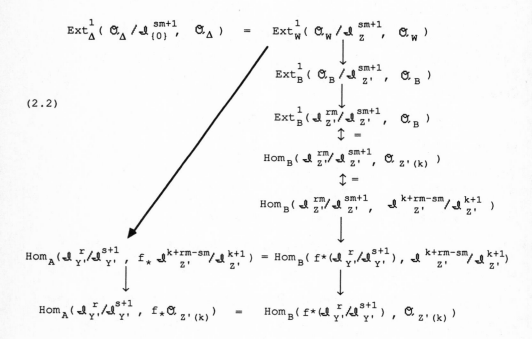

Since $\mathcal{I}_Z{}^k/\mathcal{I}_Z{}^{k+1} \approx \mathcal{O}_Z$ for all $k \geq 0$, the symbol can be computed from the diagonal arrow above using any $k \geq 0$ and $r = s$.

We next restrict this machinery to a single fibre D_y over a point $y \in Y'$, where Y' is a dense Zariski open subset of Y which lies entirely inside the smooth points of Y which are also smooth points of X, and around which there is a fibration of a neighborhood constructed as in §1. We also assume that Y' is chosen sufficiently small that f is finite in some neighborhood of each point of Y' and W is smooth at points of $f^{-1}(Y')$. Again, by making Y' smaller if necessary, we have that for $y \in Y'$ and for any $z \in f^{-1}(y)$, the component E_z of $f^{-1}(D_y)$ containing z is a disc transverse to some component of Z. Choose such a z in Z'', a component of Z for which $m = \max\{m' : f^*\mathcal{I}_Y \to \mathcal{I}_{Z''}{}^{m'}\}$. We have maps

$$H^1_{\{z\}}(E_z; \mathcal{O}_{E_z}) \xrightarrow{\approx} H^1_{\{z\}}(f^{-1}(D_y); \mathcal{O}_{f^{-1}(D_y)}) \xrightarrow{\approx} H^1_{\{0\}}(U; \mathcal{O}_U)$$

which are isomorphisms since the fibration π has reduced fibres by assumption.

We now examine the commutative diagram

$$\begin{array}{ccccc} \mathrm{Gr}_{rm} H^1_{\{0\}}(\Delta; \mathcal{O}_\Delta) & \longleftarrow & \mathrm{Gr}_{rm} H^0(\mathcal{H}^1_Z(W; \mathcal{O}_W)) & \xrightarrow{\sigma} & H^0(i^* S^r N_{Y/X}) \\ \updownarrow & & \downarrow & & \downarrow \\ \mathrm{Gr}_{rm} H^1_{\{z\}}(E_z; \mathcal{O}_{E_z}) & = & \mathcal{O}_{E_z} \otimes \mathrm{Gr}_{rm} \mathcal{H}^1_Z(W; \mathcal{O}_W) & \longrightarrow & \mathcal{O}_{E_z} \otimes i^*(S^r N_{Y/X}) \end{array}$$

and we interpret the top left and lower right groups as symbols of differential operators at 0 in Δ and at y in D_y respectively. We are then in the situation of Lemma (1.1). So the operators

$$\partial^{rm}/\partial t^{rm}$$

on Δ give sections of

$$\mathcal{O}_{E_z} \otimes i^*(S^r N_{Y/X})$$

which are just the powers of the section given by the image of

$\partial^m/\partial t^m$.

Now $i_*\mathcal{O}_Z$ is torsion-free by our assumption of generic finiteness, elements of $\mathrm{Hom}_Y(\mathcal{I}_Y{}^r/\mathcal{I}_Y{}^{r+1}, i_*\mathcal{O}_Z)$ which are r-th powers at generic points must be r-th powers everywhere. So we can conclude the following:

(2.3) **Theorem:** Suppose that a multiple of Y moves in X, that is, suppose we have a diagram (2.1) satisfying 1) - 4). Then there is a non-trivial section ρ of $\mathrm{Hom}_Y(\mathcal{I}_Y/\mathcal{I}_Y{}^2, i_*\mathcal{O}_Z)$ such that every positive power of ρ is the "symbol" of an element of

$$H^1_{\{0\}}(\Delta;\mathcal{O}_\Delta) \approx H^0(\mathcal{H}^1_Z(W;\mathcal{O}_W)).$$

(2.4) **Corollary:** Let $\phi\in \mathrm{Ext}_X{}^1(\mathcal{O}_X/\mathcal{I}_Y{}^r, \mathcal{I}_Y{}^r/\mathcal{I}_Y{}^{r+1})$ be the obstruction to splitting the sequence

$$0\longrightarrow \mathcal{I}_Y{}^r/\mathcal{I}_Y{}^{r+1}\longrightarrow \mathcal{O}_X/\mathcal{I}_Y{}^{r+1}\longrightarrow \mathcal{O}_X/\mathcal{I}_Y{}^r\longrightarrow 0.$$

Then, under the hypotheses of the theorem, there must be a non-trivial $\beta\in \mathrm{Hom}_Y(\mathcal{I}_Y/\mathcal{I}_Y{}^2, i_*\mathcal{O}_Z)$ such that, for

$$\beta^r\in \mathrm{Hom}_Y(\mathcal{I}_Y{}^r/\mathcal{I}_Y{}^{r+1}, i_*\mathcal{O}_Z)\approx \mathrm{Hom}_Y(\mathcal{I}_Y{}^r/\mathcal{I}_Y{}^{r+1}, i_*\mathcal{I}_Z{}^{rm}/\mathcal{I}_Z{}^{rm+1})$$
$$\subseteq \mathrm{Hom}_X(\mathcal{I}_Y{}^r/\mathcal{I}_Y{}^{r+1}, f_*\mathcal{O}_{Z(rm)}),$$

$\phi(\beta^r) = 0$ in $\mathrm{Ext}_X{}^1(\mathcal{O}_X/\mathcal{I}_Y{}^r, f_*\mathcal{O}_{Z(rm)})$.

Note: By filtering $f_*\mathcal{O}_{Z(rm)}$ we obtain graded quotients of the form $i_*\mathcal{I}_Z{}^n/\mathcal{I}_Z{}^{n+1}\approx i_*\mathcal{O}_Z$, so a sequence of obstructions which take their values in $\mathrm{Ext}_X{}^1(\mathcal{O}_X/\mathcal{I}_Y{}^r, i_*\mathcal{O}_Z)$.

Proof: Apply the functor

$$R\mathrm{Hom}_W(\quad , \mathcal{O}_{Z(k)})$$

to the sequence

$$0 \longrightarrow \mathcal{I}_Z^{rm}/\mathcal{I}_Z^{rm+1} \longrightarrow \mathcal{O}_W/\mathcal{I}_Z^{rm+1} \longrightarrow \mathcal{O}_W/\mathcal{I}_Z^{rm} \longrightarrow 0$$

to get a map

$$\mathrm{Hom}_W(\mathcal{I}_Z^{rm}/\mathcal{I}_Z^{rm+1}, \mathcal{O}_{Z(k)}) \longrightarrow \mathrm{Ext}_W^1(\mathcal{O}_W/\mathcal{I}_Z^{rm}, \mathcal{O}_{Z(k)}).$$

Let δ_r be the element of

$$\mathrm{Ext}_W^1(\mathcal{O}_W/\mathcal{I}_Z^{rm+1}, \mathcal{O}_W)$$

which maps to the operator $\partial^{2m}/\partial t^{2m}$ on U, and let ε_2 be the corresponding element in

$$\mathrm{Hom}_W(\mathcal{I}_Z^{rm}/\mathcal{I}_Z^{rm+1}, \mathcal{O}_{Z(k)}).$$

Since this element comes from an element of

$$\mathrm{Hom}_W(\mathcal{O}_W/\mathcal{I}_Z^{rm+1}, \mathcal{O}_{Z(k)})$$

as long as $k \geq rm$, it must go to 0 in

$$\mathrm{Ext}_W^1(\mathcal{O}_W/\mathcal{I}_Z^{rm}, \mathcal{O}_{Z(k)}).$$

Let β_r be the image of ε_r in

$$\mathrm{Hom}_X(\mathcal{I}_Y^r/\mathcal{I}_Y^{r+1}, f_*\mathcal{I}_Z^{rm}/\mathcal{I}_Z^{rm+1})) \subseteq \mathrm{Hom}_X(\mathcal{I}_Y^r/\mathcal{I}_Y^{r+1}, f_*\mathcal{O}_{Z(rm)}).$$

Then β_r must go to zero in

$$\mathrm{Ext}_X^1(\mathcal{O}_X/\mathcal{I}_Y^r, f_*\mathcal{O}_{Z(rm)})),$$

since the diagram

$$\begin{array}{ccc}
\mathrm{Hom}_B(\mathcal{I}_{Z'}^{rm}/\mathcal{I}_{Z'}^{rm+1}, \mathcal{O}_{Z'(k)}) & \longrightarrow & \mathrm{Ext}_B^1(\mathcal{O}_B/\mathcal{I}_{Z'}^{rm}, \mathcal{O}_{Z'(k)}) \\
\downarrow & & \downarrow \\
\mathrm{Hom}_A(\mathcal{I}_{Y'}^{r}/\mathcal{I}_{Y'}^{r+1}, f_*\mathcal{O}_{Z'(k)}) & \longrightarrow & \mathrm{Ext}_A^1(\mathcal{O}_A/\mathcal{I}_{Y'}^{r}, f_*\mathcal{O}_{Z'(k)})
\end{array}$$

is commutative. So β_r must go to zero in

$$\mathrm{Ext}_X^1(\mathcal{O}_X/\mathcal{I}_Y^r, f_*\mathcal{O}_{Z(rm)}).$$

But the above theorem says that β_r in

$$\mathrm{Hom}_Y(\mathcal{I}_Y^r/\mathcal{I}_Y^{r+1}, i_*\mathcal{I}_Z^{rm}/\mathcal{I}_Z^{rm+1})) \approx \mathrm{Hom}_Y(\mathcal{I}_Y^r/\mathcal{I}_Y^{r+1}, i_*\mathcal{O}_Z))$$

must be the r-th power of a non-trivial section of

$$\mathrm{Hom}_Y(\mathcal{I}_Y/\mathcal{I}_Y^2, i_*\mathcal{O}_Z)).$$

When X and Y are both smooth, we have the identification $\mathrm{Hom}_Y(\mathcal{I}_Y{}^r/\mathcal{I}_Y{}^{r+1}, i_*\mathcal{O}_Z)) = \mathrm{Hom}_Z(i^*(\mathcal{I}_Y{}^r/\mathcal{I}_Y{}^{r+1}), \mathcal{O}_Z)) =$ $H^0(i^*(S^rN_{Y/X}))$, and β_r must be the r-th power of the section of $H^0(i^*N_{Y/X})$ which is the symbol of $\partial^m/\partial t^m$.

§3. Computing the obstruction

Suppose now that we have a (reduced) connected equidimensional subvariety

$$Y \subseteq P$$

and are given a motion

(3.1)
$$\begin{array}{ccccc} Z & \xrightarrow{\subseteq} & W & \xrightarrow{\pi} & \Delta \\ {\scriptstyle i}\downarrow & & \downarrow{\scriptstyle f} & & \\ Y & \xrightarrow{\subseteq} & P & & \end{array}$$

of a multiple of Y satisfying 1) - 4) listed at the beginning of §2. As in §2, define

$$m = \max\{m': f^*\mathcal{I}_Y \subseteq \mathcal{I}_Z{}^{m'}\}.$$

Let Z# be some connected union of components of Z for which

$$f^*\mathcal{I}_Y = \mathcal{I}_Z{}^m$$

at the generic points of Z# and let Z^ denote the union of the other components of Z.

Next suppose that we have a (reduced) divisor

$$X \subseteq P$$

defined by the section

$$F$$

of a line bundle $\mathcal{L}$. Also assume that X is smooth at the generic point of each component of Y. Now suppose that

(3.2)
$$F\bullet f \in \Gamma(\mathcal{I}_{Z\#}{}^{2m}\mathcal{L} \cap \mathcal{I}_{Z\wedge}{}^{2m+1}\mathcal{L}),$$

an assumption which implies that Y and the "first non-trivial order" deformation of a multiple of Y lie in X. We will show that the vanishing of the obstruction in Corollary (2.4) is equivalent to the condition that F•f vanish to order 2m+1 along Z.

The non-trivial element

$$\beta \in \mathrm{Hom}_P(\mathcal{I}_Y/\mathcal{I}_Y^2, i_*\mathcal{O}_Z)$$

given by this motion as in §2 has the property that

$$\beta^2 \in \mathrm{Hom}_P(\mathcal{I}_Y^2/\mathcal{I}_Y^3, f_*\mathcal{O}_{Z(2m)})$$

is the image of an element

$$\gamma \in \mathrm{Hom}_P(\mathcal{O}_P/\mathcal{I}_Y^3, f_*\mathcal{O}_{Z(2m)})$$

under the natural restriction map. Notice that γ must be surjective at each generic point of Z# in order to have restriction β^2. Furthermore, γ comes from

$$\gamma' \in \mathrm{Hom}_W(\mathcal{O}_W/\mathcal{I}_Z^{2m+1}, \mathcal{O}_{Z(2m)}).$$

The assumption (3.2) on F•f tells us that

$$\gamma'(F\bullet f) \in \mathrm{Hom}_W(f^*\mathcal{L}^{-1}/\mathcal{I}_Z^{2m+1}f^*\mathcal{L}^{-1}, \mathcal{O}_{Z(2m)})$$

is actually an element of

$$\mathrm{Hom}_W(f^*\mathcal{L}^{-1}/\mathcal{I}_Z^{2m+1}f^*\mathcal{L}^{-1}, \mathcal{I}_Z^{2m}/\mathcal{I}_Z^{2m+1})$$

$$\subseteq \mathrm{Hom}_W(f^*\mathcal{L}^{-1}/\mathcal{I}_Z^{2m+1}f^*\mathcal{L}^{-1}, \mathcal{O}_{Z(2m)}).$$

To relate this situation to the computation in §2 for the obstruction to moving a multiple of Y in X, we let

$$0 \longrightarrow f_*\mathcal{O}_{Z(2m)} \longrightarrow \mathcal{J}^{\cdot}$$

be an injective resolution of $f_*\mathcal{O}_{Z(2m)}$ as an $\mathcal{O}_P$-module and apply the functor

$$\mathcal{H}om_P(\ \ ,\mathcal{J}^{\cdot})$$

to the diagram below, where $\mathcal{O} = \mathcal{O}_P$ and $\mathcal{I}_Y = \mathcal{I}_{Y/P}$:

$$\begin{array}{ccccccccc}
 & & 0 & & 0 & & 0 & & \\
 & & \uparrow & & \uparrow & & \uparrow & & \\
0 & \leftarrow & \mathcal{O}_X/\mathcal{I}_{Y/X}{}^2 & \leftarrow & \mathcal{O}_X/\mathcal{I}_{Y/X}{}^3 & \leftarrow & \mathcal{I}_{Y/X}{}^2/\mathcal{I}_{Y/X}{}^3 & \leftarrow & 0 \\
 & & \uparrow & & \uparrow & & \uparrow & & \\
0 & \leftarrow & \mathcal{O}/\mathcal{I}_Y{}^2 & \leftarrow & \mathcal{O}/\mathcal{I}_Y{}^3 & \leftarrow & \mathcal{I}_Y{}^2/\mathcal{I}_Y{}^3 & \leftarrow & 0 \\
 & & \uparrow \cdot F & & \uparrow \cdot F & & \uparrow \cdot F & & \\
0 & \leftarrow & \mathcal{L}^{-1}/\mathcal{I}_Y\mathcal{L}^{-1} & \leftarrow & \mathcal{L}^{-1}/\mathcal{I}_Y{}^2\mathcal{L}^{-1} & \leftarrow & \mathcal{I}_Y\mathcal{L}^{-1}/\mathcal{I}_Y{}^2\mathcal{L}^{-1} & \leftarrow & 0 \\
 & & \uparrow & & \uparrow & & \uparrow & & \\
 & & 0 & & 0 & & 0 & &
\end{array}$$

Call the resulting diagram **D**. From the left-hand column of **D**, we obtain the exact sequence

$$\mathrm{Hom}_P(\mathcal{O}_P/\mathcal{I}_Y{}^2, f_*\mathcal{O}_{Z(2m)}) \to \mathrm{Hom}_P(\mathcal{L}^{-1}/\mathcal{I}_Y{}^2\mathcal{L}^{-1}, f_*\mathcal{O}_{Z(2m)})$$
$$\to \mathrm{Ext}_P{}^1(\mathcal{O}_X/\mathcal{I}_{Y/X}{}^2, f_*\mathcal{O}_{Z(2m)}) \to \mathrm{Ext}_P{}^1(\mathcal{O}_P/\mathcal{I}_Y{}^2, f_*\mathcal{O}_{Z(2m)}).$$

The assumption (3.2) on $F\bullet f$ easily implies that the first map in the above sequence is the zero map.

Since β^2 goes to 0 in $\mathrm{Ext}_P{}^1(\mathcal{O}_P/\mathcal{I}_Y{}^r, f_*\mathcal{O}_{Z(2m)})$, it must, by diagram **D**, come from an element

$$\xi \in \mathrm{Hom}_P(\mathcal{L}^{-1}/\mathcal{I}_Y\mathcal{L}^{-1}, f_*\mathcal{O}_{Z(2m)})$$

which in turn must be the image of the element

$$\gamma \in \mathrm{Hom}_P(\mathcal{O}_P/\mathcal{I}_Y{}^3, f_*\mathcal{O}_{Z(2m)})$$

defined above. So if the obstruction defined in the Corollary (2.4) is to vanish for Y in X, it is necessary that

$$\gamma'(F\bullet f) \in \mathrm{Hom}_W(\mathcal{L}^{-1}/\mathcal{I}_Z{}^{2m+1}\mathcal{L}^{-1}, \mathcal{I}_Z{}^{2m}/\mathcal{I}_Z{}^{2m+1})$$

is actually zero. Since γ' is surjective at each generic point of Z#, we conclude that the obstruction vanishes if and only if

(3.3) $$F\cdot f \in \Gamma(\mathcal{I}_Z{}^{2m+1}\mathcal{L}).$$

For example, consider the case in which $P = \mathbf{P}^n$,

$$W = Z \times \Delta,$$

and

$$f: W \longrightarrow \mathbf{P}^n$$

is given by

$$\lambda = (\lambda_0, \ldots, \lambda_n),\ \lambda_i \in H^0(f^*\mathcal{O}_{\mathbf{P}^n}(1)).$$

Then by developing $F\cdot f$ in powers of t, the condition (3.3) is equivalent to saying that f gives a solution in the variables

$$\gamma_i \in H^0(i^*\mathcal{O}_{\mathbf{P}^4}(1))$$

to the equation

(3.4) $$\Sigma(\partial^2F/\partial X_i\partial X_j)\cdot\beta_i\beta_j + \Sigma(\partial F/\partial X_i)\cdot\gamma_i$$

where $\beta_i = \partial^m\lambda_i/\partial t^m$. In some examples, such as the one below, the condition (3.3) essentially reduces to (3.4) for *all* W.

To see an example of this phenomenon, we will work out in detail the example that motivated this research. We let X be the quintic hypersurface in $\mathbf{P}^4$ defined by the polynomial

(3.5) $F(X_0, \ldots, X_4) =$

$$X_0\cdot h(X_0, \ldots, X_4) + c(X_2, X_3, X_4)\cdot d(X_0, \ldots, X_4) + X_1{}^2\cdot b(X_0, \ldots, X_4),$$

where $X_0 = X_1 = c = 0$ defines a plane smooth conic Y and all other forms are, for the moment, generic of appropriate degree in the variables indicated. Geometrically the cone over Y in the hyperplane $X_0 = 0$ is tangent to X along Y. We parametrize Y via

$$\alpha = \{\alpha_i(Y_0, Y_1)\}_{i=0,\ldots,4},$$

where the α_i are of degree two in the homogeneous coordinates (Y_0, Y_1) of $Y \approx \mathbf{P}^1$. That X is smooth at points of Y is achieved by choosing h such that it does not vanish at the six zeros $\{p_j\}$ of

d on Y.

This case suggests itself for study because of interest in finding a rational curve on a threefold with trivial canonical bundle which neither contracts nor has a multiple which moves. In fact, using global analytic methods, it can be shown[1] that for generic F of the form (3.5), no multiple of C moves. The obstruction in §2 for the case $r = 2$ does not give this stronger result. However it does show that C itself does not move for generic F, and it restricts possible "higher-order" motions of a multiple of C to motions of multiples of a certain double cover of C branched at specified points.

Since X is tangent to the cone given by

$$X_o = c = 0$$

and $K_X = \mathcal{O}_X$, one easily sees from the adjunction formula and the decomposability of vector bundles on $\mathbf{P}^1$ that

$$N_{Y/X} \approx \mathcal{O}_Y(2) + \mathcal{O}_Y(-4).$$

To compute $H^0(i^*N_{Y/X})$ for any finite morphism

$$i\colon Z \longrightarrow Y,$$

we apply i^*, which is exact on sequences of vector bundles, to the following diagram with exact rows and columns:

$$\begin{array}{ccccccccc}
 & & 0 & & 0 & & & & \\
 & & \downarrow & & \downarrow & & & & \\
 & & \mathcal{O}_Y(1)^2 & = & \mathcal{O}_Y(1)^2 & & & & \\
 & & \downarrow & & \downarrow{\scriptstyle (\partial\alpha_i/\partial Y_k)\cdot} & & & & \\
0 & \longrightarrow & \mathcal{K} & \longrightarrow & \mathcal{O}_{\mathbf{P}^4}(1)^5\big|_Y & \xrightarrow{\Sigma(\partial F/\partial X_i)\cdot} & \mathcal{O}_{\mathbf{P}^4}(5)\big|_Y & \longrightarrow & 0 \\
 & & \downarrow & & \downarrow & & \updownarrow = & & \\
0 & \longrightarrow & N_{Y/X} & \longrightarrow & N_{Y/\mathbf{P}^4} & \longrightarrow & N_{X/\mathbf{P}^4}\big|_Y & \longrightarrow & 0 \\
 & & \downarrow & & \downarrow & & & & \\
 & & 0 & & 0 & & & &
\end{array}$$

Now one computes that $H^0(i^*\mathcal{K})$ consists of (β_i) such that

$$h(\lambda)\cdot\beta_0 + d(\lambda)[(\partial c/\partial X_2)(\lambda)\cdot\beta_2 + \ldots + (\partial c/\partial X_4)(\lambda)\cdot\beta_4] = 0.$$

Notice that any section (β_i) of $H^0(i^*\mathcal{K})$ has the property that $\beta_0 = 0$, since $h(\lambda)$ does not vanish at any of the six points p_j. Then

$$(\partial c/\partial X_2)(\lambda)\cdot\beta_2 + \ldots + (\partial c/\partial X_4)(\lambda)\cdot\beta_4 = 0$$

which means by the diagram just above that, if $\beta_1 = 0$, the section goes to zero in $i^*N_{Y/X}$. Thus the sections of $i^*N_{Y/X}$ are given exactly by the 5-tuples

$$(0, \beta_1, 0, 0, 0)$$

where $\beta_1 \in H^0(i^*\mathcal{O}_{\mathbf{P}^4}(1))$ is arbitrary.

Next suppose that a multiple of Y moves in $\mathbf{P}^4$. Then we must have a diagram (3.1). Suppose further that

$$F\bullet f \in \Gamma(\mathcal{I}_{Z\#}{}^{rm}(5) \cap \mathcal{I}_{Z^\wedge}{}^{rm+1}(5)).$$

Here

$$f: W \longrightarrow \mathbf{P}^4$$

is given by sections

$$\lambda = \{\lambda_i\}_{i=0,\ldots,4}$$

of $f^*\mathcal{O}_{\mathbf{P}^4}(1)$. So, by our calculations earlier in this section, if the obstruction of §2 for r = 2 is to vanish in this case, we have a non-trivial solution to the equation

$$0 \equiv h(\lambda)\cdot\lambda_0 + c(\lambda)\cdot d(\lambda) + b(\lambda)\cdot\lambda_1{}^2 \mod t^{2m+1}$$

where the λ_i are sections of $f^*\mathcal{O}_{\mathbf{P}^4}(1)$. Notice that we can write

$$\lambda_0 = t^m\cdot\beta_0, \quad \lambda_1 = t^m\cdot\beta_1, \quad \text{and} \quad c(\lambda) = t^m\cdot\varepsilon$$

for β_i sections of $f^*\mathcal{O}_{\mathbf{P}^4}(1)$, ε a section of $f^*\mathcal{O}_{\mathbf{P}^4}(2)$. So we must solve

$$0 \equiv h(\lambda)\cdot\beta_0 + \varepsilon\cdot d(\lambda) + t^m\cdot b(\lambda)\cdot\beta_1{}^2 \mod t^{m+1}.$$

In particular, we must have

$$0 \equiv h(\lambda)\cdot\beta_0 + \varepsilon\cdot d(\lambda) \mod t^m$$

so that

$$\beta_0 = t^m\cdot\gamma_0 \quad \text{and} \quad \varepsilon = t^m\cdot\sigma.$$

Then we must solve

$$0 \equiv h(\lambda)\cdot\gamma_0 + \sigma\cdot d(\lambda) + b(\lambda)\cdot\beta_1{}^2 \mod t.$$

So, we must solve

$$0 = h(\lambda)\cdot\gamma_0 + \sigma\cdot d(\lambda) + b(\lambda)\cdot\beta_1{}^2$$

in $i^*\mathcal{O}_{\mathbf{P}^2}(4)$ for α_i and β_1 sections of $i^*\mathcal{O}_{\mathbf{P}^2}(1)$ and σ a section of $i^*\mathcal{O}_{\mathbf{P}^2}(2)$ for some branched cover

$$i: Z\# \longrightarrow Y.$$

(Compare with (3.2).) The solution must be non-trivial in that $\beta_1 \neq 0$. So, for example, if $b(X_0,\ldots,X_4)$ is chosen to vanish at two of the six points p_j and also a set B consisting in four of the eight zeros of h along Y, then the map

$$i: Z\# \longrightarrow Y$$

must have even-order branching above each of the four points of B. In particular, Y itself cannot move in X. Other choices of $B|_Y$ force other branching configurations.

Reference

1. Clemens, H., "The infinitesimal Abel-Jacobi mapping and moving the $\mathcal{O}(2)+\mathcal{O}(-4)$ curve." Preprint, Univ. of Utah, 1988.

HALF-CANONICAL SURFACES IN $\mathbb{P}_4$

Wolfram Decker
Fachbereich Mathematik
Erwin-Schrödinger-Straße
D-6750 Kaiserslautern
Federal Rep. of Germany

Thomas Peternell
Mathematisches Institut
Postfach 10 12 51
D-8580 Bayreuth
Federal Rep. of Germany

Joseph le Potier
UER Mathématiques
Université de Paris VII
Place Jussieu
F-75230 Paris
France

Michael Schneider
Mathematisches Institut
Postfach 10 12 51
D-8580 Bayreuth
Federal Rep. of Germany

0. Introduction

There is a well known correspondence between subcanonical surfaces in $\mathbb{P}_4$ and rank-2 vector bundles on $\mathbb{P}_4$. The essentially only known rank-2 bundle on $\mathbb{P}_4$ - the Horrocks-Mumford bundle - corresponds to an abelian surface in $\mathbb{P}_4$ [HM].

One possibility to find more 2-bundles would be to classify smooth surfaces in $\mathbb{P}_4$ with $K_X = \mathcal{O}_X(a)$, a being a small positive integer. For a = 1 this has been done by Ballico and Chiantini [BC]. In this case X is necessarily a complete intersection.

In this paper we investigate half-canonical surfaces in $\mathbb{P}_4$ i.e. $K_X = \mathcal{O}_X(2)$. There are two types of non-degenerate complete inter- section surfaces i) an intersection of a quadric and a quintic hypersurface, ii) an intersection of a cubic and a quartic hyper- surface.

A smooth section in H(1), H the Horrocks-Mumford bundle, has as zero locus a half-canonical surface of degree 16. We show that any half-canonical surface of degree 16 in $\mathbb{P}_4$ is of this type.

We determine also the only other possibilities for the degree of half-canonical surfaces in $\mathbb{P}_4$ - deg X = 18 or deg X = 22.

Moreover the irregularity of these surfaces can be bounded by

$$1 \le q \le 4 \quad \text{resp.} \quad 5 \le q \le 9.$$

The existence of these surfaces would lead to stable 2-bundles on $\mathbb{P}_4$ whose monads can be described. In this way we can exclude the case q = 1. In view of the "conjecture" [02], [OV] that the irregularity of smooth surfaces $X \subset \mathbb{P}_4$ is bounded by 2 one would expect that there are only the three classes of half-canonical surfaces in $\mathbb{P}_4$ mentioned above.

The paper is organized as follows. In section 1 we determine the invariants (degree, $\chi(\mathcal{O}_X)$, sectional genus, ...) of smooth half-canonical surfaces in $\mathbb{P}_4$ and show that they are complete intersections if their degree is less than 16.
The rank-2 vector bundle belonging to a smooth half-canonical surface is stable precisely if X is not contained in a cubic threefold. This leads us to study in section 2 smooth surfaces on cubic hypersurfaces in $\mathbb{P}_4$. This is used to classify all half-canonical surfaces of degree 16 in $\mathbb{P}_4$. An essential tool here is the uniqueness of the Horrocks-Mumford bundle [DS].
In section 3 we estimate the irregularity of half-canonical surfaces of degree 18 and 22. This is achieved by restricting the associated stable vector bundle to a suitable plane, using Barth's restriction theorem [Ba].
In section 4 we describe the vector bundles belonging to surfaces of degree 18 or 22 by monads. In this way the case of irregularity $q = 1$ can be eliminated.
In section 5 we discuss some more properties of half-canonical surfaces in $\mathbb{P}_4$. In particular we show that the Albanese map has mostly two-dimensional image. In the final section we show $q(X) = h^1(X, N^\vee_{X/\mathbb{P}_4})$ and $h^1(X, S^\nu(N^\vee_{X/\mathbb{P}_4})) = 0$ for $\nu \geq 2$ for an arbitrary smooth surface in $\mathbb{P}_4$. We hope that these results will be useful for future work on this question.
To summarize some of our results we give the following table. Here $d = \deg(X)$, g is the sectional genus, q is the irregularity of X, $\chi = \chi(\mathcal{O}_X)$

d	χ	g	q	surface
10	15	16	0	$V_2 \cap V_5$
12	16	19	0	$V_3 \cap V_4$
16	16	25	0	section in H(1)
18	15	28	$2 \leq q \leq 4$	?
22	11	34	$5 \leq q \leq 9$	?

1. Invariants of half-canonical surfaces in $\mathbb{P}_4$

Let $X \subset \mathbb{P}_4$ be a smooth surface with $K_X = \mathcal{O}_X(2)$ and let E be the corresponding rank-2 vector bundle on $\mathbb{P}_4$ (c.f. [OSS]). There is an exact sequence

(1) $$0 \to \mathcal{O} \to E \to \mathcal{J}_X(7) \to 0.$$

Moreover we have

$$c_2(E) = \deg(X) = d.$$

Since $X \subset \mathbb{P}_4$ is smooth, one gets from the normal bundle sequence the well-known relation

$$d^2 - 10d - 5c_1(K_X)c_1(\mathcal{O}_X(1)) - 2c_1(K_X)^2 + 12\chi(\mathcal{O}_X) = 0.$$

In our case $K_X = \mathcal{O}_X(2)$ this becomes

(2) $$d(d-28) + 12\chi(\mathcal{O}_X) = 0.$$

In particular we have

(3) $$d(d-4) \equiv 0 \pmod{12}$$

and therefore d is even.

Proposition 1.1. Let $X \subset \mathbb{P}_4$ be a smooth half-canonical surface of degree d. Then

$$d \leq 22.$$

Proof. X is of general type and therefore we have $c_2(X) > 0$. From

$$0 \to T_X \to T_{\mathbb{P}_4}|X \to N_{X/\mathbb{P}_4} \to 0$$

we obtain

$$c_2(X) = c_2(\mathbb{P}_4)|X - c_2(N) - c_1(X)c_1(N).$$

Inserting $c_2(N) = d^2, c_1(N) = 7h_X, c_1(X) = -2h_X$ and $c_2(\mathbb{P}_4)|X = 10h_X^2$ we get

$$c_2(X) = (24-d)d.$$

Here of course $h_X = c_1(\mathcal{O}_X(1))$. Therefore it follows $d < 24$ and because $d \equiv 0 \pmod 2$ we finally get $d \leq 22$.

Proposition 1.2. Every smooth half-canonical surface $X \subset \mathbb{P}_4$ of degree $d \leq 14$ is a complete intersection.

Proof. Let E be the corresponding vector bundle as in (1). We distinguish two cases.

Case 1: $c_1(E)^2 - 4c_2(E) \leq 0$.

Since $c_1(E) = 7$ and $c_2(E) = d$ we get immediately $d \geq 14$. But $d = 14$ is impossible by formula (3).

Case 2: $c_1(E)^2 - 4c_2(E) > 0$.

In this case E is unstable by Barth [Ba]. Now we use theorem 4.2 of [HS] to conclude that E splits into a direct sum of line bundles provided $c_2(E) = d < 15$. Therefore X is a complete intersection.

Remark. The non-degenerate half-canonical complete intersection surfaces in $\mathbb{P}_4$ are precisely the surfaces

$$X = V_2 \cap V_5$$

and

$$X = V_3 \cap V_4 .$$

The remaining cases for the degree are

$$d = 16,\ 18,\ 20,\ 22.$$

By formula (3) the case d = 20 is impossible.
Next we estimate the geometric genus $p_g = p_g(X) = h^0(X, K_X)$.

Proposition 1.3. Let $X \subset \mathbb{P}_4$ be a smooth half-canonical surface which is not a complete intersection. Then we have

$$p_g(X) \geq 15.$$

Proof. Consider the exact sequence

$$0 \to H^0(\mathbb{P}_4, \mathcal{I}_X(2)) \to H^0(\mathbb{P}_4, \mathcal{O}_{\mathbb{P}_4}(2)) \to H^0(X, \mathcal{O}_X(2)) \to \ldots$$

If $H^0(\mathbb{P}_4, \mathcal{I}_X(2)) \neq 0$, X is contained in a quadric and therefore (cf. [O1]) X is a complete intersection.
Hence $H^0(\mathbb{P}_4, \mathcal{I}_X(2)) = 0$ and we obtain

$$p_g(X) = h^0(X, K_X) = h^0(\mathcal{O}_X(2)) \geq h^0(\mathbb{P}_4, \mathcal{O}_{\mathbb{P}_4}(2)) = 15.$$

If we denote by g the sectional genus of X (i.e. g is the genus of a smooth hyperplane section of X) we see from the adjunction formula

$$g = \frac{3d+2}{2}.$$

We summarize our knowledge on the invariants in the table

d	$\chi(\mathcal{O}_X)$	g	p_g	q
10	15	16	14	0
12	16	19	15	0
16	16	25	≥ 15	≥ 0
18	15	28	≥ 15	≥ 1
22	11	34	≥ 15	≥ 5

It is an open problem, whether the irregularity of smooth surfaces in $\mathbb{P}_4$ is bounded (for instance by 2), [O2],[OV].
We will show lateron that in case d = 16 we have q = 0 and that for d = 18 resp. d = 22 there are inequalities

$$2 \le q \le 4 \quad \text{resp.} \quad 5 \le q \le 9.$$

2. Surfaces on cubic threefolds

Let $X \subset \mathbb{P}_4$ be a smooth half-canonical surface and let E be the associated rank-2 vector bundle. The bundle E is stable precisely if X is not contained in a hypersurface of degree ≤ 3. Therefore we study first smooth surfaces on cubic threefolds V. For the case that V has only finitely many singularities see [A].

Proposition 2.1. Let $X \subset \mathbb{P}_4$ be a smooth surface of degree d and let $V = V(F)$ be a cubic hypersurface in $\mathbb{P}_4$ such that $X \cap \mathrm{Sing}(V)$ is finite. Then - after a suitable choice of coordinates -

$$V(\frac{\partial F}{\partial X_0}, \frac{\partial F}{\partial X_1}) \cap X$$

is finite.

Proof. Let $\hat{X}$ be the cone over X in $\mathbb{C}^5 \setminus \{0\}$ and set

$$\Sigma = \{(x,\lambda) \in \hat{X} \times L_{inj}(\mathbb{C}^2,\mathbb{C}^5) : d_xF \circ \lambda = 0\}.$$

Here $L_{inj}(\mathbb{C}^m,\mathbb{C}^n)$ denotes the set of injective linear maps from $\mathbb{C}^m$ into $\mathbb{C}^n$. Consider the projections

$$\begin{array}{ccc} \Sigma & \xrightarrow{q} & L_{inj}(\mathbb{C}^2,\mathbb{C}^5) \\ p \downarrow & & \\ \hat{X} & & \end{array}$$

Obviously

$$p^{-1}(X) \cong \begin{cases} L_{inj}(\mathbb{C}^2,\mathbb{C}^4) & \text{for } x \in \hat{X} \setminus \mathrm{Sing}(V) \\ L_{inj}(\mathbb{C}^2,\mathbb{C}^5) & \text{for } x \in \hat{X} \cap \mathrm{Sing}(V). \end{cases}$$

Hence

$$\dim \Sigma = 11$$

and therefore the general fibre of q is either empty or one-dimensional. Pick an element $\lambda \in L_{inj}(\mathbb{C}^2,\mathbb{C}^5)$ such that

$$\dim \{x \in \hat{X} : d_X F \circ \lambda = 0\} = 1$$

or

$$\{x \in \hat{X} : d_X F \circ \lambda = 0\} = \emptyset .$$

This implies that we can find coordinates such that

$$\dim(V(\frac{\partial F}{\partial X_o}, \frac{\partial F}{\partial X_1}) \cap X) \leq 0.$$

This proves our proposition.

Corollary 2.2.

$$\mu = \#(\mathrm{Sing}V \cap X) \leq 4d.$$

Proof. $\mathrm{Sing}(V) \cap X \subset V(\frac{\partial F}{\partial X_o}, \frac{\partial F}{\partial X_1}) \cap X$. Hence

$$\mu = \#(\mathrm{Sing}(V) \cap X) \leq \#(V(\frac{\partial F}{\partial X_o}, \frac{\partial F}{\partial X_1}) \cap X) \leq 4d$$

by Bezout`s theorem.

Proposition 2.3. Let $X \subset \mathbb{P}_4$ be a smooth non-degenerate surface of degree d contained in the cubic hypersurface $V = V(F)$. Assume X contains at most finitely many singularities of V. If we have

$$\mu = \#(\mathrm{Sing}(V) \cap X) = 4d,$$

the surface X is necessarily contained in a quadric hypersurface.

Proof. Let $\mathcal{J}$ be the ideal of $X \cap V(\frac{\partial F}{\partial X_i}, 0 \leq i \leq 4)$ in X. In 2.1 it was shown that one has in suitable coordinates

$$\mathrm{Sing}(V) \cap X \subset V(\frac{\partial F}{\partial X_o}, \frac{\partial F}{\partial X_1}) \cap X$$

and that

$$\mu = \#(\mathrm{Sing}(V) \cap X) \leq \#(V(\frac{\partial F}{\partial X_o}, \frac{\partial F}{\partial X_1}) \cap X) = 4d.$$

The equality $\mu = 4d$ therefore implies

$$\mathrm{Sing}(V) \cap X = V(\frac{\partial F}{\partial X_o}, \frac{\partial F}{\partial X_1}) \cap X.$$

Hence we obtain

$$\mathcal{J} = (\frac{\partial F}{\partial X_o}|X, \frac{\partial F}{\partial X_1}|X).$$

Now we consider the Koszul-complex

$$0 \to \mathcal{O}_X(-4) \to \mathcal{O}_X(-2)^{\oplus 2} \to \mathcal{J} \to 0.$$

Taking cohomology and using the Kodaira vanishing theorem we get

$$H^0(X,\mathcal{J}(2)) \cong H^0(X,\mathcal{O}_X)^{\oplus 2}.$$

This means, that we have in suitable coordinates

$$\frac{\partial F}{\partial X_2}|X = \frac{\partial F}{\partial X_3}|X = \frac{\partial F}{\partial X_4}|X = 0.$$

There is $i \in \{2,3,4\}$ such that $\frac{\partial F}{\partial X_i} \neq 0$. If not, $V = V(F)$ would be the union of three hyperplanes. Since X is non-degenerate this is impossible. Therefore X is contained in the quadric hypersurface $V(\frac{\partial F}{\partial X_i})$.

Corollary 2.4. Let $X \subset \mathbb{P}_4$ be a smooth, non-degenerate surface of degree d. Assume d even and that X is contained in a cubic hypersurface. If X contains only finitely many singularities of V and

$$\mu = \#(\mathrm{Sing}(V) \cap X) = 4d$$

then X is a complete intersection.

Before we are able to classify smooth half-canonical surfaces in $\mathbb{P}_4$ of degree 16 we have to recall a result from [EP].

Lemma 2.5. Let $X \subset \mathbb{P}_4$ be a smooth non-degenerate surface which is contained in the hypersurface V_k of degree k. Then

(*) $\quad c_2(N(-k)) \le (k-1)^2 \deg(X) - c_1(\det(N(-1))|D)$,

where D is the (possibly empty) one-dimensional part of $\mathrm{Sing}(V_k) \cap X$.

Corollary 2.6. Let X as in 2.5. If

$$c_2(N(-k)) = (k-1)^2 \deg(X),$$

then

$$\mathrm{Sing}(V_k) \cap X$$

is finite.

Proof. Since X is non-degenerate $\det(N(-1))$ is an ample line bundle. From (*) we conclude $c_1(\det(N(-1)|D) = 0$ and therefore $D = \emptyset$.

Proposition 2.7. Let $X \subset \mathbb{P}_4$ be a smooth (non-degenerate) half-canonical surface and let E be the associated rank-2 bundle. If X is not a complete intersection the bundle E is stable.

Proof. From the sequence

$$0 \to \mathcal{O} \to E \to \mathcal{J}_X(7) \to 0$$

we see

$$H^0(\mathbb{P}_4, E(-4)) = H^0(\mathcal{J}_X(3)).$$

Therefore E is stable precisely if X is not contained in a cubic hypersurface.

Assume that X is contained in a cubic V.

By the lemma 2.5 of Ellingsrud and Peskine we get

$$c_2(N(-3)) \leq 4d - 5c_1(\mathcal{O}_X(1)|D).$$

From

$$c_2(N(-3)) = d(d-12)$$

we obtain therefore

$$d \leq 16.$$

In case of equality d = 16 the corollary 2.6 implies that $\mathrm{Sing}(V) \cap X$ is finite and

$$\mu = \#(\mathrm{Sing}(V) \cap X) = c_2(N(-3)) = 4d.$$

By 2.4 we conclude that X is a complete intersection.

Hence X is not contained in a cubic and E is stable.

Now we are in position to classify all half-canonical surfaces in $\mathbb{P}_4$ of degree 16.

<u>Theorem 2.8.</u> The smooth half-canonical surfaces of degree 16 are precisely the smooth zero-loci of sections in H(1), where H is the Horrocks-Mumford bundle on $\mathbb{P}_4$.

<u>Proof.</u> Let H be the Horrocks-Mumford bundle on $\mathbb{P}_4$ (i.e. $c_1(H) = 5$, $c_2(H) = 10$). Take a section $\sigma \in H^0(\mathbb{P}_4, H(1))$ such that $X = V(\sigma)$ is a smooth surface. Since H(1) is generated by sections [M] a general section of H(1) has this property.

Since $c_1(H(1)) = 7$ we have

$$K_X = \mathcal{O}_X(2).$$

Moreover

$$\deg X = c_2(H(1)) = 16.$$

Conversely let $X \subset \mathbb{P}_4$ be a smooth half-canonical surface of degree d = 16. Consider the associated vector bundle E

$$0 \to \mathcal{O} \to E \to \mathcal{J}_X(7) \to 0.$$

By 2.7 and the remark following 1.2 we know that E is stable. The bundle E(-1) has Chern-classes

$$c_1(E(-1)) = 5, \quad c_2(E(-1)) = 10.$$

Since the Horrocks-Mumford bundle is the only stable 2-bundle on $\mathbb{P}_4$ with these Chern classes [DS] we get

$$E \cong H(1)$$

and theorem 2.8 is proved.

3. Estimating the irregularity

In this section we will estimate the irregularity of smooth half-canonical surfaces in $\mathbb{P}_4$. For this purpose we have to know the cohomology groups of the vector bundle associated to X.
Let

$$0 \to \mathcal{O} \to E \to \mathcal{I}_X(7) \to 0$$

be the sequence (1) and let $F = E(-4)$ be the normalized bundle. By section 2 we know that F is stable provided X is not a complete intersection.

Proposition 3.1. Let $X \subset \mathbb{P}_4$ be a smooth half-canonical surface of degree d which is not a complete intersection. If F is the normalized rank-2 bundle belonging to X we have the following table for the dimension of the cohomology groups of F

q

	0	0	0	0	0	
	$\chi+\frac{3}{2}d-35$	p_g-15	0	0	0	
(T)	0	q	$10+\frac{d}{2}-\chi$	q	0	$h^q(\mathbb{P}_4, F(p))$
	0	0	0	p_g-15	$\chi+\frac{3}{2}d-35$	
	0	0	0	0	0	

p

where $\chi = \chi(\mathcal{O}_X)$, $p_g = p_g(X)$, $q = q(X)$.

Proof. Since F is stable by 2.7 we have $h^0(F) = 0$. This gives the bottom row. By Serre duality and $F^{\vee} \cong F(+1)$ we get then the first row.

From

$$(*) \qquad 0 \to \mathcal{O}(-4) \to F \to \mathcal{J}_X(3) \to 0$$

we obtain

$$h^1(F(-k)) = h^1(\mathcal{J}_X(3-k)).$$

Since the only non linearly normal surface in $\mathbb{P}_4$ is the Veronese surface we have $H^1(\mathcal{J}_X(\nu)) = 0$ for $\nu \le 1$. Hence

$$h^1(F(-k)) = 0 \text{ for } k \ge 2.$$

To calculate $h^1(F(-1)) = h^1(\mathcal{J}_X(2))$ we consider

$$(**) \qquad 0 \to \mathcal{J}_X(2) \to \mathcal{O}_{\mathbb{P}_4}(2) \to \mathcal{O}_X(2) \to 0$$

and get

$$h^1(F(-1)) = h^1(\mathcal{J}_X(2)) = h^0(\mathcal{O}_X(2)) - h^0(\mathcal{O}_{\mathbb{P}_4}(2))$$
$$= p_g - 15.$$

In the same way we get

$$h^1(F) = h^1(\mathcal{J}_X(3)) = h^0(\mathcal{O}_X(3)) - 35$$

since X is not contained in a cubic by 2.7. Since $K_X = \mathcal{O}_X(2)$, the Kodaira vanishing theorem implies

$$h^0(\mathcal{O}_X(3)) = \chi(\mathcal{O}_X(3)).$$

Riemann-Roch shows $\chi(\mathcal{O}_X(3)) = \chi(\mathcal{O}_X)+\frac{3}{2}d$ and therefore

$$h^1(F) = \chi + \frac{3}{2}d - 35.$$

This confirms the fourth row and by duality reasons the second row. By (*), (**), $K_X = \mathcal{O}_X(2)$ and the Kodaira vanishing theorem we get

$$h^2(F) = h^2(\mathcal{J}_X(3)) = h^1(\mathcal{O}_X(3)) = 0.$$

In the same way we have

$$h^2(F(-1)) = h^2(\mathcal{J}_X(2)) = h^1(\mathcal{O}_X(2)) = h^1(K_X) = q.$$

Since the middle row is selfdual it remains to compute $h^2(F(-2))$.

As above

$$h^2(F(-2)) = h^2(\mathcal{J}_X(1)) = h^1(\mathcal{O}_X(1)).$$

Since $K_X = \mathcal{O}_X(2)$ we have

$$h^2(\mathcal{O}_X(1)) = h^0(\mathcal{O}_X(1))$$

and therefore

$$\chi(\mathcal{O}_X(1)) = 2h^0(\mathcal{O}_X(1)) - h^1(\mathcal{O}_X(1)).$$

Since X is linearly normal and nondegenerate we know that

$$h^0(\mathcal{O}_X(1)) = 5$$

and therefore by Riemann-Roch

$$h^1(\mathcal{O}_X(1)) = 10-\chi(\mathcal{O}_X(1)) = \chi- \frac{d}{2}.$$

Putting things together we find

$$h^2(F(-2)) = 10 + \frac{d}{2} - \chi$$

and 3.1 is proved.

Now we are able to give inequalities for the irregularity of X. The known inequalities for surfaces of general type [BPV] yield rather coarse estimates. The knowledge of the cohomology of the associated stable rank-2 bundle and Barth`s restriction theorem [Ba] give much better inequalities.

<u>Proposition 3.2.</u> Let $X \subset \mathbb{P}_4$ be a smooth half-canonical surface of degree d. Then we have the inequalities for the irregularity $q = q(X)$

i) $1 \le q \le 4$ for $d = 18$

ii) $5 \le q \le 9$ for $d = 22$.

<u>Proof.</u> Let F be the normalized rank-2 bundle on $\mathbb{P}_4$ belonging to X. F is stable by 2.7. Hence the restriction $F|P$ to a general plane $P \subset \mathbb{P}_4$ is stable by [Ba]. The Koszul-complex for $\mathcal{J}_P$ leads to an exact sequence

$$0 \to F(-2) \to F(-1)^{\oplus 2} \to F \to F|P \to 0$$

which we cut into two pieces

(a) $$0 \to F(-2) \to F(-1)^{\oplus 2} \to Q \to 0$$

(b) $$0 \to Q \to F \to F|P \to 0.$$

From 3.1 and (a) we conclude

$$2q(X) \le h^2(F(-2)) + h^2(Q).$$

Since $h^2(F) = 0$ we infer from (b)

$$h^2(Q) \le h^1(F|P).$$

Since $F|P$ is stable, the groups $H^0(F|P)$ and $H^2(F|P)$ are zero.

Riemann-Roch applied to $F|P$ therefore yields

$$h^1(F|P) = c_2(F)-1 = d-13.$$

Altogether we get

$$2q \le 10 + \frac{d}{2}-\chi + d-13 = \frac{3}{2}d-3-\chi.$$

For $d = 18$ we have $\chi = 15$ and hence

$$2q \le 9 \quad \text{i.e.} \quad q \le 4.$$

For $d = 22$ the value of χ is 11 and the estimate above becomes

$$2q \le 19 \quad \text{i.e.} \quad q \le 9.$$

The inequalities $q \ge 1$ resp. $q \ge 5$ follow from the fact $p_g \ge 15$ proved in 1.3.

4. Excluding q = 1

In this section it is shown that the irregularity $q = 1$ cannot occur.

Let $X \subset \mathbb{P}_4$ be a smooth half-canonical surface of degree $d = 18$ and irregularity $q = 1$. The associated normalized rank-2 bundle F on $\mathbb{P}_4$ has the following cohomology groups by 3.1

q

0	0	0	0	0	
7	0	0	0	0	
0	1	4	1	0	$h^q(\mathbb{P}_4, F(p))$
0	0	0	0	7	
0	0	0	0	0	

p

<u>Proposition 4.1.</u> There is no rank-2 vector bundle F on $\mathbb{P}_4$ with Chern classes $c_1(F) = -1$, $c_2(F) = 6$ and cohomology groups given by the table above.

<u>Proof.</u> Suppose such a bundle F exists. The Beilinson spectral sequence (cf. [OSS]) implies the existence of a self-dual monad

$$H^2(F(-3))\otimes\Omega^3(3) \xrightarrow{B} H^2(F(-2))\otimes\Omega^2(2) \xrightarrow{A} H^2(F(-1))\otimes\Omega(1)$$

$H^2(F(-2))$ carries a non-degenerate symplectic form (Given by the pairing

$$H^2(F(-2))\otimes H^2(F(-2)) \to H^4(\Lambda^2(F(-2))) = H^4(\mathcal{O}(-5)) \cong \mathbb{C}).$$

In a suitable basis this form becomes

$$Q = \left(\begin{array}{cc|cc} 0 & 1 & & \\ -1 & 0 & \multicolumn{2}{c}{0} \\ \hline & & 0 & 1 \\ \multicolumn{2}{c|}{0} & -1 & 0 \end{array}\right).$$

Choosing a basis in $H^2(F(-3))$ and the dual basis in $H^2(F(-3))^\vee \cong H^2(F(-1))$ the monad becomes

$$\Omega^3(3) \xrightarrow{B} \Omega^2(2)^{\oplus 2} \xrightarrow{A} \Omega(1)$$

with

$$A = (a_1,\ldots,a_4),\ a_i \in V$$

and $B = QA^t$. Here $\mathbb{P}_4 = \mathbb{P}(V^*)$ is the space of lines in the 5-dimensional complex vector space V and A and B operate by contraction (cf.[D]).

From $A \wedge B = 0$ we get $A \wedge QA^t = 0$ which is equivalent to

$$a_1\wedge a_2 + a_3\wedge a_4 = 0.$$

This implies

$$\dim \operatorname{span}(a_1,a_2,a_3,a_4) \le 2.$$

But this contradicts the surjectivity of the bundle map

$$A : \Omega^2(2)^{\oplus 2} \to \Omega(1),$$

cf. [D], remark 1 in section 1.

<u>Corollary 4.2.</u> There is no smooth half-canonical surface $X \subset \mathbb{P}_4$ with irregularity $q(X) = 1$.

Also in the case $q(X) > 1$ the associated normalized vector bundle F can be obtained as the cohomology of a suitable monad.

<u>Proposition 4.3.</u> Let F be a rank-2 vector bundle on $\mathbb{P}_4$ with $c_1(F) = -1$ and cohomology groups as given in the table (T). Then F is the cohomology of a monad.

$$(H^3(F(-4))\otimes\Omega^4(4)) \oplus (H^2(F(-3))\otimes\Omega^3(3))$$

$$\downarrow$$

$$(H^3(F(-3))\otimes\Omega^3(3)) \oplus (H^2(F(-2))\otimes\Omega^2(2)) \oplus (H^1(F(-1))\otimes\Omega(1))$$

$$\downarrow$$

$$(H^2(F(-1))\otimes\Omega(1)) \oplus (H^1(F)\otimes\mathcal{O}).$$

Proof. Copy the proof of Lemma 7 in [D].

Remark 4.4. For further properties of this monad see [D].
If no such monad exists (with the dimensions given bei (T)) then there is no smooth half-canonical surface $X \subset \mathbb{P}_4$ of degree 18 or 22 and the classification of half-canonical surfaces in $\mathbb{P}_4$ would be finished.
Of course it would be much more interesting if such a monad would exist providing new rank 2-bundles on $\mathbb{P}_4$.

5. The Albanese dimension

In this section we show that the Albanese dimension of a half-canonical surface $X \subset \mathbb{P}_4$ of degree 18 or 22 is mostly 2.

Proposition 5.1. Let $X \subset \mathbb{P}_4$ be a half-canonical surface and let alb(X) denote the dimension of the image of the Albanese map $X \to \mathrm{Alb}(X)$.
Then

a) if X is of degree 22; alb(X) = 2

b) if X is of degree 18 and alb(X) = 1 then q(X) = 2; in this case we have $6 \le g(F) \le 10$, where F is a general fibre of the Albanese map.

Proof. Let $f : X \to \mathrm{Alb}(X)$ be the Albanese map. Assume alb(X) = 1. It is well known that $C = f(X)$ is a smooth curve of genus $g(C) = q = q(X)$. Choose a non zero $\omega \in H^0(\Omega_C)$ having $2q - 2$ simple zeroes, no zero being a critical value of f. Then $f^*(\omega) \in H^0(X,\Omega^1_X)$ defines a monomorphism

$$\mathcal{O}_X \to \Omega^1_X$$

and we let $\mathcal{O}_X(D)$ denote its image. D is effective and nothing else than the one-dimensional part of $\{f^*(\omega) = 0\}$. Since Ω_X is semi-stable with respect to K_X by [Bo] we get

$$c_1(\mathcal{O}_X(D))c_1(K_X) \le \frac{c_1(\Omega_X)c_1(K_X)}{2}$$

and hence

$$c_1(\mathcal{O}_X(D))c_1(\mathcal{O}_X(1)) \le \deg(X).$$

Since D contains all the fibres of f over the zeroes of ω we deduce from this

(*) $$(2q-2)\deg(F) \le \deg(X),$$

F being a general fibre of f. F is connected.
From the adjunction formula and $K_X = \mathcal{O}_X(2)$ we get

(**) $$\deg(F) = g(F)-1.$$

This implies $\deg(F) \ge 4$ and in fact $\deg(F) \ge 5$ by the inequality $g(F) \le \frac{1}{2}(\deg(F)-1)(\deg(F)-2)$.

a) Assume $d = \deg(X) = 22$. Since $\deg(F) \ge 4$ we get from (*) the inequality $q \le 3$ contradicting $q \ge 5$ (proposition 1.3).

b) In case $d = 18$ we get from (*) and $\deg(F) \ge 5$ that

$$q \le 2.$$

By section 1 we know $q \ge 1$ and by 4.2 the case $q = 1$ does not occur. The inequality $6 \le g(F) \le 10$ follows immediately from $\deg(F) \ge 5$ and the relations (*) and (**).

In case $q = 2$, $\mathrm{alb}(X) = 1$ we can get more information.

<u>Proposition 5.2.</u> Let $X \subset \mathbb{P}_4$ be a half-canonical surface of degree 18 and irregularity $q = 2$. Assume $\mathrm{alb}(X) = 1$ and set $r = h^0(\mathcal{O}_F(1))-1$, where F is the general fibre of the Albanese map. Then

$$2 \le r \le 3$$

and

a) $r = 2$ precisely if $g(F) = 6$

b) $r = 3$ precisely if $g(F)$ is 9 or 10.

<u>Proof.</u> We know

$$K_F = K_X|F = \mathcal{O}_F(2).$$

Hense F is not hyperelliptic and Clifford`s theorem implies

$$r < \tfrac{1}{2}\deg(\mathcal{O}_F(1)) = \tfrac{1}{2}\deg(F).$$

Since $\deg(F) \leq 9$ by 5.1 we obtain $r \leq 4$. Clearly $r \geq 2$.

Assume $r = 2$. Then F is a plane curve and from

$$g(F) = \tfrac{1}{2}(\deg(F)-1)(\deg(F)-2)$$

and

$$\deg(F) = g(F)-1$$

we get

$$g(F) = 6.$$

On the other hand we have

$$h^0(K_F) = h^0(\mathcal{O}_F(2)) \geq 2h^0(\mathcal{O}_F(1))-1.$$

This implies

$$r + 1 \leq \frac{g+1}{2}.$$

If $g(F) = 6$ we obtain $r \leq 2$ and hence $r = 2$.

If $r = 3$, Castelnuovo's bound

$$g(F) \leq \begin{cases} (k-1)^2 & \text{for } \deg(F) = 2k \\ k(k-1) & \text{for } \deg(F) = 2k+1 \end{cases}$$

and

$$\deg(F) = g(F)-1$$

lead to $g(F) \geq 9$ and by b) of 5.1. to $g(F) \in \{9,10\}$.
It remains to show that $r = 4$ is impossible. To do this we use the following fact from [ACGH], p. 198, E-1:
Let C be a smooth curve of genus g and let L be a special line bundle (i.e. $h^1(L) > 0$) of degree d. For $r = h^0(L)-1$ the inequality

$$d \geq g - d + 2r + h^1(L^{\otimes 2})$$

holds.

We apply this for $C = F$ and $L = \mathcal{O}_F(1)$ and obtain (using again $g(F) = \deg(F)+1$)

$$r \leq \frac{\deg(F)-2}{2}.$$

Since $\deg(F) \leq 9$ by 5.1 we conclude $r \leq 3$.

6. Irregularity and the conormal bundle.

In this section we show that the irregularity $q = q(X)$ of an arbitrary non-degenerate smooth surface $X \subset \mathbb{P}_4$ equals $h^1(X,N^{\vee})$ and that all the groups $H^1(X,S^{\nu}(N^{\vee}))$ vanish for $\nu \geq 2$.

<u>Proposition 6.1.</u> Let $X \subset \mathbb{P}_4$ be a smooth surface.
Then

$$q(X) = h^1(X,N^{\vee}_{X/\mathbb{P}_4}).$$

<u>Proof.</u> From the Euler sequence

$$0 \to \Omega_{\mathbb{P}_4}|X \to \mathcal{O}_X(-1)^{\oplus 5} \to \mathcal{O}_X \to 0$$

we deduce

$$H^1(X,\Omega_{\mathbb{P}_4}|X) \cong \mathbb{C}.$$

Since $\Omega_{\mathbb{P}_4}|X$ is a negative bundle we obtain from

$$0 \to N^{\vee} \to \Omega_{\mathbb{P}_4}|X \to \Omega_X \to 0$$

the exact sequence

$$0 \to H^0(\Omega_X) \to H^1(X,N^{\vee}) \to H^1(X,\Omega_{\mathbb{P}_4}|X) \to H^1(X,\Omega_X).$$

The map $H^1(X,\Omega_{\mathbb{P}_4}|X) \to H^1(X,\Omega_X)$ is non-zero and therefore injective, $H^1(X,\Omega_{\mathbb{P}_4}|X)$ being one-dimensional. This implies

$$q = h^0(X,\Omega_X) = h^1(X,N^{\vee}).$$

<u>Remark.</u> The same argument works for smooth d-dimensional submanifolds of $\mathbb{P}_n$ provided $d \geq 2$.

<u>Proposition 6.2.</u> Let $X \subset \mathbb{P}_4$ be a smooth surface. Then

$$H^1(X,S^k(N^{\vee})) = 0 \quad \text{for } k \geq 2.$$

<u>Proof.</u> To shorten our presentation we prove the case $k = 2$ and leave the general case to the reader. The exact sequence

$$0 \to N^{\vee} \to \Omega_{\mathbb{P}_4}|X \to \Omega_X \to 0$$

gives an exact sequence

$$(*) \qquad 0 \to S^2N^{\vee} \to N^{\vee} \otimes \Omega_{\mathbb{P}_4}|X \to \Omega^2_{\mathbb{P}_4}|X \to K_X \to 0.$$

We cut (*) into two exact sequences

$$(1) \qquad 0 \to S^2N^{\vee} \to N^{\vee} \otimes \Omega_{\mathbb{P}_4}|X \to Q \to 0.$$

(2) $$0 \to Q \to \Omega^2_{\mathbb{P}_4}|X \to K_X \to 0.$$

From (2) we get $H^0(Q) = 0$ and therefore by (1) it is enough to show that $H^1(X, N^\vee \otimes \Omega_{\mathbb{P}_4}|X) = 0$. Tensoring the Euler-sequence by $N^\vee$ gives an exact sequence

$$0 \to \Omega_{\mathbb{P}_4}|X \otimes N^\vee \to \mathcal{O}_X(-1)^{\oplus 5} \otimes N^\vee \to N^\vee \to 0.$$

This sequence tells us that it is sufficient to prove $H^1(X, N^\vee(-1)) = 0$. From

$$0 \to N^\vee(-1) \to \Omega_{\mathbb{P}_4}(-1)|X \to \Omega_X(-1) \to 0$$

it follows

$$H^0(X, \Omega_X(-1)) \cong H^1(X, N^\vee(-1)).$$

But $H^0(X, \Omega_X(-1)) = 0$ due to the following general fact [BPV], p.212:

Let X be a smooth projective surface, L a line bundle on X such that $H^0(X, \Omega_X \otimes L^{-1}) \neq 0$.

Then $\kappa(X, L) \leq 1$.

In particular L cannot be ample.

This finishes the proof for k = 2. For general k use the analogue of (*)

$$0 \to S^k N^\vee \to S^{k-1} N^\vee \otimes (\Omega_{\mathbb{P}_4}|X) \to \ldots \to \Omega^k_{\mathbb{P}_4}|X \to \Omega^k_X \to 0.$$

Remarks 6.3.

i) In order to compute q, we want to relate $H^1(X, N^\vee)$ and $H^1(X, S^2 N^\vee)$. To do this we use the exact sequence

(S) $$0 \to N^\vee \otimes \det N^\vee \to S^2 N^\vee \to \mathcal{J}^2_Z \to 0$$

from [PPS], Lemma 4.3. Here Z is the zero locus of a general section $s \in H^0(X, N)$ vanishing in codimension two.

Tensoring (S) with $(\det N)^{\otimes 2}$ gives an exact sequence

$$0 \to N \to S^2 N \to \mathcal{J}^2_Z \otimes (\det N)^{\otimes 2} \to 0.$$

Tensoring with K_X and taking cohomology yields an exact sequence

$$H^0(S^2 N \otimes K_X) \to H^0(\mathcal{J}^2_Z \otimes (\det N)^{\otimes 2} \otimes K_X) \to H^1(N \otimes K_X) \to H^1(S^2 N \otimes K_X).$$

By 6.2 the last group $H^1(S^2 N \otimes K_X) \cong H^1(S^2 N^\vee)$ is zero. Hence

$$q(X) = h^1(N^\vee) = 0$$

precisely if

$$H^0(S^2N\otimes K_X) \to H^0(\mathcal{J}_Z^2\otimes(\det N)^{\otimes 2}\otimes K_X)$$

is an epimorphism.

ii) Tensor the sequence (S) by detN to obtain an exact sequence

$$0 \to N^\vee \to (S^2N^\vee)\otimes\det N \to \mathcal{J}_Z^2\otimes\det N \to 0.$$

From this we get

$$0 \to H^0(S^2N^\vee\otimes\det N) \to H^0(\mathcal{J}_Z^2\otimes\det N) \to H^1(N^\vee) \to H^1(S^2N^\vee\otimes\det N).$$

Hence $q(X) = 0$ if we have

(a) $H^1(S^2N^\vee\otimes\det N) = 0$

and

(b) $h^0(S^2N^\vee\otimes\det N) = h^0(\mathcal{J}_Z^2\otimes\det N)$.

From

(+) $N^\vee\otimes N \cong (S^2N^\vee\otimes\det N) \oplus \mathcal{O}_X$

we see that (a) is equivalent to

$$h^1(X,\mathrm{End}(N)) = h^1(\mathcal{O}_X),$$

i.e. N is rigid up to tensoring with topologically trivial line bundles. By (+) we also see that

$$H^0(X,\mathrm{End}(N)) \cong H^0(S^2N^\vee\otimes\det N) \oplus \mathbb{C}.$$

If N is a simple bundle (this is satisfied for instance if N is stable) this shows

$$H^0(X,S^2N^\vee\otimes\det N) = 0.$$

Therefore the "quasi-rigidity" of N, the "simplicity" of N and $H^0(\mathcal{J}^2_Z\otimes\det N) = 0$ would imply $q(X) = h^1(\mathcal{O}_X) = 0$.

Acknowledgements: We would like to thank L. Ein for numerous discussions and the DFG Schwerpunktprogramm "Komplexe Mannigfaltigkeiten" and the DAAD program PROCOPE for financial support.

References

[A] Aure,A.: On surfaces in projective 4-space. Preprint, Oslo 1987.

[ACGH] Arbarello,E.; Cornalba,M; Griffiths,P.A.; Harris,J.: Geometry of algebraic curves. Springer Verlag 1985.

[Ba] Barth,W.: Some properties of stable rank-2 vector bundles on $\mathbb{P}_n$. Math. Ann. 226, 125-150 (1977).

[Bo] Bogomolov,F.A.: Holomorphic tensors and vector bundles on projective manifolds. Isvestija Akad. Nauk SSR, Ser.Mat. 42, 1227-1287 (1978).

[BC] Ballico,E.; Chiantini,L.: On smooth subcanonical varieties of codimension 2 in $\mathbb{P}^n$, $n \geq 4$. Annali di Math. 135, 99-117 (1983).

[BPV] Barth,W.; Peters,C.; Van de Ven,A.: Compact complex surfaces. Springer Verlag 1984.

[D] Decker,W.: Stable rank 2 vector bundles with Chern classes $c_1 = -1$, $c_2 = 4$. Math. Ann. 275, 481-500 (1986).

[DS] Decker,W.; Schreyer,F.O.: On the uniqueness of the Horrocks-Mumford bundle. Math. Ann. 273, 415-443 (1986).

[EP] Ellingsrud,G.; Peskine,C.: Sur les surfaces lisses de $\mathbb{P}_4$. Preprint 1987.

[HM] Horrocks,G.; Mumford,D.: A rank 2 bundle on $\mathbb{P}^4$ with 15,000 symmetries. Topology 12, 63-81 (1973).

[HS] Holme,A.; Schneider,M.: A computer aided approach to codimension 2 subvarieties of $\mathbb{P}_n$, $n \geq 6$. Crelle`s J. 357, 205-220 (1985).

[M] Manolache,N.: Syzygies of abelian surfaces embedded in $\mathbb{P}^4(\mathbb{C})$. Crelle`s J. 384, 180-191 (1988).

[O1] Okonek,C.: Flächen vom Grad 8 im $\mathbb{P}^4$. Math. Z. 191, 207-223 (1986).

[O2] Okonek, C.: On codimension 2 submanifolds in $\mathbb{P}^4$ and $\mathbb{P}^5$. Mathematica Gottingensis 50 (1986).

[OSS] Okonek,C.; Spindler,H.; Schneider,M.: Vector bundles on complex projective spaces. Progress in Math. 3, Birkhäuser (1980).

[OV] Okonek,C.; Van de Ven,A.: Vector bundles and geometry: Some open problems. Mathematica Gottingensis n° 17 (1988).

[PPS] Peternell,T.; le Potier,J.; Schneider,M.: Vanishing theorems, linear and quadratic normality. Invent. math. 87, 573-586

GROUPES DE POINTS DE P^2:

CARACTERE ET POSITION UNIFORME.

Ph. Ellia
C.N.R.S. U.A. 168
Département de Mathématiques
Université de Nice
Parc Valrose
06034 Nice Cedex (Fr).

Ch. Peskine
C.N.R.S. U.A. 213
Université de Paris VI
Tour 45-46, 5e étage
4, Place Jussieu
75252 Paris Cedex 05 (Fr).

Quelles sont les relations entre la postulation d'un groupe de points plan et ses propriétés géométriques? Il n'y a évidemment pas de réponse exhaustive à cette question. La littérature fourmille de réponses partielles, le plus souvent suffisantes pour les applications en vue. La classification des courbes gauches est un champ fréquent d'application.

On sait par exemple que si un groupe de points, E, est section plane générale d'une courbe intégre (resp. lisse) de $\mathbb{P}^3$, alors:

(1) E vérifie le principe de position uniforme [H].

(2) le caractère numérique de E est sans lacunes; de plus tous les caractères sans lacunes sont ainsi obtenus [GP].

Ce dernier énoncé détermine les fonctions de Hilbert des sections planes générales des courbes intègres (resp. lisses) de l'espace. Ce résultat a été récemment retrouvé ([MR]). En fait, et sans le dire, ces auteurs caractérisent les fonctions de Hilbert des groupes de points vérifiant le principe de position uniforme : ce sont les mêmes que celles des sections planes générales des courbes intègres de l'espace (donc celles dont le caractère numérique est sans lacune). C'est en cherchant à redémontrer ce résultat par des méthodes propres au caractère que nous avons été amenés à interpréter la présence d'une lacune dans le caractère de E par une décomposition naturelle de E (cf la proposition) contredisant le principe de position uniforme (cf corollaire 1).

Le lecteur a bien compris que ce n'est pas dans ces pages qu'il trouvera des résultats nouveaux. Pour le conforter dans cette opinion, nous présentons une nouvelle démonstration d'un résultat sans doute classique (corollaire 2). Plus que des résultats, nous avons voulu exposer ici un point de vue qui nous paraît utile.

Notations: On désigne par $\mathbb{P}^2$ le plan projectif, Proj($K[x_o,x_1,x_2]$), sur un corps K algébriquement clos, de caractéristique quelconque. Si E est un groupe de points (non nécéssairement réduit) de $\mathbb{P}^2$, rappelons que le cône projetant, A, de E est l'anneau gradué $K[x_o,x_1,x_2]/I$, où I est l'idéal gradué des formes s'annulant sur E. La fonction de Hilbert de E est $H(E, k) := \text{rang } A_k$. Si la droite, L, d'équation $x_2 = 0$ ne rencontre pas E, l'application naturelle $K[x_o,x_1] \to A$ induit pour A une structure de module gradué de type fini sur R $= K[x_o,x_1]$. Une résolution minimale de A comme R-module gradué de type fini sera de la forme:

$$0 \to \oplus_o^{s-1} R[-n_i] \to \oplus_o^{s-1} R[-i] \to A \to 0 \quad (*).$$

De plus si la droite L est générale on a:

(i) s est le degré minimal d'une courbe contenant E.

(ii) $n_i \geq s$ pour $i = 0, ..., s-1$.

(iii) $\deg(E) - H(E, n) = \Sigma_{o \leq i \leq s-1} [(n_i-n-1)_+ - (i-n-1)_+]$;

en particulier : $\deg(E) = \Sigma_{o \leq i \leq s-1} [(n_i-i)]$.

Il est clair que la suite (n_i) ne dépend que de la fonction H(E, n) et qu'elle la caractérise. Rappelons la

Définition ([GP]) : La suite d'entiers $(n_o,...,n_{s-1})$, $n_o \geq ... \geq n_{s-1}$, est le caractère numérique de E.

Proposition: Soit E un groupe de points plan, de cône projetant A et de caractère $(n_o,...,n_{s-1})$. Si $n_{t-1} > n_t + 1$, il existe une courbe T de degré t telle que:

(i) si $E' = T \cap E$ et si $f = 0$ est l'équation de T, le cône projetant de E' est $A/fA = A'$ et son caractère $(n_o,...,n_{t-1})$.

(ii) si E" est le groupe de points résiduel de E' dans E par rapport à T (donc de cône projetant $A'' \cong Af$), le caractère de E" est $(m_o, ..., m_{s-t-1})$ avec $m_i = n_{t+i} - t$.

Dém: Appliquons le foncteur dualisant $\text{Hom}_R(\ . \ , R[-2])$ à la suite exacte (*). On trouve une résolution minimale, comme R-module, du A-module dualisant, que nous noterons Ω_A:

$$0 \to \oplus_o^{s-1} R[i-2] \to \oplus_o^{s-1} R[n_i-2] \to \Omega_A \to 0 \quad (**).$$

Le R-module Ω_A a donc un systéme minimal de générateurs $(\alpha_o, ..., \alpha_{s-1})$ avec $\deg(\alpha_j) = -n_j + 2$. Si $x_2.\alpha_j = \Sigma_{o \leq k \leq s-1} a_{kj} \alpha_k$, on a : $a_{kj} \in R_{nk-nj+1}$, ce qui montre $a_{kj} = 0$ pour $j < t$ et $k \geq t$. Le sous R-module M de Ω_A engendré par $(\alpha_o, ..., \alpha_{t-1})$ est donc un A-module. Dégageons alors l'énoncé suivant:

Lemme: Si le sous R-module de Ω_A engendré par $(\alpha_0, \ldots, \alpha_{t-1})$ est un A-module, la conclusion de la proposition est vraie.

Dém: Considérons la matrice $P = (\delta_{mn}x_2 - a_{mn})$, $0 \leq m,n \leq s-1$, à coefficients dans $K[x_0,x_1,x_2]$. La courbe, S, d'équation $\det(P) = 0$ a degré s et contient E. Mais P est de la forme:

$$P = \begin{pmatrix} N & 0 \\ P' & \end{pmatrix} \quad \text{où N est la matrice carrée } (\delta_{mn}x_2 - a_{mn}),\ 0 \leq m,n \leq t-1.$$

La courbe T d'équation $\det(N) = 0$ a degré t et est contenue dans S. Posons $E' := T \cap E$. Soit A' le cône projetant de E'.
Montrons $M = \Omega_{A'}$, où $\Omega_{A'}$ est un A'-module dualisant. On a $M \subset \Omega_{A'}$ car $\Omega_{A'} = \mathrm{Hom}_A\ (A/\det(N), \Omega_A)$ et il est clair que $\det(N).\alpha_j = 0$ pour $j = 0, \ldots, t-1$. D'autre part si $\mathfrak{m}$ désigne l'idéal (x_0, x_1) de R, on a par construction une application injective de R-modules : $M/\mathfrak{m}M \to \Omega_A/\mathfrak{m}\Omega_A$. Cette application se factorise à travers $M/\mathfrak{m}M \to \Omega_{A'}/\mathfrak{m}\Omega_{A'}$ qui est donc nécessairement injective. Par suite les générateurs minimaux , $\alpha_0, \ldots, \alpha_{t-1}$, de M sont aussi générateurs minimaux de $\Omega_{A'}$ (comme R-modules). Mais comme $E' \subset T$ avec $\deg(T) = t$, le caractère de E' est au plus de longueur t et $\Omega_{A'}$ a au plus t générateurs minimaux, ce qui démontre non seulement $M = \Omega_{A'}$ mais aussi que E' n'est pas contenu dans une courbe de degré $< t$. De plus le caractère de E' est $(n_0, \ldots, n_{t-1})$.
Soit $\mathcal{B}$ le noyau de la surjection naturelle $A \to A'$.
Comme $M = \Omega_{A'} = \mathrm{Hom}_A\ (A', \Omega_A)$ et comme $\mathrm{Hom}_A\ (\ .\ , \Omega_A)$ est un foncteur dualisant on a un diagramme commutatif:

$$\begin{array}{ccccccccc}
 & & 0 & & 0 & & 0 & & \text{(D.1)} \\
 & & \downarrow & & \downarrow & & \downarrow & & \\
0 & \to & \oplus_0^{t-1} R[i-2] & \to & \oplus_0^{t-1} R[n_i-2] & \to & \Omega_{A'} & \to & 0 \\
 & & \downarrow & & \downarrow & & \downarrow & & \\
0 & \to & \oplus_0^{s-1} R[i-2] & \to & \oplus_0^{s-1} R[n_i-2] & \to & \Omega_A & \to & 0 \\
 & & \downarrow & & \downarrow & & \downarrow & & \\
0 & \to & L & \to & \oplus_t^{s-1} R[n_i-2] & \to & \mathrm{Hom}(\mathcal{B}, \Omega_A) & \to & 0 \\
 & & \downarrow & & \downarrow & & \downarrow & & \\
 & & 0 & & 0 & & 0 & &
\end{array}$$

Comme $\mathrm{Hom}_A(\mathcal{B}, \Omega_A)$ est de profondeur un, le R-module L est libre et la première suite verticale est scindée; on a donc $L = \oplus_t^{s-1} R[i-2]$. En dualisant le diagramme précédent on obtient:

$$\begin{array}{ccccccccc} & & 0 & & 0 & & 0 & & \\ & & \downarrow & & \downarrow & & \downarrow & & \\ 0 & \to & \oplus_t^{s-1} R[-n_i] & \to & \oplus_t^{s-1} R[-i] & \to & \mathcal{B} & \to & 0 \\ & & \downarrow & & \downarrow & & \downarrow & & \\ 0 & \to & \oplus_0^{s-1} R[-n_i] & \to & \oplus_0^{s-1} R[-i] & \to & A & \to & 0 \\ & & \downarrow & & \downarrow & & \downarrow & & \\ 0 & \to & \oplus_0^{t-1} R[-n_i] & \to & \oplus_0^{t-1} R[-i] & \to & A' & \to & 0 \\ & & \downarrow & & \downarrow & & \downarrow & & \\ & & 0 & & 0 & & 0 & & \end{array} \qquad (D.2)$$

où les deux suites verticales de gauche sont scindées. Montrons enfin que l'idéal $\mathcal{B}$ de A est principal engendré par la classe de det(N) dans A. Considérons la matrice à coefficients dans R dont les colonnes sont les coordonnées des classes (dans A) des éléments det(N), x_2.det(N), ..., x_2^{s-t-1}.det(N), par rapport au systéme minimal de générateurs du R-module $\mathcal{B}$ (rangés par degrés decroissants). Elle est triangulaire à termes diagonaux de degrés 0. Si elle n'est pas inversible, il existe $j \le s-t-1$ et $a_i \in R_{j-i}$ tels que la classe de $x_2^j.\det(N) + \Sigma a_i.\det(N).x_2^i$ dans A soit nulle, et E est contenu dans une courbe de degré $j+t < s$ ce qui est impossible. On a donc bien $\mathcal{B} = \det(N).A = A''[-t]$ où A" est le cône projetant du groupe de points E" résiduel de E' dans E par rapport à T. Le lemme et la proposition sont démontrés.

Définition: Un groupe de points plan, E, vérifie le principe de position uniforme si pour tout sous groupe de points, F, de E et tout entier k: $H(F, k) = \min(\deg(F), H(E, k))$.

Corollaire1: Tout groupe de points plan vérifiant le principe de position uniforme a un caractère sans lacunes.

Dém: Soit E un groupe de points de caractère $(n_i)_{0 \le i \le s-1}$. Supposons qu'il existe $t < s$ tel que $n_{t-1} > n_t + 1$ et montrons que le groupe de points E' décrit dans la proposition contredit le principe de position uniforme. Comme E' est contenu dans une courbe de degré t, il suffit de vérifier $\deg(E') \ge h^0(\mathcal{O}_{\mathbb{P}^2}(t))$. Mais le caractère de E' est $(n_0, ..., n_{t-1})$ avec $n_{t-1} \ge n_t + 2 \ge s + 2 \ge t + 3$. Donc:

$$\deg(E') \ge \sum_0^{t-1} (t+3-i) = t(t+3) - t(t-1)/2 = t(t+7)/2 \ge (t+2)(t+1)/2.$$

Corollaire 2: Soit E un groupe de points plan de degré d. Soit $\tau = \max \{n, H(E,n) < d\}$. Soit s un entier tel que $s \leq d/s$ et que $\tau \geq s - 3 + d/s$. L'une des conditions suivantes est vérifiée:

(i) E est intersection complète d'une courbe de degré s et d'une courbe de degré d/s et $\tau = s - 3 + d/s$.

(ii) il existe t, avec $0 < t < s$, et un sous groupe de points E' de E contenu dans une courbe de degré t tel que : $t[\tau + (5-t)/2] \geq \deg(E') \geq t[\tau - t + 3]$.

Dém: Soient $(n_o,...)$ le caractère de E et t le plus grand entier tel que $(n_o, ..., n_{t-1})$ est connexe. Remarquons d'abord que $t \leq s$.

En effet $n_o = \tau + 2 \geq s - 1 + d/s$ et $\sum_o^{s-1} (s - 1 + d/s - 2i) = d$. Considérons alors le groupe de points E' de caractère $(n_o, ..., n_{t-1})$ qui est soit E soit le sous groupe décrit dans la proposition. Les inégalités $\tau + 2 \geq n_i \geq \tau + 2 - i$ démontrent: $\sum_o^{t-1} (\tau + 2 - i) \geq \deg(E') \geq \sum_o^{t-1} (\tau + 2 - 2i)$ soit $t[\tau + (5-t)/2] \geq \deg(E') \geq t[\tau - t + 3]$.

Il reste à traiter le cas t = s. Dans ce cas les inégalités $d \geq \deg(E') \geq \sum_o^{s-1} (n_i-i) \geq$ $\geq \sum_o^{s-1} (s - 1 + d/s - 2i) = d$ entrainent E = E' et $n_i = s - 1 + d/s - i$ pour $i = 0, ..., s-1$. Posons $s' = d/s$. Soit A le cône projetant de E. Considérons la résolution minimale, comme R-module, d'un A-module dualisant Ω_A:

$$0 \to \oplus_o^{s-1} R[i-2] \to \oplus_o^{s-1} R[s+s'-1-i] \to \Omega_A \to 0$$

Le R-module Ω_A a donc un systéme minimal de générateurs $(\propto_o, ..., \propto_{s-1})$ avec $\deg(\propto_i) = 1 + i - s - s'$. Considérons la matrice, à coefficients dans R, dont les colonnes sont les coordonnées de $x_2^j \propto_o$ par rapport à $(\propto_o, ..., \propto_{s-1})$. C'est une matrice triangulaire à termes diagonaux de degrés 0.

Si l'un de ces termes diagonaux est nul, disons celui correspondant à la colonne d'ordre t, il est clair que, si t est minimal, le sous R-module de Ω_A engendré par $\propto_o, ..., \propto_{t-1}$ est identique au sous R-module engendré par $\propto_o$, $x_2 \propto_o, ..., x_2^{t-1} \propto_o$ et que ce sous R-module est un A-module. Utilisant le lemme, il existe un sous groupe de points E' de E de caractère $(n_o, ..., n_{t-1})$, donc de degré $\sum_o^{t-1} (n_i-i) = \sum_o^{t-1} (\tau + 2 - 2i) = t(\tau - t + 3)$, contenu dans une courbe de degré t.

Si tous les termes diagonaux sont inversibles la matrice est inversible et $\Omega_A = A.\propto_o = A[s+s'-1]$. Dans ce cas A est un anneau de Gorenstein de dimension

un quotient de $K[x_0, x_1, x_2]$, donc une intersection complète d'après un théorème de Serre. Mais la résolution minimale:
$0 \to \oplus_o^{s-1} R[-s'-i] \to \oplus_o^{s-1} R[-i] \to A \to 0$, exprime bien que E est l'intersection complète annoncée.

Remarques: (i) Illustrons le corollaire 2 par un exemple bien connu:
Si $\tau \geq d/3$ le groupe de points vérifie l'une des conditions suivantes:
(1) il y a $\tau + 2$ points alignés (comptés avec multiplicités)
(2) il y a $2\tau + 2$ ou $2\tau + 3$ points sur une conique (comptés avec multiplicités)
(3) $\tau = d/3$ et le groupe de points est intersection complète d'une cubique et d'une courbe de degré τ (éventuellement dégénérées).

(ii) Nous n'avons rien dit de la stratification du schéma de Hilbert par la postulation. Ce point de vue est traité dans [C] où Coppo obtient indépendamment des résultats voisins des nôtres.

(iii) Dans [D] Davis présente un résultat analogue à notre proposition.

Bibliographie:

[C] Coppo, M.A. : Thèse en préparation, Université de Nice.
[D] Davis,E: "0-dimensional subschemes of $\mathbb{P}^2$:lectures on Castelnuovo's function" Queen's Papers in pure & applied Math. 76 (1986)
[GP] Gruson,L.-Peskine,Ch. : "Genre des courbes de l'espace projectif" Lectures Notes in Math. 687, Springer (1978)
[H] Harris,J. : "The genus of space curves" Math. Ann. 249, 191-204 (1980)
[MR] Maggioni,R -Ragusa,A. :"The Hilbert function of generic plane sections of curves in $\mathbb{P}^3$" Invent. Math. 91, 253-255 (1988)

On singular Del Pezzo varieties

Takao FUJITA
Department of Mathematics, College of Arts and Sciences
University of Tokyo
Komaba, Meguro, Tokyo, 153 Japan

Introduction

By a *Del Pezzo variety* we mean a pair (V, L) of a variety V and an ample line bundle L on V such that

1) V has only Gorenstein (but possibly non-normal) singularities,
2) the dualizing sheaf ω_V is $\mathcal{O}_V((1 - n)L)$ for $n = \dim V$, and
3) $H^q(V, tL) = 0$ for any integers q, t with $0 < q < n$.

We have $\Delta(V, L) = 1$ for any Del Pezzo variety (see (0.3)). So we have a complete classification in case V is smooth (see **[F3]** and **[F4]**). In **[F5]** we studied the case in which L is very ample (but V may have singularity). In this paper we consider the general case.

In §1 we establish an existence theorem of a ladder of (V, L). From this it follows that L is very ample if and only if $L^n \geq 3$. Thus, we can apply the theory in **[F5]** in most cases. The cases in which L^n is small can be treated by ad hoc method. In §3, as an application, we study the exceptional divisor of a contraction of an extremal ray on a smooth 4-fold which is mapped to a point.

I would like to express my hearty thanks to the members of the organizing committee who gave me the opportunity to attend this conference.

§0. Preliminaries

(0.0) Throughout this paper we use the notation and convention as in **[F3]**, **[F4]**, **[F5]**. In particular, vector bundles and the locally free sheaves of their sections are not distinguished notationally.

Tensor products of line bundles are denoted additively, while we use multiplicative notation for products in Chow rings. The line bundle determined by a Cartier divisor D is denoted by $[D]$. Given a morphism $f\colon X \longrightarrow Y$ and a line bundle L on Y, we denote f^*L by L_X, or sometimes just by L when confusion is impossible or harmless.

(0.1) Let L be a line bundle on a variety V with $n = \dim V$. Then $\chi(V, tL) = \sum_{q=0}^{\infty}(-1)^q h^q(V, tL)$ is a polynomial of degree $\leq n$ of the form $\sum_{j=0}^{n}\chi_j t^{[j]}/j!$, where χ_j's are integers, $t^{[0]} = 1$ and $t^{[j]} = t(t+1)\cdots(t+j-1)$. Then $\chi_n = L^n$. We set $g(V, L) = 1 - \chi_{n-1}$, which will be called the *sectional genus* of (V, L).

The Δ-genus of (V, L) is defined by $\Delta(V, L) = n + L^n - h^0(V, L)$.

(0.2) If V has only Gorenstein singularities, the line bundle corresponding to the dualizing sheaf ω is called the canonical bundle of V, and is denoted by K. Then $(K + (n-1)L)L^{n-1} = 2g(V, L) - 2$.

To see this, take a very ample line bundle A on V such that $B = K + A$ is also very ample. Let D be a general member of $|A|$. We have an exact sequence $0 \longrightarrow \mathcal{O}_V(tL - A) \longrightarrow \mathcal{O}_V(tL) \longrightarrow \mathcal{O}_D(tL) \longrightarrow 0$. So $\chi(V, tL) = \chi(V, tL - A) + \chi(D, tL_D)$. Moreover $\chi(D, tL_D) \sim at^{n-1}$ for $a = L^{n-1}A/(n-1)!$ modulo terms of degree $< n - 1$. Similarly we have $\chi(V, B - tL) \sim \chi(V, -tL) + bt^{n-1}$ for $b = (-L)^{n-1}B/(n-1)!$. On the other hand $\chi(B - tL) = \chi(K + A - tL) = (-1)^n\chi(tL - A)$ by Serre duality. Combining these observations and computing the coefficients of t^{n-1} we obtain $(K + (n-1)L)L^{n-1} = -2\chi_{n-1}$.

Thus, in particular, $g(V, L) = 1$ for any Del Pezzo variety.

(0.3) Conversely, a polarized variety (V, L) is a Del Pezzo variety if $g(V, L) = 1$ and if $H^q(V, tL) = 0$ for any integers t, q with $0 < q < n = \dim V$. Moreover, $\Delta(V, L) = 1$ in this case.

For a proof, see [**F4**; (5.7.5)].

§1. Existence of a ladder

(1.1) From now on, throughout in this paper, everything is defined over an algebraically closed field of characteristic zero.

The main result of this section is the following

(1.2) **Theorem.** *Let (V, L) be a polarized variety such that $n = \dim V$, $\Delta(V, L) = 1$ and $g(V, L) < L^n = d$. Suppose that V has only Cohen Macaulay singularities. Then (V, L) has a ladder, namely, a sequence $V = V_n \supset V_{n-1} \supset \cdots \supset V_1$ of subvarieties V_j of V such that each V_{j-1} is a Cartier divisor on V_j whose linear equivalence class is the restriction of L to V_j.*

Proof. Note that $|L|$ has only finitely many base points by [F1; Theorem 1.9]. First we consider the case $n = 2$.

Let $\pi: M \longrightarrow V$ be a desingularization of V such that $\pi^*|L| = E + \Lambda$ for some effective divisor E and a linear system Λ without base points. So Λ defines a morphism $\rho: M \longrightarrow \mathbb{P}$ such that $\mathcal{O}[\Lambda] = \rho^*\mathcal{O}_{\mathbb{P}}(1)$. We may assume that π is "minimal", that means, there is no curve Z in M such that $Z \simeq \mathbb{P}^1$, $Z^2 = -1$ and both $\pi(Z)$ and $\rho(Z)$ are points.

Let W be the image of ρ. If $\dim W > 1$, any general member H of Λ is irreducible. Then π_*H is an irreducible member of $|L|$. It is generically smooth and hence reduced since V is locally Cohen Macaulay. Thus (V, L) has a ladder in this case.

We should study the case $\dim W = 1$ and $w = \deg(W) > 1$. We will derive a contradiction.

Let X be a general fiber of $f: M \longrightarrow W$. Then, numerically, $[\Lambda] \sim wX$. So $d = L(E + wX) = wLX \geq w$, while $\dim \Lambda = h^0(V, L) - 1 = 2 + d - \Delta(V, L) - 1 = d$ and $0 \leq \Delta(W, \mathcal{O}(1)) \leq 1 + w - (1 + \dim \Lambda) = w - d$. Hence $w = d$, $LX = 1$ and $\Delta(W, \mathcal{O}(1)) = 0$. So $W \simeq \mathbb{P}^1$.

Since $1 = LX = (E + H)X = EX$ for $H = [\Lambda]$, the prime decomposition of E is of the form $E = \sum \mu_i E_i$ with $\mu_0 = 1$, $E_0X = 1$ and $E_iX = 0$ for $i \neq 0$. We set $e = -E_0^2$. Since $L(L - E_0 - H) = L(E - E_0) = 0$, we have $0 \geq (L - E_0 - H)^2 = d - e$ by the index theorem. Hence $e \geq d \geq 2$.

Let K be the canonical bundle of M. We claim $KE_i \geq 0$ for any i. Note that $E_i^2 < 0$ since $LE_i = 0$. So E_i must be a (-1)-curve if $KE_i < 0$. Then $i \neq 0$, $E_iX = 0$ and $f(E_i)$ is a point. This contradicts the minimality of π. Thus we prove the claim, which implies $KE \geq KE_0 = e - 2$.

Assume that $KX \geq 0$. Then $KL = K(E + wX) \geq e - 2$ and hence $2g(M, L) - 2 = (K + L)L \geq d + e - 2 \geq 2d - 2$. Since $g(V, L) \geq g(M, L)$ this contradicts the hypothesis $g(V, L) < d$.

Thus we conclude $KX < 0$. Then $X \simeq \mathbb{P}^1$ and $KX = -2$. We further claim that f is a $\mathbb{P}^1$-bundle. Indeed, if there is a singular fiber Y of f, it contains either two (-1)-curves or a (-1)-curve of multiplicity two. Since $LY = LX = 1$, we have $LZ = 0$ for one of them, Z. But this is impossible by the minimality of π.

Now we easily see that $E = E_0$ and E is a section of f. So M is a Hirzebruch surface Σ_e. Moreover $d = e$ since $0 = LE = E^2 + wEX$. Note that $g(M, L_M) = 0 = \Delta(M, L_M)$.

Set $\mathcal{F} = \pi_*\mathcal{O}_M$ and let $\mathcal{C}$ be the cokernel of the natural injective homomorphism $\mathcal{O}_V \longrightarrow \mathcal{F}$. Set $v = \pi(E)$, the unique base point of $|L|$. Let $\mathcal{E}$ be the free sheaf $\mathcal{O}_V(H^0(V, L))$. Then we have natural homomorphisms $\mathcal{E} \longrightarrow \mathcal{O}_V[L]$, $\mathcal{E} \otimes [-L] \longrightarrow \mathcal{O}_V$ and $\mathcal{E}[-L]_M \longrightarrow \mathcal{O}_M$. Since the last one defines the linear system $E + |H|$, its cokernel is $\mathcal{O}_E$. Therefore $\mathrm{Coker}(\pi_*\mathcal{E}[-L] \simeq \mathcal{F} \otimes \mathcal{E}[-L] \longrightarrow \mathcal{F})$ is a subsheaf of $\pi_*\mathcal{O}_E$, which is of length one. This implies $\ell(\mathcal{F}/\mathfrak{m}\mathcal{F}) = 1$ for the maximal ideal $\mathfrak{m}$ of $\mathcal{O}_V$ defining the point v. Thus $\mathcal{O}_V/\mathfrak{m} \simeq \mathcal{F}/\mathfrak{m}\mathcal{F}$ and hence $\mathcal{C} = \mathfrak{m}\mathcal{C}$. So $v \notin \mathrm{Supp}(\mathcal{C})$ by Nakayama's Lemma.

If $\dim(\mathrm{Supp}(\mathcal{C})) < 1$, then V is normal off finite points. So the singular points are finite. Therefore V is normal by Serre's criterion. This contradicts $0 = \Delta(M, L_M) < \Delta(V, L) = 1$. Thus we have a curve Γ in $\mathrm{Supp}(\mathcal{C})$.

We claim $L\Gamma \geq d$. To see this, take a curve Y in M such that $\pi(Y) = \Gamma$. Then $Y \cap E = \emptyset$ since $v \notin \Gamma$. So $YX > 0$ and Y meets a general member of $|L|_M = E + |H|$ at points on d different fibers of f. On the other hand, points on different fibers off E are separated by $\pi^*|L|$, and hence are mapped to different points on V. Therefore Γ meets a general member of $|L|$ at d points. Thus we prove the claim.

Now we have $\chi(\mathscr{C}[tL]) \sim rt$ modulo constant terms in t, where r is an integer with $r \geq d$. Since $\chi(M, tL_M) \sim \chi(\mathscr{F}[tL]) = \chi(V, tL) + \chi(\mathscr{C}[tL])$, we obtain $g(V, L) = g(M, L_M) + r \geq d$. This contradicts the hypothesis and we finish the proof in case $n = 2$.

The case $n > 2$ is rather easy. Let $\pi: M \longrightarrow V$, $\pi^*|L| = E + \Lambda$ and $f: M \longrightarrow W \subset \mathbb{P}$ be as before. Since $\pi(E) = \mathrm{Bs}|L|$ is a finite set, $LE = 0$ in the Chow ring of M. We denote the pull-back of $\mathcal{O}_{\mathbb{P}}(1)$ by H. Then $L = E + H$ in $\mathrm{Pic}(M)$ and $LH^{n-1} = L^2H^{n-2} = \cdots = d > 0$. This implies $\dim W \geq n - 1 > 1$. Hence any general member S of Λ is irreducible and so is $D = \pi_*S$. Similarly as before, D is locally Macaulay and reduced. We have $g(D, L_D) = g(V, L)$ and $\Delta(D, L_D) \leq \Delta(V, L) = 1$. If $\Delta(D, L_D) = 1$, the induction hypothesis applies. If $\Delta(D, L_D) \leq 0$, then L_D is very ample (cf. [**F1**] or [**F4**]). In either case (D, L_D) has a ladder, hence so does (V, L). Q. E. D.

(1.3) Before giving applications, we exhibit an example showing that the hypothesis $g < d$ is indispensable.

Take a polarized manifold (W, H) such that $\dim W = n - 1$, $H^{n-1} = d$ and $\Delta(W, H) = 0$. For example let (W, H) be a scroll over $\mathbb{P}^1$. Note that H is very ample and $g(W, H) = 0$. Set $\mathscr{E} = \mathcal{O}_W \oplus \mathcal{O}_W[2H]$ and let $p: P = \mathbb{P}_W(\mathscr{E}) \longrightarrow W$ be the $\mathbb{P}^1$-bundle associated to $\mathscr{E}$. Let H_ζ be the tautological line bundle $\mathcal{O}_P(1)$ and set $H_\alpha = p^*H$. Then the unique member S of $|H_\zeta - 2H_\alpha|$ is a section of p. Let B_1 be a general member of $|(2b - 1)H_\zeta|$ for some positive integer b. Then B_1 is smooth and connected, and $S \cap B_1 = \emptyset$. Since $B = B_1 + S \in |2F|$ for $F = bH_\zeta - H_\alpha$, there is a finite double covering $f: M \longrightarrow P$ with branch locus B such that $f_*\mathcal{O}_M \simeq \mathcal{O}_P \oplus \mathcal{O}_P(-F)$. We have $f^*S = 2E$ for some prime divisor E on M and $E \simeq S \simeq W$. Moreover, since $[S]_S = -2H$, we have $[E]_E = -H$. So there is a birational morphism $\pi: M \longrightarrow V$ contracting E to a normal point. This singularity is formally isomorphic to the vertex of the projective cone over (W, H). Hence it is rational and Cohen Macaulay. We infer that $L = E + f^*H_\alpha$ comes from $\mathrm{Pic}(V)$ since $L_E = 0$. This pair (V, L) is an example with the desired property if $b > 1$.

To see this, note first that $2L_M = f^*H_\zeta$ is generated by global sections. So, in order to prove the ampleness of L_V, it suffices to show $LC > 0$ for any curve C in V. Let Z be the proper transform of C on M. Then Z is not contained in E. So $EZ \geq 0$ and $LC = (E + H_\alpha)Z \geq 0$. The equality holds only if $EZ = H_\alpha Z = 0$. But the latter implies that $f(Z)$ is contained in a fiber X of p. Since $S \cap X$ is a simple point, $f^{-1}(X)$ is irreducible and $Z = f^{-1}(X)$. This contradicts $EZ = 0$. Thus we conclude $LC > 0$, proving the ampleness.

We next show $L^n = d$ and $g = g(V, L) = (b - 1)d$. Indeed, since $L_E = 0$, we have $L^n = L^{n-1}H_\alpha = \cdots = LH_\alpha^{n-1} = EH_\alpha^{n-1} = H^{n-1}\{W\} = d$. To calculate g, we infer $g = g(M, L_M)$ since V is normal. So $2g - 2 = (K + (n - 1)L)L^{n-1}$ for the canonical bundle K of M. In view of $K = f^*(K^P + F) = (K^W - 2H_\zeta + 2H_\alpha + F)_M = (K^W + (b - 2)H_\zeta + H_\alpha)_M$ for the canonical bundle K^W of W, we infer $KL^{n-1} = (2b - n - 1)d - 2$ since $H_\zeta L^{n-1} = 2L^n = 2d$, $H_\alpha L^{n-1} = L^n = d$ and $K^W L^{n-1}\{M\} = 2^{1-n}K^W H_\zeta^{n-1} = 2^{2-n}K^W H_\zeta^{n-1}\{P\} = 2^{2-n}K^W s_{n-2}(\mathcal{E})\{W\} = 2^{2-n}K^W(2H)^{n-2} = -2 - (n - 2)d$. Thus we obtain $g = (b - 1)d$ by easy computation.

Now we claim $\Delta(V, L) = 1$ unless $b = 1$. Indeed, $h^0(V, L) = h^0(M, E + H_\alpha) \geq h^0(M, H_\alpha) \geq h^0(P, H_\alpha) = h^0(W, H) = n - 1 + d$. Thus $\Delta(V, L) \leq 1$. If $\Delta < 1$, we must have $g = 0$, which is impossible when $b > 1$. Hence $\Delta = 1$. When $b = 1$, we have in fact $\Delta(V, L) = 0$.

Alternately, when $b > 1$, we can directly check that E is the fixed component of $|L_M|$, since $E\Gamma = 1$ for any general fiber Γ of $\rho: M \longrightarrow W$ and Γ is a curve of arithmetic genus $b - 1$.

Thus, (V, L) satisfies the assumptions of the theorem (1.2) except $g < d$. But it does not have a ladder if $d > 1$ and $b > 1$. Indeed, when $n = 2$, any member of $|L_M|$ is of the form $E + \rho^* D$ for some divisor D of degree d on $W \simeq \mathbb{P}^1$. So the member $\pi_*(E + \rho^* D)$ is not prime. Thus there is no ladder. When $n > 2$, for any general member Y of $|L|$, (Y, L_Y) is a polarized variety of the same type as (V, L) constructed from a hyperplane section of W. Therefore we can find a sequence of subvarieties of V until dimension two, but never till dimension one.

(1.4) If V is smooth at each base point of $|L|$, we need not assume $g < d$ in (1.2). For a proof, see [**F2**] or [**F4**; (3.4)].

In the example (1.3), $\pi(E)$ is the unique base point of $|L|$ and this is a singular point of V. It is an ordinary double point when $d = 2$ and W is a hyperquadric.

(1.5) **Corollary.** *Let* (V, L) *be a polarized variety as in (1.2). Then* $g = g(V, L) > 0$ *and*

1) *any ladder of* (V, L) *is regular, which means, the restriction map* $H^0(V_j, L) \longrightarrow H^0(V_{j-1}, L)$ *is surjective for each* j,
2) $\mathrm{Bs}|L| = \emptyset$ *if* $d = L^n \geq 2$,
3) $g = 1$ *and* L *is simply generated if* $d \geq 3$, *and*
4) L *is quadratically presented if* $d \geq 4$.

For a proof, see [**F2**; Theorem 4.1] or [**F4**; (5.2) & (5.3)]. Here, "simply generated" means that the graded algebra $\oplus_{t\geq 0} H^0(V, tL)$ is generated by $H^0(V, L)$. In particular L is very ample in this case. If furthermore V is defined by equations of degree two under the embedding given by $|L|$, L is said to be quadratically presented.

(1.6) **Corollary.** *Let* (V, L) *be as above and assume* $d \geq 3$. *Then* $H^q(V, tL) = 0$ *for any* q, t *with* $0 < q < n$.

For a proof, see [**F4**; (3.8)]. Thus, in this case, (V, L) is a Del Pezzo variety by (0.3).

(1.7) **Corollary.** *If in addition* $d = 3$, V *is a hypercubic. If* *d = 4, V is a complete intersection of two hyperquadrics.*

§2. Classification of Del Pezzo varieties

(2.0) Throughout this section (V, L) is a Del Pezzo variety with $\dim V = n$ and $d = L^n$. So $g(V, L) = \Delta(V, L) = 1$.

(2.1) (V, L) has a ladder. Indeed, (1.2) applies if $d > 1$. If $d = 1$, any general member D of $|L|$ is irreducible and reduced since $L^{n-1}D = 1$. So one obtains a ladder.

For any ladder $\{V_j\}$ of (V, L), each (V_j, L) is also a Del Pezzo variety.

(2.2) When $d = 1$, (V, L) is a weighted hypersurface of degree 6 in the weighted projective space $\mathbb{P}(3, 2, 1, \cdots, 1)$.

For a proof, we use the induction on n. In case $n = 1$, the assertion is well-known since V is of arithmetic genus one. In case $n > 1$, we take a ladder and use Mori's argument **[M]** (or see **[F2**; §2]).

(2.3) When $d = 2$, (V, L) is a weighted hypersurface of degree 4 in the weighted projective space $\mathbb{P}(2, 1, \cdots, 1)$. Moreover there is a finite double covering $f: V \longrightarrow \mathbb{P}^n$ branched along a hypersurface of degree 4 such that $L = f^*\mathcal{O}_{\mathbb{P}}(1)$.

The first assertion is proved similarly as (2.2). The second assertion follows from (1.5; 2).

(2.4) When $d \geq 3$, L is very ample by (1.5). If $d = 3$ or 4, then (1.7) applies. If $d \geq 5$, we can use the theory in **[F5]**. In particular, if V is not a cone over another variety, its structure is very precisely described. But this is not always the case. Indeed:

(2.5) If (V, L) is a Del Pezzo variety, then the projective cone (C, H) over (V, L) is also a Del Pezzo variety.

To see this, note that V is identified with a general member of $|H|$. Then $L = H_V$ and $g(C, H) = g(V, L)$. So (1.6) and (0.3) apply.

(2.6) This trouble can be avoided by the following

Lemma. *Suppose that* $d \geq 5$ *and that* V *has only local complete intersection singularities. Then* V *is not a cone.*

Proof. Let D be a general member of $|L|$. We will derive a contradiction assuming that V is a cone over D. Indeed, since the vertex of V is local complete intersection, (D, L_D) must be a global complete intersection. This cannot be the case since (D, L_D) is a Del Pezzo variety of degree $d \geq 5$.

(2.7) **Corollary.** *Let* (V, L) *be as above. Then* V *is normal unless* $n \leq 3$.

Proof. If V is not normal, [**F5**; (2.7)] applies. The classification table there shows that the singularity of V is local complete intersection only if it is of type (N^1) or (C^1) with $n = 1$, $\mathbb{P}^1(N^1, H(C^1))$ or $\mathbb{P}^1(N^1, Q(C^1))$ with $n = 2$, or $\mathbb{P}^2(N^1, Q(C^1))$ with $n = 3$. In particular $n \leq 3$.

(2.8) Following [**F5**], we will describe precisely the structure of such a non-normal Del Pezzo threefold (V, L) with local complete intersection singularities.

We have $h^0(V, L) = d + 2$ and V is embedded in $P \simeq \mathbb{P}^{d+1}$ by $|L|$. The singular locus R of V is a plane in P. For any point v on R, the cone $W = v * V$ over V with vertex v is a 4-fold of degree $d - 2$. Taking three times general hyperplane sections of W we get a Veronese curve M of degree $d - 2$ (V is of type (c) in the terminology of [**F5**]). Moreover $W = M * R$, that is, the generalized cone over M with R being the set of vertices (see [**F5**; (6)]).

Let $\tilde{P}$ be the blowing-up of P along R and let $\tilde{W}$ be the proper transform of W. Then $\tilde{W} \simeq \mathbb{P}_M((d - 2)H_\beta \oplus \mathcal{O} \oplus \mathcal{O} \oplus \mathcal{O})$ where H_β is the line bundle on $M \simeq \mathbb{P}^1$ of degree one. The tautological line bundle is the pull-back of $\mathcal{O}_P(1)$ via $\tilde{W} \longrightarrow W \subset P$, which will be denoted by H_α. Thus $f: \tilde{W} \longrightarrow M$ is a $\mathbb{P}^3$-bundle.

The unique member D of $|H_\alpha - f^*(d - 2)H_\beta|$ is the exceptional divisor of $\pi: \tilde{W} \longrightarrow W$. Moreover $\pi(D) = R$, $D \simeq M \times R$ and the second projection is the restriction of π.

The proper transform $\hat{V}$ of V is a divisor on $\hat{W}$. It is a member of $|H_\alpha + f^*2H_\beta|$. If we choose homogeneous coordinates $(\beta_0:\beta_1)$ of M and $(\alpha_0:\alpha_1:\alpha_2)$ of R suitably, then the divisor $Q = \hat{V} \cap D$ on D is defined by the equation $\alpha_0\beta_0{}^2 + \alpha_1\beta_0\beta_1 + \alpha_2\beta_1{}^2 = 0$. Hence $f_Q\colon Q \longrightarrow M$ is a $\mathbb{P}^1$-bundle, while $\pi_Q\colon Q \longrightarrow R$ is a finite double covering with branch locus defined by $\alpha_1{}^2 = 4\alpha_0\alpha_2$. This implies $Q \simeq M \times \mathbb{P}^1_\sigma$ and $\pi^*H_{\alpha|Q} = H_\beta + H_\sigma$, where H_σ is the pull-back of $\mathcal{O}(1)$ of $\mathbb{P}^1_\sigma$. For proofs, see (7), (8), (c.1), (c.3) in **[F5]**.

The map $\hat{V} \longrightarrow V$ is finite and is the normalization of V. Moreover $\hat{V}$ is a $\mathbb{P}^2$-bundle over M. Since $\hat{V} \in |H_\alpha + 2H_\beta|$, we have an exact sequence $0 \longrightarrow \mathcal{O}(-2) \longrightarrow \mathcal{O}(d - 2, 0, 0, 0) \longrightarrow \mathcal{E} \longrightarrow 0$ on M such that $\mathbb{P}(\mathcal{E}) \simeq \hat{V}$. By the above observation about Q, we infer that the cokernel of the composition $\mathcal{O}(-2) \longrightarrow \mathcal{O}(d - 2, 0, 0, 0) \longrightarrow \mathcal{O}(0, 0, 0)$ is isomorphic to $\mathcal{O}(1, 1)$. From this we obtain $\mathcal{E} \simeq \mathcal{O}(d - 2, 1, 1) = (d - 2)H_\beta \oplus H_\beta \oplus H_\beta$.

(2.9) Unlike the normal case below, there is no upper bound of d in the above case (2.8), where V is assumed not to be a cone.

When V is normal and singular, **[F5**; (2.9)] applies unless V is a cone. In particular $d \leq 8$ (resp. 6, 6, 5) if $n = 2$ (resp. 3, 4, 5). The case $n \geq 6$ does not occur if $d \geq 5$.

When V is smooth and $d \geq 5$, we have $n \leq 6$. Moreover $d \leq 9$ (resp. 8, 6, 5, 5) if $n = 2$ (resp. 3, 4, 5, 6). See **[F3]**.

§3. Applications

Here we recall and will improve a few results in **[B2]**, making some necessary supplements.

(3.1) Let X be a smooth 4-fold whose canonical bundle $K = K_X$ is not nef. Let $\phi\colon X \longrightarrow Y$ be a contraction of an extremal ray. Here, as in **[B2]**, we will be interested in the case where ϕ is birational and contracts an effective divisor E to a point.

(3.2) In the above case there is an ample line bundle L on X such that $K_{|E} = -aL_E$ and $[E]_E = -bL_E$ for some positive integers a, b. We have $a = b = 1$ unless $\Delta(E, L_E) = 0$. Moreover, when $\Delta = 0$, E is isomorphic to either $\mathbb{P}^3$ or a (possibly singular) hyperquadric. For a proof, see the beginning of the proof of Prop. 2.4 in [B2].

(3.3) In the sequel we study the case $a = b = 1$. Then (E, L_E) is a Del Pezzo variety by [B1; Lemma 2.3] and (0.3). Therefore we can apply the results in §2. We use (1.2) instead of Prop. 2.3 in [B2], which is not always true (see (1.3)).

(3.4) When E is normal, we can classify (E, L_E) by the method in [B2]. So, from now on, we assume that E is not normal.

We further assume $d \geq 5$, since otherwise (2.2), (2.3) or (1.7) applies.

Thus (2.7) applies and (E, L_E) is of the type (2.8). In particular the singular locus R of E is isomorphic to $\mathbb{P}^2$.

Let $\mu\colon \tilde{X} \longrightarrow X$ be the blow-up of X along R and let Z be the exceptional divisor over R. By the universality of the blowing-up, the proper transform $\tilde{E}$ of E on $\tilde{X}$ is isomorphic to the abstract blowing-up of E along the subscheme R of E. This applies to $\tilde{V}$ in (2.8) too. Therefore $\tilde{E} \simeq \tilde{V}$. Furthermore $Z \cap \tilde{E} \simeq Q \simeq \mathbb{P}^1_\beta \times \mathbb{P}^1_\sigma$. Thus $[Z]_Q$ is the normal bundle of Q in $\tilde{E}$, which is $[D]_Q = H_\sigma + (3 - d)H_\beta$ in (2.8).

Let $\mathcal{N}^\vee$ be the conormal bundle of R in X. Then $Z \simeq \mathbb{P}_R(\mathcal{N}^\vee)$ and $[-Z]_Z$ is the tautological line bundle, which will be denoted by H_ζ. We have $H_\zeta H_\alpha\{Q\} = -(H_\sigma + (3 - d)H_\beta)(H_\sigma + H_\beta) = d - 4$. On the other hand this is equal to $H_\zeta H_\alpha[\tilde{E}]_Z\{Z\}$, while $\tilde{E} = \mu^*E - 2Z$ in $\mathrm{Pic}(\tilde{X})$. So $\tilde{E}_Z = -bH_\alpha + 2H_\zeta$ when $[E]_R = \mathcal{O}_R(-b)$. Actually we have $b = 1$ in case (3.3).

Suppose that $K_{X|E} = -aL_E$. Then $K_{X|R} = \mathcal{O}_R(-a)$ and $c_1(\mathcal{N}^\vee) = \mathcal{O}_R(3 - a)$. So $H_\zeta{}^2H_\alpha\{Z\} = 3 - a$ and $d - 4 = H_\zeta H_\alpha[\tilde{E}]\{Z\} = H_\zeta H_\alpha(2H_\zeta - bH_\alpha) = 6 - 2a - b$.

In our particular case we have $a = b = 1$ and hence $d = 7$. Moreover $\tilde{E} \simeq \mathbb{P}_M(5H_\beta \oplus H_\beta \oplus H_\beta)$ by (2.8).

Remark. It is uncertain whether this case $d = 7$ does really occur or not.

(3.5) Any way, thus we have shown that $d \leq 8$ in case (3.3). This improves upon Prop. 2.4 in [B2].

Bibliography

[B1] M. Beltrametti, On d-folds whose canonical bundle is not numerically effective, according to Mori and Kawamata, to appear in Annali Mat. Pura e Appl.

[B2] M. Beltrametti, Contractions of non-numerically effective extremal rays in dimension 4, in Proc. Conf. on Alg. Geom. Berlin 1985, pp. 24-37, Teubner Text zur Math. 92, 1987.

[F1] T. Fujita, On the structure of polarized varieties with Δ-genera zero, J. Fac. Sci. Univ. of Tokyo, **22** (1975), 103-115.

[F2] T. Fujita, Defining equations for certain types of polarized varieties, in Complex Analysis and Algebraic Geometry, pp.165-173, Iwanami, Tokyo, 1977.

[F3] T. Fujita, On the structure of polarized manifolds of total deficiency one, I, II and III, J. Math. Soc. Japan **32** (1980), 709-725, ibid., **33** (1981), 415-434 & **36** (1984), 75-89.

[F4] T. Fujita, On polarized varieties of small Δ-genera, Tôhoku Math. J. **34** (1982), 319-341.

[F5] T. Fujita, Projective varieties of Δ-genus one, in Algebraic and Topological Theories —— to the memory of Dr. Takehiko MIYATA, pp. 149-175, Kinokuniya, 1985.

[M] S. Mori, On a generalization of complete intersections, J. Math. Kyoto Univ. **15** (1975), 619-646.

Note. Here I would like to correct an error in [**F5**, (2.11)]. There I claimed "the possible type of singularities are subgraphs of Dynkin diagram ··· ", but this is not true. I should have written "the possible type of singularities are graphs whose corresponding root systems are subsystems of the root system of the Dynkin diagram ··· ".

I would like to thank Dr. T. Urabe who pointed out this mistake.

ABELIAN SURFACES IN PRODUCTS OF PROJECTIVE SPACES.

Klaus Hulek
Mathematisches Institut, Universität Bayreuth
Postfach 10 12 51, D-8580 Bayreuth
Federal Republic of Germany

0. Introduction

It is well known that every abelian surface in $\mathbb{P}_4$ has necessarily degree ten. To prove the existence of such surfaces is a much harder problem [HM], [R], [L], [HL]. In this brief note we shall investigate the existence of abelian surfaces in $\mathbb{P}_2 \times \mathbb{P}_2$ and $\mathbb{P}_1 \times \mathbb{P}_3$. Again this problem falls into two parts. Here we shall show that in both cases there is only one possible class for which an abelian surface can exist. This follows mostly from the self-intersection formula, but not entirely. As a consequence it follows that the only abelian surfaces in $\mathbb{P}_2 \times \mathbb{P}_2$ are the obvious ones, namely products of cubic curves. For $\mathbb{P}_1 \times \mathbb{P}_3$ we exhibit a possible candidate. The proof that such a surface really exists, however, needs methods which are a lot more subtle than the rather elementary considerations of this note. We shall not pursue this here, but hope to come back to it at some future point. Although our results are quite easy to prove, there does not seem to be a reference in the literature.

1. Preliminaries

If $Z = X \times Y$ is a product we denote the canonical projections by p and q respectively:

$$\begin{array}{ccc} & Z = X\times Y & \\ {}^{p}\swarrow & & \searrow^{q} \\ X & & Y \end{array}$$

If $\mathcal{L}$ and $\mathcal{M}$ are line bundles on X resp. Y we set

$$\mathcal{L} \boxtimes \mathcal{M} := p^*\mathcal{L} \otimes q^*\mathcal{M}.$$

In particular, if $X = \mathbb{P}_k$ and $Y = \mathbb{P}_m$ we set

$$\mathcal{O}(a,b) := \mathcal{O}_{\mathbb{P}_k}(a) \boxtimes \mathcal{O}_{\mathbb{P}_m}(b).$$

We denote the classes of $\mathcal{O}_Z(1,0)$, resp. $\mathcal{O}_Z(0,1)$ in $H^2(Z,\mathbb{Z})$ by h_1 resp. h_2.

Lemma 1.1: Let C be a curve. Then the product $C \times \mathbb{P}_2$ does not contain an abelian surface X unless C is an elliptic curve and $X = C \times D$ where D is a smooth cubic curve.

Proof. First assume $g(C) \geq 2$. Then the assertion is obvious since the projection $X \to C$ must be surjective and this would imply the existence of a non-constant 1-form on X.

Case 1: $g(C) = 0$. Then $C = \mathbb{P}_1$ and

$$\mathcal{O}_{\mathbb{P}_1 \times \mathbb{P}_2}(X) = \mathcal{O}(a,b)$$

for some $a,b > 0$. By the adjunction formula

$$\omega_X = \mathcal{O}_X(a-2,b-3).$$

Since X is abelian $\omega_X = \mathcal{O}_X$ and this implies

$$(ah_1+bh_2)((a-2)h_1+(b-3)h_2)h_i = 0 \qquad (i = 1,2)$$

i.e.

$$b(b-3) = 0 \qquad (i = 1)$$
$$a(b-3)+b(a-2) = 0 \qquad (i = 2).$$

It follows that $b = 3$ and $a = 2$. On the other we get from

$$0 \to \mathcal{O}(-2,3) \to \mathcal{O} \to \mathcal{O}_X \to 0$$

an exact sequence

$$H^1(\mathcal{O}) \to H^1(\mathcal{O}_X) \to H^2(\mathcal{O}(-2,-3)).$$

By Serre-duality and the Künneth formula

$$h^2(\mathcal{O}(-2,-3)) = h^1(\mathcal{O}) = 0.$$

Hence $h^1(\mathcal{O}_X) = 0$, a contradiction.

Case 2: $g(C) = 1$. Here our argument is very similar. By [Ha,p.292]

$$\operatorname{Pic}(C \times \mathbb{P}_2) = \operatorname{Pic} C \times \operatorname{Pic} \mathbb{P}_2$$

i.e. we can write

$$\mathcal{O}_{C \times \mathbb{P}_2}(X) = \mathcal{L} \boxtimes \mathcal{O}(b)$$

for some $b > 0$ and $\mathcal{L} \in \operatorname{Pic} C$. Since

$$\omega_{C \times \mathbb{P}_2} = \mathcal{O}_C \boxtimes \mathcal{O}(-3)$$

the adjunction formula gives

$$\omega_X = \mathcal{L} \boxtimes \mathcal{O}(b-3)|X.$$

Let $a = \deg \mathcal{L}$. then arguing as before we find

$$b(b-3) = 0$$
$$ab + a(b-3) = 0$$

i.e. $b = 3$, $a = 0$. Since

$$H^\circ(\mathcal{L} \boxtimes \mathcal{O}(3)) = H^\circ(\mathcal{L}) \otimes H^\circ(\mathcal{O}(3))$$

it follows that $\mathcal{L} = \mathcal{O}_C$ and the assertion is then obvious.

Finally we recall the <u>self-intersection formula</u> from [F,p.103]. For any regular embedding $i : X \to Z$ of codimension d with normal bundle $N_{X/Z}$:

$$i^* i_*[\alpha] = c_d(N_{X/Z}) \cap [\alpha]$$

for all $\alpha \in A_*(X)$.

In particular, if X is a smooth surface in a 4-manifold Z then

$$[X]^2 = c_2(N_{X/Z}) \cap [X].$$

2. <u>Abelian surfaces in $\mathbb{P}_2 \times \mathbb{P}_2$</u>

In this section we shall prove

<u>Proposition</u> 2.1: <u>Every abelian surface</u> X <u>in</u> $\mathbb{P}_2 \times \mathbb{P}_2$ <u>is of the form</u> $X = C \times D$ <u>where</u> C <u>and</u> D <u>are smooth cubics</u>.

<u>Proof.</u> We shall first apply the self-intersection formula. The class of X is of the form

$$[X] = \alpha h_1^2 + \beta h_2^2 + \gamma h_1 h_2$$

with integers $\alpha, \beta, \gamma \geq 0$. From the normal bundle sequence

$$0 \to T_X \to T_{\mathbb{P}_2 \times \mathbb{P}_2}|X \to N_{X/\mathbb{P}_2 \times \mathbb{P}_2} \to 0$$

and the fact that T_X is trivial one finds

$$c(N_{X/\mathbb{P}_2\times\mathbb{P}_2}) = c(T_{\mathbb{P}_2\times\mathbb{P}_2}|X)$$

$$= (1+3h_1+3h_1^2)(1+3h_2+3h_2^2).[X].$$

Hence

$$c_2(N_{X/\mathbb{P}_2\times\mathbb{P}_2}) = 3\alpha + 3\beta + 9\gamma.$$

Since

$$[X]^2 = \gamma^2 + 2\alpha\beta$$

the self-intersection formula implies

$$(1) \qquad 3\alpha + 3\beta + 9\gamma = \gamma^2 + 2\alpha\beta.$$

Claim 1: (i) $\alpha = 0$ or $\alpha \geq 6$

(ii) $\beta = 0$ or $\beta \geq 6$.

Clearly it is enough to prove (i). Assume that $\alpha > 0$. Then projection onto the second factor gives a surjective map

$$q|_X : X \to \mathbb{P}_2$$

of degree α. Since an abelian surface does not contain curves with negative self-intersection it follows that $q|_X$ is finite. Hence the Nakai-Moishezon criterion implies that

$$\mathcal{O}_X(0,1) = (q|_X)^* \mathcal{O}_{\mathbb{P}_2}(1)$$

is ample. By Kodaira vanishing $h^1(\mathcal{O}_X(0,1)) = h^2(\mathcal{O}_X(0,1)) = 0$ and Riemann-Roch gives

$$h^\circ(\mathcal{O}_X(0,1)) = \tfrac{1}{2}[\mathcal{O}_X(0,1)]^2 = \tfrac{1}{2}[X]h_2^2 = \tfrac{\alpha}{2}.$$

Since $h^\circ(\mathcal{O}_X(0,1)) \geq h^\circ(\mathcal{O}_{\mathbb{P}_2}(1)) = 3$ we find $\alpha \geq 6$.

Claim 2: $\alpha = 0$ or $\beta = 0$.

Assume $\alpha,\beta > 0$. From (1) we get

$$\gamma^2-9\gamma = \alpha(3-\beta) + \beta(3-\alpha).$$

By Claim 1 we have $\alpha,\beta \geq 6$, hence

$$\gamma^2-9\gamma \leq -36$$

On the other hand $\gamma^2-9\gamma \geq -\frac{81}{4}$ for all $\gamma \in \mathbb{R}$, a contradiction.

In order to prove the proposition we can now assume $\alpha = 0$ the case $\beta = 0$ being analogous. In this case projection onto the second factor gives a map

$$X \to D \subset \mathbb{P}_2$$

where D is a (possibly singular) curve. Let $\nu : \tilde{D} \to D$ be the normalization map. Let D_o be the smooth part of D and let X_o be the open set of X which lies over D_o. Since $id \times \nu$ is an isomorphism away from the singularities of D we can consider X_o to be a subset of $\mathbb{P}_2 \times \tilde{D}$. Let $\tilde{X}$ be its Zariski-closure. Then we have a commutative diagram

$$\begin{array}{ccc} \tilde{X} & \to & \mathbb{P}_2 \times \tilde{D} \\ \downarrow g & & \downarrow id \times \nu \\ X & \to & \mathbb{P}_2 \times D. \end{array}$$

By construction g is finite and birational. Since X is smooth it follows from [S, theorem 5, p. 115] that g is an isomorphism. By lemma 1.1 it follows that $\tilde{D}$ is an elliptic curve and that $\tilde{X} = C \times \tilde{D}$ where C is a smooth cubic. But then D must already have been smooth and $X = C \times D$ as claimed.

3. Abelian surfaces in $\mathbb{P}_1 \times \mathbb{P}_3$

Here we prove

Proposition 3.1: *If X is an abelian surface in $\mathbb{P}_1 \times \mathbb{P}_3$ then its class is of the form $[X] = 8h_1h_2 + 6h_2^2$. Every abelian surface in $\mathbb{P}_1 \times \mathbb{P}_3$ is necessarily isogenous to a product.*

Proof. Let the class of X be

$$[X] = \alpha h_1 h_2 + \beta h_2^2.$$

Then $\alpha, \beta \geq 0$ and the interpretation of α and β is as follows: If $\alpha > 0$ the projection q induces a map

$$\bar{q} := q|_X \ X \to \bar{X} \subset \mathbb{P}_3$$

onto a surface $\bar{X}$. Then

$$\alpha = \deg \bar{q} \cdot \deg \bar{X}.$$

Note that as in the proof of proposition 2.1 the map $X \to \bar{X}$ must be finite. Since $\mathbb{P}_3$ does not contain abelian surfaces it follows that $\beta > 0$. the map

$$\bar{p} := p|_X : X \to \mathbb{P}_1$$

is surjective and the fibres are space curves of degree β.

As before we want to make use of the self-intersection formula. From

$$\begin{aligned} c(N_{X/\mathbb{P}_1\times\mathbb{P}_3}) &= c(T_{\mathbb{P}_1} \times T_{\mathbb{P}_3}|X) \\ &= (1+2h_1)(1+4h_2+6h_2^2+4h_2^3)|[X] \end{aligned}$$

we get

$$\begin{aligned} c_2(N_{X/\mathbb{P}_1\times\mathbb{P}_3}) &= (8h_1h_2+6h_2^2).[X] \\ &= 6\alpha + 8\beta. \end{aligned}$$

Since

$$[X]^2 = 2\alpha\beta$$

we find

$$(2) \qquad 2\alpha\beta = 6\alpha + 8\beta.$$

Since $\beta > 0$ this also shows that $\alpha > 0$.

<u>Claim</u> 1: $\alpha \geq 8$.

Since $\alpha > 0$ the projection q induces a finite map

$$\bar{q} : X \to \bar{X} \subset \mathbb{P}_3.$$

By lemma 1.1 $\bar{X}$ cannot be a plane, hence $\bar{X}$ spans $\mathbb{P}_3$. Since $\bar{q}$ is finite the line bundle

$$\mathcal{O}_X(0,1) = \bar{q}^*\mathcal{O}_{\mathbb{P}_3}(1)$$

is ample. By Kodaira vanishing and Riemann-Roch this shows

$$h^\circ(\mathcal{O}_X(0,1)) = \tfrac{1}{2}[\mathcal{O}_X(0,1)]^2 = \tfrac{1}{2}\, h_2^2.[X] = \tfrac{\alpha}{2}.$$

Since

$$h^\circ(\mathcal{O}_X(0,1)) \geq h^\circ(\mathcal{O}_{\mathbb{P}_3}(1)) = 4$$

we find $\alpha \geq 8$.

<u>Claim</u> 2: $4 \leq \beta \leq 6$.

We can rewrite (2) as

$$(3) \qquad \alpha(6-\beta) = \beta(\alpha-8).$$

Since $\alpha \geq 8$ we get $\beta \leq 6$. Also from (3) resp. (2) we find that $\beta \leq 2$

implies $\alpha < 0$ and $\beta = 3$ leads to a contradiction. Hence $4 \leq \beta \leq 6$.

This leaves us with the following possibilities for α and β

$$(\alpha,\beta) = (16,4),(10,5) \text{ or } (8,6).$$

We have to exclude the first two possibilities. To do this we consider the fibration

$$\bar{p} : X \to \mathbb{P}_1.$$

For each $t \in \mathbb{P}_1$ the fibre X_t is a space curve of degree $\beta = 4$ or 5. Since every morphism from $\mathbb{P}_1$ to X is constant it follows that every fibre X_t is reduced and irreducible. Moreover for a general fibre $\omega_{X_t} = \mathcal{O}_{X_t}$ by the adjunction formula, i.e. $\bar{p}$ is an elliptic fibration. We have already seen that this fibration has no multiple fibres. By Kodaira's classification of singular fibres [BPV, p.151] it can, therefore, not have any singular fibres at all. But then the canonical bundle formula [BPV, Corollary (12.3), p.162] implies

$$\omega_X = \bar{p}^* \mathcal{O}_{\mathbb{P}_1}(-2) = \mathcal{O}_X$$

a contradiction.

Hence the only remaining possibility is $(\alpha,\beta) = (8,6)$. It remains to show that X is isogenous to a product. In order to see this we look again at the fibration

$$\bar{p} : X \to \mathbb{P}_1.$$

As before the general fibre must be smooth with trivial canonical bundle. It is either an elliptic sextic curve or the union of two plane cubics. In any case X contains an elliptic curve and by Poincare's theorem on complete reducibility X is isogenous to a product.

3.2 Remark: As we have seen in the proof of proposition 3.1 the general fibre of the projection

$$\bar{p} : X \to \mathbb{P}_1$$

is either an elliptic sextic curve or a union of two plane cubics. The first case can again be excluded using the canonical bundle formula. In the second case we can use Stein factorisation to get an elliptic fibration

$$\tilde{p} : X \to C$$

over an elliptic curve C whose fibres are all smooth plane cubics. In particular X is a fibre bundle over C.

3.3 <u>Problem</u>: Do abelian surfaces X in $\mathbb{P}_1 \times \mathbb{P}_3$ with $[X] = 8h_1h_2 + 6h_2^2$ exist?

We want to conclude this section with an example which seems to be a good candidate for an abelian surface which can be embedded in $\mathbb{P}_1 \times \mathbb{P}_3$. For this purpose let E be an elliptic curve. Let P_o be a point of order 2 and let Q_o be a point of order 4 with $2Q_o = P_o$. The natural projections

$$\pi_1 : \quad E \to E_1 := E/\langle P_o\rangle$$

$$\pi_2 : \quad E \to E_2 := E/\langle Q_o\rangle$$

have degree 2 resp. 4. We set

$$X := E_1 \times E_2 .$$

The map

$$\pi = (\pi_1, \pi_2) : E \to X$$

is an embedding since $\ker \pi_1 \cap \ker \pi_2 = \{0\}$. Moreover

$$E.E_1 = \deg \pi_2 = 4$$

$$E.E_2 = \deg \pi_1 = 2.$$

We set

$$H_2 := E + E_1 .$$

Note that

$$H_2^2 = 8, \quad H_2.E_2 = 3.$$

The following statements are easy to check:

(1) The linear system $|H_2|$ is base point free and defines a map

$$\bar{q} : X \to \mathbb{P}_3 .$$

(2) Under the map $\bar{q}$ the translates of the elliptic curves E_2 are mapped isomorphically to smooth plane cubics.

Finally let

$$p' : E_1 \to \mathbb{P}_1$$

be a map of degree 2. This induces a map

$$\bar{p} : X \to \mathbb{P}_1 .$$

3.4 <u>Problem:</u> Can one choose $\bar{p}$ such that the map

$$(\bar{p},\bar{q}) : X \to \mathbb{P}_1 \times \mathbb{P}_3$$

is an embedding?

It is easy to check that the construction is such that all numerical conditions are fulfilled.

Acknowledgement: The author would like to thank the DFG for support under grant HU 337/2-1.

References

[BPV] Barth,W., Peters,C., Van de Ven,A.: Compact complex surfaces. Springer Verlag 1984.

[F] Fulton, W.: Intersection theory. Springer Verlag 1984.

[H] Hartshorne, R.: Algebraic Geometry, Springer Verlag 1977.

[HM] Horrocks, G., Mumford, D.: A rank 2 bundle on $\mathbb{P}^4$ with 15,000 symmetries. Topology 12, 63-81 (1973).

[HL] Hulek,K., Lange,H.: Examples of abelian surfaces in $\mathbb{P}_4$. J. Reine Angew. Math. 363, 201-216 (1985).

[L] Lange, H.: Embeddings of Jacobian surfaces in $\mathbb{P}^4$. J. Reine Angew. Math. 372, 71-86 (1986).

[R] Ramanan, S.: Ample divisors on abelian surfaces. Proc. London Math. Soc. 51, 231-245 (1985)

[S] Shafarevich, I.R.: Basic algebraic geometry. Springer Verlag 1977.

EMBEDDED PROJECTIVE VARIETIES OF SMALL INVARIANTS.III

Paltin Ionescu
University of Bucharest, Department
of Mathematics, str. Academiei 14,
70109 Bucharest , ROMANIA

Introduction

Several years ago we have started a program aiming at a classification of embedded smooth projective varieties (over $\mathbb{C}$) following the values of their numerical invariants, assumed to be small enough (see [8], [9], [10], [11], [12]). Although we were primarily interested in the classification according to the degree d, consideration of other invariants (namely the sectional genus g and the Δ-genus Δ) became necessary. The basic tool of our investigation was the adjunction mapping, whose properties were recently understood completely (cf. [24], [25], [14]). We gradually found out the following limitations, inherent to the method employed: $d \leq 8$ (cf. [13]) $g \leq 7$, $\Delta \leq 5$ (cf.[14]). On the other hand, the classification problem naturally splits into two parts. The first task is to obtain a maximal list; secondly, each case has to be investigated in order to decide whether or not it really occurs. Thus, for $d \leq 6$ the list was effective, cf.[8], [9]. For d=7 (see [10], [11]) the existence of four types was left open, while for d=8 the undecided cases were more numerous (cf.[12]). This paper, which is the last in this series, settles the existence problem in all these situations. Thus, the list given in [11] for d=7 turns out to be effective, while from the list given in [12] for d=8 three types have to be excluded (see the table below). We have thus completed the classification of smooth projective varieties up to degree 8. In contrast to [11] where we used mainly ad hoc methods, this time we took the opportunity to present systematically the few general methods available for proving the existence of embedded manifolds.
Finally, let us point out those cases which, in the meantime, were settled by other authors, namely: A.Buium (cf.[2]) first proved the existence of a certain surface of degre 8 in $\mathbb{P}^5$; C. Okonek ([20], [21]) proved the existence of a certain 3-fold of degree 7 in $\mathbb{P}^5$ and of two types of surfaces of degree 8 in $\mathbb{P}^4$; finally, as we can judge from [15], recently J. Alexander showed the existence of a rational surface with d=8, g=5 in $\mathbb{P}^4$, a seemingly subtle case.

The reader may consult [4], [11], [14], [21] for further references on the subject.

Acknowledgement. I am indebted to C.Bănică for some useful conversation. Special thanks are due to I. Coandă for many helpful discussions on vector bundles.

Conventions. Basically we employ the same definitions and notations as in the first two parts [11] and [12]. Let us recall from [12] that the term "linear fibration" used in [11] was replaced by "scroll"; thus, the term "scroll" from [11] became "scroll over a curve". For convenience we recall some of the notations:

- X $\mathbb{P}^n_{\mathbb{C}}$ is a smooth, connected, linearly normal and non-degenerate closed subvariety; dim $X=r$, codim $X=s$, degree of $X=d$.
- H is a (smooth) hyperplane section of X.
- g is the sectional genus of X.
- Δ is the Δ-genus of X.

- $q=h^1(\mathcal{O}_X)$
- $E^\vee$ denotes the dual of a vector bundle E.
- $T_X(\Omega^1_X)$ is the tangent (cotangent) bundle of X.
- ω_X or $\mathcal{O}_X(K)$ is the canonical bundle of X.
- $p_g = h^0(\omega_X)$
- $D_1=D_2$ (resp. $D_1\equiv D_2$) denotes linear equivalence (resp. numerical equivalence) of divisors.
- If $Y\subset X$ is a subvariety, $D|_Y$ denotes restriction of a divisor (class).
- I denotes the sheaf of ideals of Y.
- A smooth projective variety is also called a manifold.

The following table presents the list of (linearly normal, non-degenerate) submanifolds $X\subset\mathbb{P}^n_{\mathbb{C}}$ of degree 8. Notation $\sigma_Z:X\to Y$ means that X is the blowing-up of Y with center Z; E denotes the exceptional locus of σ_Z. L is a line in $\mathbb{P}^2$.

s	r	Abstract structure of X	H or $\mathcal{O}_X(H)$
	1	$\mathbb{P}^1$	$\mathcal{O}(8)$
7	2-8	scroll over $\mathbb{P}^1$	
	1	g=1	
6	2	$-\mathbb{P}^1 \times \mathbb{P}^1$ $-\sigma_P : X \to \mathbb{P}^2$	$\mathcal{O}(2,2)$ $\sigma^*(3L)-E$
	3	$\mathbb{P}^3$	$\mathcal{O}(2)$
	1	g=2	
	2	$-\sigma_{P_o,\ldots,P_4} : X \to \mathbb{P}^2$ - scroll over an elliptic curve	$\sigma^*(4L)-2E_o-E_1-\ldots-E_4$ $e=0$ $H \equiv C_o+4F$ $e=2$ $H \equiv C_o+5F$
5	3	$X \subset \mathbb{P}^1 \times Q^3$ as a hyperplane section, $Q^3 \subset \mathbb{P}^4$ the hyperquadric	
	4	$\mathbb{P}^1 \times Q^3$	Segre embedding
	1	g=3	
	2	$-\ \sigma_{P_1,\ldots,P_8} : X \to F_e$ $e \le 3$ $-\ \sigma_{P_1,\ldots,P_8} : X \to \mathbb{P}^2$ $-f : X \to \mathbb{P}^2$ double covering	$\sigma^*(H_e)-E_1-\ldots-E_8$ $H_e = 2C_o+(4+e)F$ $\sigma^*(4L)-E_1-\ldots-E_8$ $H=-2K$
4	3	- scroll over an elliptic curve $-\ \mathbb{P}^1 \times \mathbb{P}^3 \cap Q^6$, $Q^6 \subset \mathbb{P}^7$ a hyperquadric $-f : X \to Z \subset \mathbb{P}^1 \times \mathbb{P}^3 \subset \mathbb{P}^7$ double covering, Z a hyperplane section of $\mathbb{P}^1 \times \mathbb{P}^3$ $-\ \mathbb{P}(E)$, E rank-2 vector bundle on $\mathbb{P}^2$, given by $0 \to \mathcal{O}_{\mathbb{P}^2} \to E \to I^{(4)}_{\{P_1,\ldots,P_8\}} \to 0$	tautological
	4	$f : X \to \mathbb{P}^1 \times \mathbb{P}^3$ double covering, discriminant divisor $D \in \lvert\mathcal{O}(2,2)\rvert$	$f^*\mathcal{O}(1,1)$
	2	- scroll, $e=-2$, $g=2$ $-\ C \times \mathbb{P}^1$, $C \subset \mathbb{P}^2$ curve of degree 4 - geometrically ruled elliptic surface, $e=-1$	$H \equiv C_o+3F$ Segre embedding $H \equiv 2C_o+F$
	1	g=4	

s	r	Abstract structure of X	H or $\mathcal{O}_X(H)$
3	2	$-\sigma_{P_1,\dots,P_{10}}: X\to Q^2\subset \mathbb{P}^3$	$\sigma^*(3H_Q)-E_1-\dots-E_{10}$
		$-\sigma_{P_1,\dots,P_4}: X\to S\subset \mathbb{P}^3$ S cubic surface	$\sigma^*(2H_S)-E_1-\dots-E_4$
		$-\sigma_{P_1,\dots,P_{12}}: X\to F_e$, $e\leq 4$	$\sigma^*(H_e)-E_1-\dots-E_{12}$ $H_e=2C_o+(5+e)F$
	3	$\mathbb{P}(E)$, E rank-2 vector bundle on the quadric Q, given by $0\to\mathcal{O}_Q\to E\to I^{(3,3)}_{\{P_1,\dots,P_{10}\}}\to 0$	tautological
	1	g=5	
	2	K3 surface	
	≥3	complete intersection (2,2,2)	
	1	g=5	
	2	$\sigma_{P_o,\dots,P_{10}}: X\to\mathbb{P}^2$	$\sigma^*(7L)-E_o-2E_1-\dots-2E_{10}$
	1	g=6	
	2	$-\sigma_p: X\to S$, S K3 surface	
		$-\sigma_{P_1,\dots,P_{16}}: X\to\mathbb{P}^2$	$\sigma^*(6L)-E_1-\dots-E_{12}-$ $-2E_{13}-\dots-2E_{16}$
2	1	g=7	
	2	$f_{\lvert K\rvert}: X\to\mathbb{P}^1$, X minimal, elliptic, q=0	
	3	$f_{\lvert H+K\rvert}: X\to\mathbb{P}^1$ with fibres complete intersections (2,2)	
	≥1	complete intersections (2,4)	
1	≥1		

1. The Mumford-Fujita criterion

The following result due to Mumford [18] and Fujita [3] generalizes the familiar fact that on a curve of genus g, a divisor of degree $\geq 2g+1$ is very ample.

Theorem A (Mumford-Fujita). *Let H be an ample divisor on a smooth, projective variety X. Assume that $|H|$ has finitely many base-points, $\Delta \leq g$ and $d \geq 2\Delta+1$. Then H is very ample (and $\Delta = g$).*

(1.1) Corollary. *If $q(X)=0$, $|H|$ is ample with finitely many base-points and $d \geq 2g+1$, then H is very ample.*

Indeed, one may find a reduced, irreducible curve C got by intersecting dim X-1 generic members of $|H|$. We have $\Delta(X,H)=\Delta(C, H|_C) \leq g = g(C)$, so the Theorem applies.

(1.2) Let $f: X \to \mathbb{P}^1 \times \mathbb{P}^3$ be a double covering ramified along a smooth member of $|\mathcal{O}_{\mathbb{P}^1 \times \mathbb{P}^3}(2,2)|$ and let $H \in |f^* \mathcal{O}_{\mathbb{P}^1 \times \mathbb{P}^3}(1,1)|$. Then H is very ample and $d=8$, $g=\Delta=3$. Note that the analogous case of a double covering of $\mathbb{P}^1 \times \mathbb{P}^2$, having $d=6$, $g=\Delta=2$ was treated in [11], Prop.7.1.

(1.3) Proposition. *Let c_1, c_2 be two integers such that $c_1 \geq 4$, $c_1^2/4 < c_2 \leq 3(c_1-1)$ and $4c_2 \neq c_1^2+4$. Then there exists a 3-dimensional scroll over $\mathbb{P}^2$ with invariants $d=c_1^2-c_2$ and $g=\frac{1}{2}(c_1-1)(c_1-2)$.*

Proof. There is an ample and spanned stable rank-2 vector bundle E on $\mathbb{P}^2$ with Chern numbers c_1, c_2. This follows from [17] Prop.7.6 and Prop.6.5. If $X=\mathbb{P}(E)$ and $H \in |\mathcal{O}_X(1)|$, we find $d=c_1^2-c_2$, $g=1/2(c_1-1)(c_1-2)$. The result follows from (1.1) since $d \geq 2g+1$. In particular, taking $c_1=4$, $9 \geq c_2 \geq 6$, we find $g=3$, $7 \leq d \leq 10$.
We need a modification of the Mumford-Fujita criterion to cover also the case $d=2g$. It is given by the following.

(1.4) Proposition. *Let $|H|$ be an ample linear system without base-points on a smooth projective variety X. In particular there is a smooth curve C, got by intersecting dim X-1 generic members in $|H|$. Assume that:*

i) $q(X)=0$, $d=2g$;

ii) $|H|_C - K_C| = \emptyset$; *as we shall see, this condition ensures that $H|_C$ is very ample, embedding C into a certain projective space, say $\mathbb{P}^a$;*

iii) *the restriction map $H^0(\mathcal{O}_{\mathbb{P}^a}(2)) \to H^0(\mathcal{O}_C(2))$ is onto. Then H is very ample.*

Proof. A result due to Iitaka [7] shows that $H|_C$ is indeed very ample. On the other hand, by [3] Prop.1.10, the map $H^o(\mathcal{O}_C(H))\otimes H^o(\mathcal{O}_C(tH))\to H^o(\mathcal{O}_C((t+1)H))$ is onto for $t\geq 2$. Condition iii) ensures it is onto also for $t=1$, so the proof of the Mumford-Fujita criterion applies.

Now we are going to use (1.4) for proving the existence of a 3-fold in $\mathbb{P}^6$ having invariants $d=8$, $g=4$, which is a scroll over the quadric Q (see [12]). First we need the following:

(1.5) Lemma. A smooth curve in $\mathbb{P}^4$ having $d=8$ and $g=4$ is either arithmetically normal, or hyperelliptic.

Proof. By [3], Prop.1.10, it is arithmetically normal if the map $H^o(\mathcal{O}_{\mathbb{P}^4}(2))\to H^o(\mathcal{O}_C(2))$ is onto. Assuming the contrary, using the sequence:

$$0 \to I_C(2) \to \mathcal{O}_{\mathbb{P}^4}(2)\to\mathcal{O}_C(2)\to 0$$

we find that there are three hyperquadrics Q_1,Q_2,Q_3 (with linearly independent equations) containing C.

The intersection $Q_1\cap Q_2$ must be reducible, since otherwise C would be the complete intersection of Q_1,Q_2,Q_3.

As C is non-degenerate, it must be contained in a (non-degenerate) surface of degree 3 in $\mathbb{P}^4$. Such a surface is either a cone over $\mathbb{P}^1$, which is easily seen to be impossible (cf. Lemma (6.3) below), or a scroll over $\mathbb{P}^1$. In this last case, using the notations of [6] Ch.V, we get $C\varepsilon|2C_o+6f|$, so C is hyperelliptic and we are done.

(1.6) Proposition. There is a rank-2 vector bundle E on the quadric Q such that, if we let $X=\mathbb{P}(E)\xrightarrow{\pi} Q$ and H corresponds to the tautological bundle of X, the following hold:

i) $c_1(E)\varepsilon|\mathcal{O}_Q(3,3)|$, $c_2(E)=10$;

ii) E restricted to each line from $|\mathcal{O}_Q(1,0)|$ or $|\mathcal{O}_Q(0,1)|$ is of the form $\mathcal{O}(1)\oplus\mathcal{O}(2)$; |H| is ample and base-points free;

iii) $H^o(E(-1,-1))=0$.

(1.7) Assuming for the moment the truth of (1.6), let us see how it applies to give the desired example. Indeed, from i) it follows that $d(H)=8$, $g(H)=4$. On X we have two linear systems |U| and |V| corresponding to the pull-backs by π of the two systems of generators on Q. We have $K=-2H+U+V$, or $-K-H=H-U-V$. Since we have $H^1(\mathcal{O}_X(-U-V))=0$, we find easily that (1.6) iii) implies that (1.4) ii) holds. By the first part in (1.6) ii), a curve got by intersecting two generic members in |H| is mapped isomorphically by π to a curve of the linear system $|\mathcal{O}_Q(3,3)|$; thus it is non-hyperelliptic! Now the result follows combining (1.5) and (1.4).

We divide the proof of (1.6) in three steps.

Step 1. Consider $P_1,\dots,P_6$ points "in general position" on Q (as the reader will notice, the "general position" assumptions concern the linear systems $|\mathcal{O}_Q(a,b)|$, with $a,b\leq 2$). Consider rank-2 vector bundles E constructed by Serre's method from extensions of type

$$(+)\qquad 0\to \mathcal{O}_Q(1,1)\to E\to I_{\{P_1,\dots,P_6\}}(2,2)\to 0.$$

We get $c_1(E)\in|\mathcal{O}_Q(3,3)|$, $c_2(E)=10$ and E restricted to any line from $|\mathcal{O}_Q(1,0)|$ or $|\mathcal{O}_Q(0,1)|$ is $\mathcal{O}(1)\oplus\mathcal{O}(2)$ (by the assumption of "general position" such a line passes through at most one point P_i). One also gets $h^0(E(-1,-1))=1$, $H^1(E)=0$ and $H^1(E(0,-1))=0$. Now, for any $V\in|\pi^*\mathcal{O}_Q(0,1)|$, $\mathcal{O}_V(H)$ is spanned by global sections. By the exact sequence

$$0\to\mathcal{O}_X(H-V)\to\mathcal{O}_X(H)\to\mathcal{O}_V(H)\to 0,$$

we find that $\mathcal{O}_X(H)$ is spanned. We shall prove that $\mathcal{O}_X(H)$ is ample by using a criterion due to Gieseker ([5], Prop.2.1). Since $\mathcal{O}_X(H)$ is spanned, it will be enough to show that for any irreducible curve C on Q, $E|_C$ has no quotient isomorphic to $\mathcal{O}_C$ (or that the dual $\check{E}|_C$ has no non-zero global sections). Moreover, we need a convenient presentation for E, obtained as follows. Since the restriction of $E(-1,-1)$ to the lines of $|\mathcal{O}_Q(1,0)|$ and $|\mathcal{O}_Q(0,1)|$ is $\mathcal{O}\oplus\mathcal{O}(1)$, we find that $E(-1,-1)$ is a quotient of two bundles of the form $\mathcal{O}(a_1,0)\oplus\mathcal{O}(a_2,0)\oplus\mathcal{O}(a_3,0)$ and $\mathcal{O}(0,b_1)\oplus\mathcal{O}(0,b_2)\oplus\mathcal{O}(0,b_3)$, respectively. Recalling that $h^0(E(-1,-1))=1$ and computing the Chern classes we find out the following two pairs of possible types of presentations of E:

$(*)\quad 0\to\mathcal{O}_Q(-4,0)\to\mathcal{O}_Q(-2,1)\oplus\mathcal{O}_Q(0,1)\oplus\mathcal{O}_Q(1,1)\to E\to 0,$

$(*')\quad 0\to\mathcal{O}_Q(0,-4)\to\mathcal{O}_Q(1,-2)\oplus\mathcal{O}_Q(1,0)\oplus\mathcal{O}_Q(1,1)\to E\to 0,$

$(**)\quad 0\to\mathcal{O}_Q(-4,0)\to\mathcal{O}_Q(-1,1)\oplus\mathcal{O}_Q(-1,1)\oplus\mathcal{O}_Q(1,1)\to E\to 0,$

$(**')\quad 0\to\mathcal{O}_Q(0,-4)\to\mathcal{O}_Q(1,-1)\oplus\mathcal{O}_Q(1,-1)\oplus\mathcal{O}_Q(1,1)\to E\to 0.$

Let $C\in|\mathcal{O}_Q(a,b)|$ be an irreducible, reduced curve. We may assume that $a>0$, $b>0$ (if, for instance, $a=0$, it follows $b=1$, so $E|_C$ is ample by ii)). There are four cases to consider, according to the possible combinations of the presentations above. Suppose that $E|_C$ is not ample and take a surjection $E|_C\to\mathcal{O}_C$.

(1) Assume that $(*)$ and $(*')$ occur. Then $\mathcal{O}_C(1,1)$ is mapped to zero in $\mathcal{O}_C$.

$\mathcal{O}_C(0,1)$ is either mapped to zero, or trivial (in which case $a=0$).

Thus we may assume that $\mathcal{O}_C(-2,1)\simeq\mathcal{O}_C$, giving $a=2b$. Working similarly with $(*')$ it follows $b=2a$, so $a=b=0$, which is absurd.

(2) Assume the presentations are (*) and (**'). From (*) we deduce as before that $a=2b$. From (**') it follows that $\mathcal{O}_C(-1,1)$ has a non-zero global section. The exact sequence:

$$0\to\mathcal{O}_Q(-1-2b,1-b)\to\mathcal{O}_Q(-1,1)\to\mathcal{O}_C(-1,1)\to 0$$

gives $b=1$. Thus $C\in|\mathcal{O}_Q(2,1)|$. But now recall that E was given by (+). We have $c_1(E|_C)=9$ and, by assumption of "general position", C contains at most 5 of the points P_i. Moreover $C\cong\mathbb{P}^1$, so we have the exact sequence:

$$0\to\mathcal{O}_C(3+\alpha)\to E|_C\to\mathcal{O}_C(6-\alpha)\to 0,$$

with $0\leq\alpha\leq 5$, showing that $E|_C$ is ample.

(3) The case (**), (*') is completely similar.

(4) Finally, suppose the presentations to be (**), (**'). As above $\mathcal{O}_C(1,-1)$ and $\mathcal{O}_C(-1,1)$ must have non-zero global sections. It follows that $a=b=1$, $C\cong\mathbb{P}^1$, $c_1(E|_C)=6$. Again by the assumption of generality C may contain at most three of the points P_i. The exact sequence

$$0\to\mathcal{O}_C(2+\alpha)\to E|_C\to\mathcal{O}_C(4-\alpha)\to 0,$$

with $0\leq\alpha\leq 3$ shows that $E|_C$ is ample.
Up to now we have constructed bundles E satisfying i) and ii) of (1.6). But, as already remarked, $h^o(E(-1,-1))=1$, so iii) clearly fails. For the moment, observe that the restriction of $E(0,-2)$ to the members of $|\mathcal{O}_Q(1,0)|$ is $\mathcal{O}\oplus\mathcal{O}(-1)$. Therefore $E(0,-2)$ has a sub-line bundle of type $\mathcal{O}_Q(a,0)$. Thus E is an extension of line bundles and computing Chern classes we find:

(++) $$0\to\mathcal{O}_Q(-4,2)\to E\to\mathcal{O}_Q(7,1)\to 0.$$

<u>Step II</u>. Now consider bundles given by extensions of the form

$$0\to\mathcal{O}_Q(0,1)\to E'\to I_{\{P_1,\ldots,P_7\}}(3,2)\to 0,$$

where $P_1,\ldots,P_7$ are points "in general position". It follows that $h^o(E'(-1,-1))=0$. Moreover, the restriction of E' to any member of $|\mathcal{O}_Q(1,0)|$ is $\mathcal{O}(1)\oplus\mathcal{O}(2)$. So, by the preceding argument, E' too may be written as an extension of line bundles, as in (++).

<u>Step III</u>. As remarked in [19], the (indecomposable) rank-2 bundles which may be given by an extension as in (++) are parametrized by a projective space $\mathbb{P}^{19}$ and there is also a "versal" bundle on $Q\times\mathbb{P}^{19}$, inducing on various fibres all bundles which may be given as

in (++). It is easy to see that from $c_1(E) \in |\mathcal{O}_Q(3,3)|$, $c_2(E)=10$ and $h^o(E)=7$ it follows that E is indecomposable. Now, by Step I, since $h^1(E)=0$, there is an open, non-empty set of $\mathbb{P}^{19}$ corresponding to ample bundles E such that the tautological bundle on $\mathbb{P}(E)$ is spanned. By Step II and semicontinuity there is an open, non-empty set of bundles with $h^o(E(-1,-1))=0$. Thus we may find bundles satisfying all conditions of (1.6). Finally we remark that, conversely, if X is a scroll over Q with d=8, g=4, X is isomorphic to $\mathbb{P}(E)$, where E is a bundle fulfilling all conditions of (1.6).

(1.8) Inspired by [4] we shall prove here the existence of a hyperquadric fibration of dimension 4 with d=7, g=3 (cf. [11]). Let $Y=\mathbb{P}(E)$, where $E=\mathcal{O}_{\mathbb{P}^1} \oplus \mathcal{O}_{\mathbb{P}^1} \oplus \mathcal{O}_{\mathbb{P}^1}(1) \oplus \mathcal{O}_{\mathbb{P}^1}(1) \oplus \mathcal{O}_{\mathbb{P}^1}(1)$.

Let $H \in |\mathcal{O}_Y(1)|$ and denote by F a fibre of the projection $\pi: Y \to \mathbb{P}^1$.

Let $S=\mathbb{P}(\mathcal{O}_{\mathbb{P}^1} \oplus \mathcal{O}_{\mathbb{P}^1}) \subset Y$ be the embedding corresponding to the surjection $E \to \mathcal{O}_{\mathbb{P}^1} \oplus \mathcal{O}_{\mathbb{P}^1}$. We have $H|_S \in |\mathcal{O}_S(0,1)|$ and $(2H+F)|_S \in |\mathcal{O}_S(1,2)|$. Since the linear system $|2H+F|$ is base-points free, we may choose a smooth member $X \in |2H+F|$ such that $X \cap S$ is irreducible. Now we claim that $H|_X$ is ample. If not, since $|H|$ is base-points free, we may find a curve $C \subset X$ such that $(H.C)=0$. But C has to be a section for π and it follows that $C \subset S$, so $C=X \cap S$. This is absurd since $(H.C)=(H.X.S)= =(H|_S.X|_S)_S=1$. We get $(H|_X)^4=7$, $g(H|_X)=3$, $q(X)=0$, so (1.1) applies and $(X, H|_X)$ has the desired properties.

Next we want to investigate projections of a manifold from one of its points. Let X' be a manifold in $\mathbb{P}^N$ with hyperplane section H' and invariants $d'=d(H')$, $g'=g(H')$. A <u>line</u> of X' is a curve $C \subset X'$ such that $(H'.C)=1$. If P is a point on X', let $\sigma: X \to X'$ denote the blowing-up at P, $E=\sigma^{-1}(P)$, $H=\sigma^*(H')-E$. We have $d=d(H)=d'-1$, $g=g(H)=g'$, $\Delta(X,H)=\Delta(X',H')$.

(1.9) <u>Proposition</u>. <u>Assume that</u> $q(X')=0$ <u>and</u> P <u>is not contained in any line of</u> X'. <u>Then</u>

i) <u>If</u> $d' \geq 2g'+2$, H <u>is very ample on</u> X;

ii) <u>If</u> $d'=2g'+1$ <u>and</u> (X',H') <u>is not a hyperquadric fibration</u>, H <u>is very ample if and only if</u> $|-K-(r-2)H|=\emptyset$, <u>where</u> $r=\dim X$.

<u>Proof</u>. The proposition is a consequence of (1.1) and (1.4), but the following "elementary" argument may also be given. We present details for ii), i) being similar and simpler. We have to show that $|H|$ separates points and tangent vectors. Once we identify elements of $|H|$

with hyperplane sections of X' through $\mathbb{P}$, the problem is reduced to proving the same property for $\mathcal{O}_D(H)$, for some smooth member $D\in|H|$. This is seen by using Bertini's theorem (here it is important that there are no lines through P), the exact sequence

$$0 \to \mathcal{O}_X \to \mathcal{O}_X(H) \to \mathcal{O}_D(H) \to 0,$$

and the fact that $q(X)=0$.
Inductively we are reduced to showing the very ampleness of $H|_C$, for a smooth curve C got by intersecting $r-1$ members of $|H|$. The theory of the adjunction mapping (see [11], Section 1) gives that $|K'+(r-1)H'|$ is base-points free (otherwise there is a line through each point of X') and either $K'+(r-1)H'=0$, or the adjunction mapping has at least a two-dimensional image. As $\sigma^*(K'+(r-1)H')=K+(r-1)H$, it follows $H^1(\mathcal{O}_X(-K-(r-1)H))=0$ by the vanishing theorem. Combining with $|-K-(r-2)H|=\emptyset$, we find inductively that $|H|_C-K_C|=\emptyset$, so we may apply Iitaka's result quoted in (1.4).

Let us see some applications:

(1.10) Take $X'=\mathbb{P}^2$, $H'\in|\mathcal{O}(a)|$, $a\geq 2$. Let p be the maximal number of projections from generic points of X', allowed by (1.9). We find $p\geq 3a-2$ for $a\geq 4$. One knows classically that for $a=2$, $p=1$ and for $a=3$, $p=6$.

(1.11) Take $X'=Q$, the quadric, $H'\in|\mathcal{O}_Q(3)|$. We have $d'=18$, $g'=4$. We can project from α generic points, where $\alpha\leq 10$. For $\alpha=10$, remark that $\dim|-K'|=8$, so we get $|-K|=\emptyset$ by choosing the ten points generically. In particular the surface thus obtained has $d=8$, $g=4$.

(1.12) Take X' to be a cubic in $\mathbb{P}^3$, $H\in|\mathcal{O}(2)|$. It follows that we may project from four generic points, the resulting surface having $d=8$, $g=4$. Here the maximal number of projections allowed is five, cf. [11] Prop.8.1.

(1.13) Let X' be the Del Pezzo surface of [11] Th.4.1 iv), $H'=-2K'$ and $P\in X'$ any point. Then, keeping the notations above, H is very ample on X. Indeed, $d'=8$, $g'=3$, $q=0$ and there are no lines on X'.

(1.14) Now take X' to be the Del Pezzo surface having $H'=-3K'$, $d'=9$, $g'=4$. This surface cannot be projected from any of its points. Indeed, we have $\dim|-K'|=1$, so $|-K|\neq\emptyset$ for any position of the point P.

(1.15) Take $X'=\mathbb{P}^3$, $H'\in|\mathcal{O}(2)|$, $P\in X'$. Then we may project one time.

2. A Bertini-type theorem for vector bundles

Theorem B (Kleiman [16]). Let Y be a smooth, projective variety and E a vector bundle of rank a on Y spanned by global sections. Take an integer $b \le a$ such that $\dim Y < 2(a-b+2)$. Then the dependency locus of b generic global sections of E is either empty or smooth, of pure codimension a-b+1 in Y.

Remark. Actually, the proof of the above is simpler and somewhat different from that of [16] (where one has to assume that E(-1) is spanned) since we are in characteristic 0 and generic smoothness holds.

(2.1) Take $Y=\mathbb{P}^4$, $E=\Omega^1_Y(2)$, $b=3$. The dependency locus of three generic sections of E must be non-empty, since otherwise we find an exact sequence of the form

$$0 \to \mathcal{O}_Y^{\oplus 3} \to \Omega^1_Y(2) \to \mathcal{O}_Y(3) \to 0$$

This is absurd since $H^0(\Omega^1_Y(1))=0$. Thus, by Theorem B, there is a smooth surface X (which must be connected since $Y=\mathbb{P}^4$) such that its ideal sheaf I_X has a resolution:

$$0 \to \mathcal{O}_Y^{\oplus 3} \to \Omega^1_Y(2) \to I_X(3) \to 0. \tag{5}$$

Dualising we get:

$$0 \to \mathcal{O}_Y(-3) \to T_Y(-2) \to \mathcal{O}_Y^{\oplus 3} \to \omega_X(2) \to 0.$$

Now it is easy to see that X must be the Veronese surface of degree 4 in $\mathbb{P}^4$.

(2.2) (C.Okonek [20]). Take $Y=\mathbb{P}^5$, $E=\Omega^1_Y(2)$, $b=4$.

The above procedure yields the existence (left open in [11])of a 3-fold of degree 7 in $\mathbb{P}^5$ which is a scroll over the cubic surface of $\mathbb{P}^3$.

3. The results of Peskine-Szpiro for the case of codimension two

Recall that an arithmetically Cohen-Macaulay submanifold of codimension two in $\mathbb{P}^n$ is a complete intersection if $n \ge 6$. For $n \le 5$ one has:

Theorem C (Peskine-Szpiro [22]). Let $m \ge 2$, a_i $(i=1,\dots,m-1)$, b_i $(i=1,\dots,m)$ be given positive integers such that

$\sum_{i=1}^{m-1} a_i = \sum_{i=1}^{m} b_i$ and $a_i > b_j$ for any i, j. If $n \le 5$, there is some submanifold $X \subset \mathbb{P}^n$, of codimension two, such that I_X has the resolution:

$$0 \to \bigoplus_{i=1}^{m-1} \mathcal{O}(-a_i) \to \bigoplus_{i=1}^{m} \mathcal{O}(-b_i) \to I_X \to 0.$$

Next, starting with some given submanifold Y of codimension two in $\mathbb{P}^n$, one can try to find a new one, say X, such that the union of X and Y is the complete intersection of two hypersurfaces. One says that X and Y are "linked". We have:

Theorem D (Peskine-Szpiro [22]).

i) Assume that X and Y are linked by the complete intersection of two hypersurfaces of degree a and b. If

$$0 \to E \to F \to I_Y \to 0$$

is a locally free resolution for I_Y, a resolution for I_X is given by:

$$0 \to F^\vee(-a-b) \to E^\vee(-a-b) \oplus \mathcal{O}(-a) \oplus \mathcal{O}(-b) \to I_X \to 0,$$

ii) If $n \le 5$, a, b, e are integers such that $a, b \ge e$ and $I_Y(e)$ is spanned by global sections, there are generic forms $u \in H^0(I_Y(a))$, $v \in H^\vee(I_Y(b))$ and a smooth subvariety $X \subset \mathbb{P}^n$ such that X and Y are linked by the complete intersection of the hypersurfaces given by u and v.

(3.1) Combining Theorem C and Theorem D i) we treated in [11] and [12] the following codimension two cases: $d=5$, $g=2$; $d=6$, $g=3$; $d=7$, $g=5,6$ and $d=8$, $g=7$.

(3.2) C. Okonek (see [21]) used Theorem D ii) for proving the existence of the two types of surfaces in $\mathbb{P}^4$ having $d=8$, $g=6$ and $p_g=0$ or 1. For convenience we give here a simplified version of his argument. Start with the Veronese surface in $\mathbb{P}^4$, denoted by Y. From (5) we see that $I_Y(3)$ is spanned. By Theorem D we may link Y to a (smooth) surface X_1 by two forms of degree 3 and 4 respectively, such that I_{X_1} has the resolution:

(6) $\quad 0 \to T(-6) \to \mathcal{O}^{\oplus 4}(-4) \oplus \mathcal{O}(-3) \to I_{X_1} \to 0$; in particular $I_{X_1}(4)$ is spanned.

Twisting (6) by $\mathcal{O}(4)$ and dualising, we get:

$$0 \to \mathcal{O}(-4) \to \mathcal{O}^{\oplus 4} \oplus \mathcal{O}(-1) \to \Omega^1(2) \to \omega_{X_1}(1) \to 0.$$

Now it is easy to see that X_1 has invariants $d=8$, $g=6$, $p_g=0$. Similarly we find an X_2 which is linked to X_1 by two forms of degree 4. From (6) we get:

$$0\to\mathcal{O}^{\oplus 4}\oplus\mathcal{O}(-1)\to\Omega^1(2)\oplus\mathcal{O}^{\oplus 2}\to I_{X_2}(4)\to 0 \text{ and dualising}$$

$$0\to\mathcal{O}(-4)\to T(-2)\oplus\mathcal{O}^{\oplus 2}\to\mathcal{O}^{\oplus 4}\oplus\mathcal{O}(1)\to\omega_{X_2}(1)\to 0.$$ It follows that X_2 has invariants $d=8$, $g=6$, $p_g=1$.

4. Reider's theorem on surfaces

The following remarkable result gives an efficient way of proving the very ampleness of a linear system on a surface.

Theorem E (I.Reider [23]). Let X be a smooth, projective surface and H a divisor on X such that:

i) H-K is nef;

ii) $(H-K)^2\geq 9$;

iii) there is no effective divisor E on X such that either

$(H-K.E)=0$, $(E^2)=-1,-2$ or

$(H-K.E)=1$, $(E^2)=0,-1$ or

$(H-K.E)=2$, $(E^2)=0$ or

$H-K\equiv 3E$, $(E^2)=1$.

Then H is very ample.

(4.1) Let X be a geometrically ruled elliptic surface with invariant $e=-1$ and let $H\equiv 2C_o+F$ (notations as in [6] Ch.V). Then H is very ample. Indeed we have $H-K\equiv 4C_o$, so it is ample (cf.[6] loc cit). The remaining conditions in Theorem E are obvious. The existence of this type of surfaces of degree 8 was first proved by Buium in [2].

(4.2) Let X be a geometrically ruled surface over a curve of genus 2 with invariant $e=-2$ and let $H\equiv C_o+3F$. Then H is very ample. Indeed, $H-K\equiv 3C_o-F$ is ample and condition iii) in Theorem E is easily verified. We also remark that, conversely, any surface of degree 8 which is a scroll over a curve of genus 2 is necessarily of this type.

5. Scrolls over a curve of genus ≤1

The following is classical and easy. It was included only for perspective.

(5.1) Proposition. A scroll of dimension r and degree d over $\mathbb{P}^1$ has $\Delta=g=0$; the existence of such a scroll with given invariants d,r is equivalent to the numerical condition $d\geq r$. They are all obtained as linear sections of the Segre embedding of $\mathbb{P}^1 \times \mathbb{P}^{d-1}$ in $\mathbb{P}^{2d-1}$.

(5.2) Proposition. A scroll of dimension r and degree d over an elliptic curve C has $g=1$, $\Delta=r$; the existence of such a scroll with given invariants d, r is equivalent to the numerical condition $d>2r$.

Proof. The equality $\Delta=r$ was proved in [11] Prop.3.11. Assume that X exists in $\mathbb{P}^n$, with $n=h^0(\mathcal{O}_X(H))-1$. Since $\Delta=r$ it follows $d=n+1$. If $d\leq 2r$, we would get $r-(n-r)\geq 1$ and Barth's Theorem (cf [1]) gives $q(X)=0$. This is a contradiction, so necessarily $d>2r$. Conversely, assume this last condition holds. First we show that for any integers $a,b>0$, there exists an ample vector bundle E on C having rank b and $c_1(E)=a$. If $b=1$ this is obvious. Assume we have already found an ample rank$(b-1)$ bundle F on C with $c_1(F)=a$. By Riemann-Roch there is some non-split exact sequence of the form

$$0\to\mathcal{O}_C\to E\to F\to 0.$$

By a result due to Gieseker (see [5], Th.2.2), E is ample and, obviously, $c_1(E)=a$, $rk(E)=b$. The above argument shows that we may find an ample vector bundle E_1 on C with $rk(E_1)=r$, $c_1(E_1)=d-2r$ and an exact sequence

$$0\to\mathcal{O}_C\to E_1\to F\to 0.$$

Take some $L\in Pic(C)$ with $c_1(L)=2$. Let $E=E_1\otimes L$, $X=\mathbb{P}(E)\overset{\pi}{\to}C$. Since we have $c_1(E)=d$ it will be enough to prove that $\mathcal{O}_X(1)$ is very ample on X. For any two points $P,Q\in C$ we let $L_1=L\otimes\mathcal{O}_C(-P-Q)\in Pic^0(C)$. We get $H^1(E\otimes\mathcal{O}_C(-P-Q))=H^1(E_1\otimes L_1)=H^0(E_1^\vee\otimes L_1^\vee)=0$ since E_1 is ample and L_1 has degree 0. Thus we have $H^1(\mathcal{O}_X(1)\otimes\pi^*\mathcal{O}_C(-P-Q))=0$ and the result is a consequence of the following simple lemma.

(5.3) Lemma (cf.[2] Lemma 3.4). Let X be a manifold and $\pi:X\to C$ a morphism onto some smooth curve. Let $X_P=\pi^{-1}(P)$ for $P\in C$. If M is an invertible sheaf on X such that $M|_{X_P}$ is very ample and $H^1(M\otimes\mathcal{O}_X(-X_P-X_Q))=0$ for any $P,Q\in C$, then M is very ample.

6. The effective list of manifolds of degree 7 and 8

(6.1) Theorem. The list of manifolds of degree 7 given in [11] is effective.

For a proof, apply (5.2), (1.3), (1.8) and (2.2) to show the existence of the four types left undecided in [11].

(6.2.) Consider now the case when (X, H) is a hyperquadric fibration over $\mathbb{P}^1$ having invariants $d=8$, $g=\Delta=4$ (cf.[12]). Examples with dim $X=2$ are got by taking divisors of type $(4,2)$ on $\mathbb{P}^1 \times \mathbb{P}^2$, embedded Segre into $\mathbb{P}^5$. Next we prove that the cases dim $X \geq 3$ are not possible. We start with a useful remark valid for hyperquadric fibrations $\varphi : X \to C$ of dimension r, over any base curve C. Let us first introduce the following notations: Q for a fibre of φ, $E=\varphi_*(\mathcal{O}_X(H))$, $Y=\mathbb{P}(E) \xrightarrow{\pi} C$, F for a fibre of π and $\mathcal{O}_Y(L)=:\mathcal{O}_Y(1)$. Then we claim that $\mathcal{O}_Y(L)$ is spanned by global sections. Indeed, consider the exact sequence

(7) $$0 \to \mathcal{O}_X(H-Q) \to \mathcal{O}_X(H) \to \mathcal{O}_Q(H) \to 0.$$

Since Q is a hyperquadric in $\mathbb{P}^r$, the restriction map $H^0(\mathcal{O}_X(H)) \to H^0(\mathcal{O}_Q(H))$ is surjective, so we get
$h^0(\mathcal{O}_X(H-Q)) = h^0(\mathcal{O}_X(H)) - r - 1$. It follows that
$h^0(\mathcal{O}_Y(L-F)) = h^0(\mathcal{O}_Y(L)) - r - 1$ and the exact sequence
$0 \to \mathcal{O}_Y(L-F) \to \mathcal{O}_Y(L) \to \mathcal{O}_F(L) \to 0$ shows that $\mathcal{O}_Y(L)$ is spanned, since $\mathcal{O}_F(L)$ is so for any fibre F.
Now return to our case when dim $X=3$, $d=2g=8$, $C=\mathbb{P}^1$.
We find $H^1(\mathcal{O}_X(H))=0$ and, using (7), $H^1(\mathcal{O}_X(H-Q))=0$. Thus we must have $E \cong \bigoplus_{i=1}^{4} \mathcal{O}(a_i)$, with $a_i \geq 0$. Moreover, we get $X \in |2L+2F|$ and $c_1(E) = \sum_{i=1}^{4} a_i = 3$.
Since at least one a_i is zero, the map $\Psi_{|L|} : Y \to \mathbb{P}^6$ maps Y onto a cone of degree 3. Thus X is contained in such a cone. Passing to hyperplane sections and using Bertini's theorem we find that some smooth sectional curve of X lies on a two-dimensional cone of degree 3. This contradicts the following lemma.

(6.3) Lemma. Let C be a smooth curve of degree d contained in a surface of degree b which is a cone with vertex P over some smooth curve. Then:

i) if $P \notin C$, b divides d,

ii) if $P \in C$, b divides $d-1$.

For a proof, blow-up P and compute intersection numbers on the resulting geometrically ruled surface.

(6.4) Consider now surfaces in $\mathbb{P}^4$ of degree 8 with $g=5$ and $q=1$ which are hyperquadric fibrations (cf[12]).

We show that they cannot exist. Using the notations introduced in (6.2) we find $(L^3)=4$, $X \in |2L+\pi^*B|$ for some degree-zero divisor on the

elliptic curve C. As remarked in (6.2) we have a morphism $\Psi_{|L|}:Y\to\mathbb{P}^4$. Since $(L^3)=4$, either $\Psi_{|L|}$ maps Y birationally onto a hypersurface of degree 4, say Z, or X is contained in a hyperquadric. This last possibility is absurd since otherwise X would be a complete intersection. Next we show that Z is a cone. Indeed, since the fibres F are mapped to planes, it is enough to find a curve contracted by $\Psi_{|L|}$. If there are no such curves, L is ample and it follows $H^1(E)=H^1(\mathcal{O}_X(H))=0$. But we find $h^1(\mathcal{O}_X(H))=1$. Now, if T is a curve such that $(L.T)=0$, we have $T\cap X=\emptyset$ (because $L|_X=H$ is very ample we cannot have $T\subset X$).
As a consequence, no divisor on Y is contracted by $\Psi_{|L|}$. Indeed, if, say $D\equiv\alpha L+\beta F$ is contracted, it follows:

$$0=(D.L^2)=4\alpha+\beta$$

and, since $D\cap X=\emptyset$, $(D.X.F)=2\alpha=0$, so $\alpha=\beta=0$ which is absurd. Thus we proved that if $S\in|L|$ is a generic member, the map induced by restricting $\Psi_{|L|}$ is a finite, birational morphism between S and a certain surface of degree 4 in $\mathbb{P}^3$. Since S is a geometrically ruled elliptic surface, using the notations of [6] Ch.V, we have $L|_S\equiv C_0+bf$, $4=2b-e$ and $b-e=(C_0+bf.C_0)\geq 2$.

Moreover, $e\geq -1$ implies that necessarily $e=0$, $b=2$, so $b-e=2$ and C_0 is mapped two to one onto some line. From this we deduce that there are infinitely many pairs of fibres F mapped to pairs of planes intersecting in a line. If we take a hyperplane of $\mathbb{P}^4$ containing such a pair of planes, its intersection with Z containes a certain quadric, besides the two planes. Taking its pullback by $\Psi_{|L|}$, we find a geometrically ruled elliptic surface mapped birationally to a quadric, which is clearly absurd. It should be pointed out that this class of surfaces was first excluded by Okonek in [21] by a completely different argument.

(6.5) Consider now the rational surfaces of $\mathbb{P}^4$ having $d=8$, $g=5$ (cf.[12]). As we understood from [15], recently J. Alexander proved the existence of this type of surfaces. This seems to be a rather subtle case, since none of the methods described so far can be applied.

Now, looking over the maximal list proposed in [12] for degree 8 and using (5.2), (1.3), (1.2), (4.2), (1.14), (1.7), (1.12), (6.2), (6.4), (6.5) and (3.2) we see that the following result was proved.

(6.6) Theorem. The effective list of manifolds of degree 8 is as given in the table following the introduction.

References

1. Barth, W. Transplanting cohomology classes in complex-projective space. Amer. J. Math.92(1970), 951-967.
2. Buium, A. On surfaces of degree at most 2n+1 in $\mathbb{P}^n$, in Proceedings of the Week of Algebraic Geometry, Bucharest 1982, Springer Lect. Notes Math., 1056(1984).
3. Fujita, T. Defining equations for certain types of polarized varieties, Complex Analysis and Algebraic Geometry, Tokyo, Iwanami, (1977), 165-173.
4. Fujita, T. Classification of polarized manifolds of sectional genus two, Preprint.
5. Gieseker, D. P-ample bundles and their Chern classes, Nagoya Math.J.43, (1971), 91-116.
6. Hartshorne, R. Algebraic Geometry, Springer (1977).
7. Iitaka, S. Algebraic Geometry: an introduction to birational geometry of algebraic varieties, Springer (1982).
8. Ionescu, P. An enumeration of all smooth, projective varieties of degree 5 and 6, INCREST Preprint Series Math., 74, (1981).
9. Ionescu, P. Variétés projectives lisses de degrés 5 et 6, C.R.Acad. Sci.Paris, 293, (1981), 685-687.
10. Ionescu, P. Embedded projective varieties of small invariants, INCREST Preprint Series Math., 72 (1982).
11. Ionescu, P. Embedded projective varieties of small invariants in Proceedings of the Week of Algebraic Geometry, Bucharest 1982, Springer Lect.Notes Math., 1056 (1984).
12. Ionescu, P. Embedded projective varieties of small invariants II, Rev. Roumaine Math. Pures Appl.,31 (1986), 539-544.
13. Ionescu, P. Varieties of small degree, An.St.Univ. A.I.Cuza, Iassy, 31 s.I.a(1985), 17-19.
14. Ionescu, P. Ample and very ample divisors on surfaces, Rev. Roumaine Math. Pures Appl., 33(1988), 349-358.
15. Katz, S. Hodge numbers of linked surfaces in $\mathbb{P}^4$, Duke Math. J., 55(1987), 89-95.
16. Kleiman, S. Geometry on grassmannians and applications to splitting bundles and smoothing cycles, Publ.Math.IHES, 36 (1969), 281-297.
17. Le Potier, J. Stabilité et amplitude sur $\mathbb{P}_2(\mathbb{C})$, in Vector bundles and differential equations, Proceedings,Nice, 1979, Progress in Math.7, Birkhäuser.
18. Mumford, D. Varieties defined by quadratic equations, in Questions on algebraic varieties (CIME Varenna 1969) Ed.Cremonese, Roma 1970.
19. Newstead P.E. and Schwarzenberger R.L.E. Reducible vector bundles on a quadric surface, Proc. Camb. Phil. Soc.,60, (1964), 421-424.
20. Okonek, C. Über 2-codimensionale Untermannigfaltigkeiten vom Grad 7 in $\mathbb{P}^4$ und $\mathbb{P}^5$, Math.Z., 187(1984), 209-219.
21. Okonek,C. Flächen vom Grad 8 im $\mathbb{P}^4$, Math.Z., 191 (1986), 207-223.
22. Peskine, C. and Szpiro, L. Liaison des variétés algébriques I, Inv.Math. 26(1974) 271-302.
23. Reider, I. Vector bundles of rank 2 and linear systems on algebraic surfaces, Ann. Math.127 (1988), 309-316.
24. Serrano, F. The adjunction mapping and hyperelliptic divisors on a surface, J. reine angew. Math. 381 (1987), 90-109.
25. Sommese, A. and Van de Ven, A. On the adjunction mapping, Math. Ann. 278 (1987), 594-603.

ON THE EXISTENCE OF SOME SURFACES

Elvira Laura Livorni
Dipartimento di Matematica,Università degli Studi de L'Aquila
Via Roma Pal.Del Tosto, 67100 L'Aquila, Italia

To my children Luca and Fabio

Introduction

The problem of classifying algebraic, projective, surfaces with small projective invariants i.e. degree or sectional genus is an old problem. It was started by Picard and Castelnuovo, see references.

Roth in [33], [34], [35], [36] and in [37], gave a birational classification of connected, smooth, algebraic, projective surfaces with sectional genus less than or equal to six. Classically, the adjunction process was introduced by Castelnuovo and Enriques [12] to study curves on ruled surfaces. Recently, after Sommese and Van de Ven study of the adjunction mapping, see [39], [40], [41], the problem of giving a biregular classification of smooth, connected, algebraic, projective surfaces with either small degree or small sectional genus has been studied again by various authors, see references.

We started the study of such surfaces while we were Sommese's student in Notre Dame in 1981. The main tool for the identification of the numerical projective invariants and for the description of a minimal pair of (X,L) were the iterated adjunction mappings, for the definition see [5]. Actually it turned out, see [7], that we really need to iterate the adjunction mapping only for $g=g(L) \geq 8$.

The reason for writing this paper is that after Reider 's results, [31], it has been possible to answer to the often subtle question if the pairs (X,L) determined in the previous papers, see references, do really exist i.e. if the line bundle L on X is very ample. We like to call those surfaces "the candidate surfaces".

Using Biancofiore's results, see [3], [4], and Buium's results, [9], we have been able to answer for most of the cases when the Kodaira dimension $\kappa(X)=-\infty$. When $\kappa(X)\geq 0$, unfortunately Reider's method doesn't help us. In this case we haved used again Buium's results in [9] but there are still open some very interesting cases for example the existence of elliptic surfaces either with $\kappa(X)=0$ or $\kappa(X)=1$. See cases 9, 14, 17, 20, 21 and 23 in §4. It was our intention to quote all the works of mathematician who gave contribution to this nice classification started by the italian classical school. We apologize if we have forgotten someone.

The organization of the paper is as follows:
In §0, we collect background material and explain the conditions we have to impose on the points that we have to blow up on a minimal model in order to guarantee the very ampleness of L. See (0.3),(0.11) and (0.19). In §1, we determine the existence of surfaces with $g \leq 7$ and whose minimal model is $\mathbf{P}^2$. In §2, we determine the existence of surfaces with $g \leq 7$ and whose minimal model is a rational ruled surface. In §3, we determine the existence of surfaces with $g \leq 7$ and whose minimal model is an irregular ruled surface. In §4, we determine the existence of surfaces with $g \leq 7$ and whose minimal model has non-negative Kodaira dimension.

I would like to express my gratitude to Andrew J. Sommese for his constant encouragement to continue in my work although the hard job of being a mother.

I woul like to thank A.Lanteri for useful conversation regarding surfaces with $\kappa(X)=1$.

§0 Background Material.

(0.1) Let X be an analytic space. We let $\mathcal{O}_X$ denote its structure sheaf and $h^{i,0}(X)=\dim H^i(X,\mathcal{O}_X)$. If X is a complex manifold, we let K_X denote its canonical bundle.

(0.2) Let X be a smooth, connected, projective surface. Let D be an effective Cartier divisor on X. Denote by [D] the holomorphic line bundle associated to D. If L is a holomorphic line bundle on X, we write $|L|$ for the linear system of Cartier divisors associated to L. Of course if $|L|$ is non empty, then $[D]=L$ for $D\in|L|$. Let E be a second holomorphic line bundle on X. Then $L\cdot E$ denotes the evaluation of the cup product, $c_1(L)\wedge c_2(E)$ on X, where $c_1(L)$ and $c_2(E)$ are the Chern classes of L and E respectively. If $D\in|L|$ and $C\in|E|$, it is convenient to write $D\cdot C=D\cdot E=L\cdot C=L\cdot E$. We let $g=g(L)=(L\cdot L+K_X\cdot L+2)/2$, which is called the adjunction formula. If there is a smooth $D\in|L|$, then $g=g(L)=h^{1,0}(D)$. We let $q=\dim H^1(X,\mathcal{O}_X)$ and $p_g=\dim H^2(X,\mathcal{O}_X)$.

(0.3) Definition: Let $p_1,\dots,p_s$ be a finite set of points of $\mathbf{P}^2$. We say that $p_1,\dots,p_s$ are in general position if no three are collinear and no ten lie on a conic, and furthermore after any finite sequence of admissible transformations, the new set of s points also has no three collinear. It is easy to check that if the points satisfy this definition then:

1) No six of them lie on a conic
2) No eight of them, one double, lie on a rational cubic
3) No nine of them, three double, lie on a rational quartic
4) No nine of them, six double, lie on a rational quintic.

This definition is an easy consequence of the definition in [15,pg.409 ex.4.15]

(0.4) Let $m\in\mathbf{N}$ and X be a smooth surface. We write $\mathcal{D}_m$ for the set of all divisors $E\subseteq X$ such that $E\neq 0$ and mE is effective.

(0.5) Definition: Let L be a line bundle on a smooth, connected, projective surface X. Let $M=L\otimes K_X^{-1}$. A divisor E on X is said to be a Reider divisor if $E\in\mathcal{D}_1$ and $(M-E)\cdot E\leq 2$.

(0.6) Let $\pi:\mathbf{P}^2_s\to\mathbf{P}^2$ be the blowing-up morphism with center $p_1,\dots,p_s\in\mathbf{P}^2$ and let P_j be the corresponding exceptional curves on $\mathbf{P}^2_s$. Then a line bundle L on $\mathbf{P}^2_s$ is of the form

$L=\pi^*(\mathcal{O}_{\mathbf{P}^2}(\delta))\otimes_{1\le j\le s}[P_j]^{-m_j}$. We write $L\equiv\delta l-\sum_{1\le j\le s}m_jP_j$, where $l\in|\pi^*(\mathcal{O}_{\mathbf{P}^2}(1))|$. Let $M=L\otimes K_X^{-1}$. Then $M\equiv(\delta+3)l-\sum_{1\le j\le s}(m_j+1)P_j$.

(0.7) Definition: Let $\mathbf{P}^2_s$, L and M be as in (0.6). A divisor E on $\mathbf{P}^2_s$ is said to be a test divisor for the pair $(\mathbf{P}^2_s,L)$ if it satisfies the following conditions:

(i) $E\in D_1$; (ii) $g(E)\ge 0$; (iii) $E\equiv yl-\sum_{1\le j\le s}\alpha_jP_j$ where $1\le y\le(\delta+2)/2$ and

$$\operatorname{Max}\{0,(m_j+2y-\delta-2)/2\}\le\alpha_j\le\begin{cases}1 & \text{if } y=1\\ \operatorname{Min}\{y-1,(m_j+1)/2\} & \text{if } y\ge 2.\end{cases}$$

We write E_y for the test divisors of the pair $(\mathbf{P}^2_s,L)$.

(0.7.1) Remark: $m_j=L\cdot P_j$; $\alpha_j=E_y\cdot P_j$.

(0.8) Theorem: Let $\mathbf{P}^2_s$, L and M be as in (0.6). Assume that $M^2\ge 9$ and that $2\ge m_1\ge\ldots\ge m_s$. Then L is very ample if for any y such that $1\le y\le(\delta+2)/2$ and for any $D\in|\pi^*(\mathcal{O}_{\mathbf{P}^2}(y)|$ we have:

$\sum_{j\in\Lambda}m_j\le y(\delta+3-y)-3$; where $\Lambda=[j\in[1,\ldots,s]/p_j\in D]$.

Proof: See [3,Theorem (2.3)]

(0.9) Theorem: Let $\mathbf{P}^2_s$, L and M be as in (0.6). Let E_y be as in (0.7). Assume that: (i) $\delta\ge 0$; (ii) $M^2\ge 9$; (iii) if $E^2_y<0$ then $(M-E_y)\cdot E_y\ge 3$; (iv) if $E^2_y=0$ then $M\cdot E_y\ne 1,2$; (v) if $E^2_y=1$ then $M\equiv 3E_y$. Then L is very ample.

Proof: See [3,Theorem (2.2)]

(0.10) Corollary: Let $\mathbf{P}^2_s$, L and M be as in (0.6). Let E_y be as in (0.7). If $y=[(\delta+3)/2]$ then $E_y\cdot(M-E_y)\ge 3$ if one of the following is satisfied:

1) $M^2\ge 10$; 2) $\delta+3$ is even; 3) At least one of the points p_j, $1\le j\le s$, has even multiplicity.

Proof: See [3, Lemma (2.1.2)].

(0.11) Definition: Let $p_1,\ldots,p_s\in\mathbf{P}^2$. Let $\mathbf{P}^2_s$ and L be as in (0.6). We say that $p_1,\ldots,p_s$ are in general position with respect to L if for any $D\in|\mathcal{O}_{\mathbf{P}^2}(r)|$ such that:

1) D is irreducible and reduced; 2) $1\le r\le(\delta+2)/2$; 3) $\mu_j(D)\le(m_j+1)/2$, j=1,...,s, then $1/2\sum_{1\le j\le s}\mu_j(D)(\mu_j(D)+1)\le r(r+3)/2$, where $\mu_j(D)$ denotes the multiplicity of D at p_j. This definition can be found in [3, (3.0)].

(0.12) Let X be a geometrically ruled surface. Let $\pi:X_s\to X$, be the blowing up morphism with center $p_1,...,p_s\in X$ and let P_j be the corresponding exceptional curves on X_s. Then a line bundle L on X_s is of the form $L=[\pi^*(\mathcal{L})-\sum_{1\le j\le s}m_jP_j]$, where $\mathcal{L}$ is the sublinear system on X corresponding to L. In [8] and in [4], it was studied the very amplness of $\mathcal{L}$. If we let $L\equiv aC_0+bf-\sum_{1\le j\le s}m_jP_j$ then $M=L-K_X\equiv(a+2)C_0+(b-2q+2+e)f+\sum_{1\le j\le s}(m_j+1)P_j$.

(0.13) Theorem: Let X_s, L and M be as in (0.12). Assume s=0 and $a\ge1$.
(a) Let $e\ge0$. If $b\ge ae+2q+1$ then L is very ample. Moreover if $q\le2$ then L is very ample if and only if $b\ge ae+2q+1$.
(b) Let $e<0$. If $b\ge(ae/2)+2q+1+\text{Max}\{ae/2,-3/2\}$, then L is very ample. On the other hand if L is very ample and either q=1, $a\ge1$ and e=-1 or q=2, a=1 and $e<0$, then $b\ge(ae/)2+2q+1+\text{Max}\{ae/2,-3/2\}$.

Proof: See [4, Corollary (1.3.3) and (1.4.2)]

(0.14) Definition: Let X, X_s and L be as in (0.12). We say that L satisfies the property P_1 if for any $E\equiv f-\sum_{1\le j\le s}\alpha_jP_j$ such that $E\in D_1$ and $0\le\alpha_j\le1$, then $L\cdot E=a-\sum_{1\le j\le s}\alpha_jm_j\ge1$. See [4, (1.2.3)].

(0.14.1) Remark: The property P_1 is a necessary condition for the very ampleness of L.

(0.15) Definition: Let X, X_s and L be as in (0.12). Let E be a divisor on X_s. We say that E is a test divisor for the pair (X_s,L), if the following conditions are satisfied:

1) $E\in D_1$; 2) $g(E)\ge0$; 3) $E\equiv xC_0+yf-\sum_{j=1,...,s}\alpha_jP_j$; where:
$1\le x\le(a+1)/2$,

$$1/4(2b-4q+3+(2x-a)e))\ge y\ge\begin{cases}0 & \text{if } (x,e)\in\Gamma_0\\ xe/2 & \text{if } (x,e)\in\Gamma_1\end{cases}$$

$$\text{Min}\{x,(m_j+1)/2\}\ge\alpha_j\ge\begin{cases}\text{Max}\{0,m_j-a-b+2q-e+2y-3\} & \text{if } (a+2-2x,e)\in\Gamma_0\\ \text{Max}\{0,m_j-a-b+2q-e+2y-3+(a+2-2x)e/2\} & \text{if } (a+2-2x,e)\in\Gamma_1\end{cases}$$

$$\text{where } \Gamma_0 = \begin{cases} \{(x,e) \text{ such that } x=0,1\ e\geq -q\} \\ \{(x,e) \text{ such that } 2\leq x\leq (a+2)/2,\ e\geq 0\} \end{cases} \quad \text{and} \quad \Gamma_1=\{(x,e) \text{ such that } 2\leq x\leq (a+4)/2,\ -q\leq e\leq 0\}.$$

We write $E_{x,y}$ for the test divisors of the pair (X_s,L). See [4, §1].

(0.16) Definition: Let X, X_s and L be as in (0.12). Let $q=0$. We say that L satisfies the property P_2 if for any $E\equiv C_0-\sum_{1\leq j\leq s}\alpha_jP_j$ such that $E\in D_1$ and $0\leq\alpha_j\leq 1$, then $L\cdot E=b-ae-\sum_{1\leq j\leq s}\alpha_jm_j\geq 1$. See [4, (2.0.3)].

(0.16.1) Remark: The property P_2 is a necessary condition for the very ampleness of L.

(0.17) Definition: Let X, X_s and L be as in (0.12). Let $q=0$. Let E be a divisor on X_s. We say that E is a test divisor for the pair (X_s,L), if the following conditions are satisfied:

1) $E\in D_1$; 2) $g(E)\geq 0$; 3) $E\equiv xC_o+yf-\sum_{1\leq j\leq s}\alpha_jP_j$

where :

$1\leq x\leq (a+1)/2$,

$$1/4(2b+3+(2x-a)e)\geq y\geq \begin{cases} \text{Max}\{xe,1\} & \text{if either } x=1 \text{ or } x\geq 2 \text{ and } e\leq 2 \\ xe+2-e & \text{if } x\geq 2 \text{ and } e\geq 2. \end{cases}$$

$$\text{Min}\{x,(m_j+1)/2\}\geq\alpha_j\geq \begin{cases} \text{Max}\{0,m_j-a-b-e+2y-3\} & \text{if } (a+2-2x,e)\in\Gamma_0 \\ \text{Max}\{0,\ (m_j-a-b-e+2y-3+(a+2-2x)e/2\} & \text{if } (a+2-2x,e)\in\Gamma_1 \end{cases}$$

$$\text{where } \Gamma_0 = \begin{cases} \{(x,e) \text{ such that } x=0,1\ \} \\ \{(x,e) \text{ such that } 2\leq x\leq (a+2)/2\} \end{cases} \quad \text{and} \quad \Gamma_1=\{(x,e) \text{ such that } 2\leq x\leq (a+4)/2, e=0\}.$$

We write $E_{x,y}$ for the test divisors of the pair (X_s,L). See [4, §2].

(0.18) Theorem: Let X, X_s, L and M be as in (0.12). Assume that:

1) $a\geq 0$; 2) $M^2\geq 9$; 3) L satisfies property P_1 (P_1 and P_2 if $q=0$); 4) $(M-E_{x,y})\cdot E_{x,y}\geq 3$ for any test divisor $E_{x,y}$ such that $E^2_{x,y}<0$. Then L is very ample unless there is a test divisor $E_{x,y}$ such that either $E^2_{x,y}=0$ and $M\cdot E_{x,y}=1,2$ or $E^2_{x,y}=1$ and $M\equiv 3E$.

Proof: See [4, (1.2.5) and (2.0.6)].

(0.19) Definition: Let X, L, X_s and L be as in (0.12). Let $q=0$. We say that the points $p_j\in X$,

$1 \le j \le s$, are in general position respect to L, if for any divisor $D \equiv xC_0 + yf$ of X such that:
1) D is irreducible and reduced; 2) $0 \le x \le (a+1)/2$; 3) $0 \le y \le ((-ae/2)+b+(x+1)e+2)/2$
4) $\mu_j(D) \le (m_j+1)/2$, $1 \le j \le s$, then
$1/2 \sum_{1 \le j \le s} \mu_j(D)\,(\mu_j(D)+1) \le h^0(D)-1$, where $\mu_j(D)$ is the multiplicity of D at p_j.

(0.19.1) Remark: If e=0,1, this definition coincide with (0.11).

(0.20) Theorem: Let (X,L) be a pair consisting of a smooth, connected, projective surface X and a very ample line bundle L on it. Suppose $h^0(L) \ge 5$. Then $d(d-5)-10(g-1)+12\chi(\mathcal{O}_X) \ge 2K_X \cdot K_X$.

Proof: See [1, Lemma (0.11)]

(0.21) Theorem: Let (X,L) be the pair consisting of a smooth, connected, projective surface X and a very ample line bundle L on it. Let $2h^0(L)-4 \le d \le 3h^0(L)-8$. Then $p_g \le d-2h^0(L)+5$. Furthermore X is a Castelnuovo surface if and only if $L \cdot K_X = 3(d-2h^0(L)+4)$.

Proof: See Buium [9, Lemma (1.4)].

(0.22) Theorem: Let (X,L) be the pair as in (0.21). If X is a Castelnuovo surface in $\mathbf{P}^n$, $n=h^0(L)-1$, then X is minimal. Furthermore X is arithmetically normal and q=0.

Proof: See Buium [9, Lemma (1.10)].

(0.23) Corollary: Let (X,L) be the pair consisting of a smooth, connected, projective surface X and a very ample line bundle L on it. Suppose d=9, $h^0(L)=6$, g=7, then X is a minimal Castelnuovo surface with q=0 and $p_g=2$.

Proof: From (0.21) it follows that $p_g \le 2$ and furthermore that X is a Castelnuovo surface. Thus $p_g=2$. Applying now (0.22) we have q=0.

§1. Existence of surfaces whose minimal model is $\mathbf{P}^2$.

(1.0) Let L be a very ample line bundle on a smooth, connected, projective surface $\mathbf{P}^2{}_s$. Suppose $g \le 7$. The main purpose of this section is to exhibit the existence of some of "the candidate surfaces" described in [5] and in [7] and precisely those whose minimal model is $\mathbf{P}^2$. The situation is described in the following table.

Instructions to understand the table:

1) We have included in our table the birational invariant $\gamma_0=d-c_1^2$ so it will be easy to identify the surfaces quoted in the tables II,III,IV in [5] for $g \leq 6$ and in the table in [7] for g=7. All the other invariants of the surfaces can be found in the mentioned tables. We like to remind that t_0 denotes the number of exceptional lines that we have to contract on $\mathbf{P}^2_s$ in order to get a minimal reduction. 2) In A we list the numbers of points blowing up which on $\mathbf{P}^2$ we get surfaces $\mathbf{P}^2_s$ whose existence was proved either by the italian classical geometers or by [2], [27], [28], [29]. 3) In B we list the numbers of points blowing up which on $\mathbf{P}^2$ we get surfaces $\mathbf{P}^2_s$ whose existence follows assuming that the points $p_1,\dots,p_s$ satisfy (0.3). 4) In C we list the numbers of points blowing up which on $\mathbf{P}^2$ we get surfaces $\mathbf{P}^2_s$ whose existence follows assuming that the points $p_1,\dots,p_s$ satisfy (0.11). (We use this only in the cases 8 and 19). 5) In D we list the numbers of points blowing up which on $\mathbf{P}^2$ we get surfaces $\mathbf{P}^2_s$ with $M^2 \leq 8$ and so we are not able to say anything about their existence. 6) We write

$|L|$ for the sublinear system of $\mathcal{O}_{\mathbf{P}^2}(\delta)$ associated to L. So when we write $|L|=\delta l - \sum_{1\leq j\leq s} m_j p^{k_j}$, we mean the sublinear system of $\mathcal{O}_{\mathbf{P}^2}(\delta)$ which passes through m_j points of multiplicity k_j.

	g	γ_0	d	$h^0(L)$	$\|L\|$	A	B	C	D
	0	-8	1	3	l	$t_0=0$			
	0	-5	4	6	$2l$	$t_0=0$			
	1	0	9,...,3	10,...,4	$3l-t_0p$	$0\leq t_0\leq 6$			
	3	7	16,...,6	15,...,5	$4l-t_0p$	$0\leq t_0\leq 10$			
	3	6	8,7	7,6	$6l-7p^2-t_0p$	$t_0=0,1$			
	4	8	9	7	$9l-8p^3$	$t_0=0$			
1)	4	9	12,...,7	10,...,5	$6l-6p^2-t_0p$	$t_0=5$	$0\leq t_0\leq 4$		
2)	5	10	9,8	6,5	$7l-10p^2-t_0p$	$t_0=0,1$			
3)	5	12	16,...,9	13,...,6	$6l-5p^2-t_0p$		$0\leq t_0\leq 6$		$t_0=7$
4)	6	15	20,...,8	16,...,5	$6l-4p^2-t_0p$	$t_0=12$	$0\leq t_0\leq 9$		$t_0=10,11$
5)	6	13	14,...,9	10,...,6	$9l-7p^3-p^2-t_0p$		$0\leq t_0\leq 3$		$t_0=4,5$
6)	6	13	13,...,9	9,...,6	$7l-9p^2-t_0p$		$t_0=0,1,2$		$t_0=3,4$
7)	6	12	11,10,9	7,6	$9l-6p^3-4p^2-t_0p$		$t_0=0$		$t_0=1,2$
8)	6	11	10,9	6	$10l-10p^3-t_0p$			$t_0=0$	$t_0=1$
9)	6	10	9	6,5	$13l-10p^4$				$t_0=0$
10)	6	16	25,...,9	21,...,6	$5l-t_0p$	$0\leq t_0\leq 12$	$t_0=13$		$t_0=14,15,16$
11)	7	12	11,10	7,6	$15l-6p^5-4p^4-t_0p$				$t_0=0,1$
12)	7	12	10	6	$12l-5p^4-6p^3-t_0p$				$t_0=0$
13)	7	18	24,...,10	19,...,6	$6l-3p^2-t_0p$		$0\leq t_0\leq 11$		$t_0=12,13,14$
14)	7	16	18,...,10	13,...,6	$9l-7p^3-t_0p$		$0\leq t_0\leq 5$		$t_0=6,7,8$
15)	7	16	17,...,10	12,...,6	$7l-8p^2-t_0p$		$0\leq t_0\leq 4$		$t_0=5,6,7$
16)	7	15	16,...,9	11,...,5	$12l-8p^4-t_0p$		$t_0=0,1,2$		$3\leq t_0\leq 7$
17)	7	15	15,...,9	10,...,5	$12l-6p^3-3p^2-t_0p$		$0\leq t_0\leq 3$		$t_0=4,5,6$
18)	7	14	12,11,10	7,6	$9l-5p^3-6p^2-t_0p$		$t_0=0$		$t_0=1,2$
19)	7	13	12,11,10	7,6	$12l-6p^4-4p^3-t_0p$			$t_0=0$	$t_0=1,2$
20)	7	13	11,10	7,6	$12l-7p^4-p^3-3p^2-t_0p$				$t_0=0,1$
21)	7	13	11,10	7,6	$10l-9p^3-2p^2-t_0p$				$t_0=0,1$

How we obtain the results described in the above table.

(1.1) In the cases 9,11,12, 20, 21, $M^2 \leq 8$ for every value of t_0. Therefore we don't know if in these cases L is very ample. In the cases 13,14 and 15 using Castelnuovo's Inequality and [15,pg.434], we see that if d=9, which seemed to be possible in [5] and [7], then $h^0(L)=6$ and we get a contradiction by (0.23).

(1.2) Consider now the cases (not contained in (1.1)) in which $\mathbf{P}^2{}_s$ is obtained by blowing up only simple or double points i.e. the cases 1,2,3,4,6,10,13,15. We have $\delta=5,6,7$. In order to apply (0.8) we need to compute the following values:

	y	1	2	3	4
	$\Sigma\alpha_j \leq$	2	5	9	s
	$\delta=5, y(\delta+3-y)-3$	4	9	12	_
Case 10 $t_0/M^2 \leq 8$; $t_0=14,15,16$	$\Sigma\alpha_j m_j \leq$	2	5	9	_
	$\delta=6, y(\delta+3-y)-3$	5	11	15	17
Case 1 $t_0/M^2 \leq 8$; $t_0=5$	$\Sigma\alpha_j m_j \leq$	4	10	12,...,15	12,...,16
Case 3 $t_0/M^2 \leq 8$; $t_0=7$	$\Sigma\alpha_j m_j \leq$	4	10	10,...,14	10,...,16
Case 4 $t_0/M^2 \leq 8$; $t_0=10,11,12$	$\Sigma\alpha_j m_j \leq$	4	8,9	8,...,13	8,...,17
Case 13 $t_0/M^2 \leq 8$; $t_0=12,13,14$	$\Sigma\alpha_j m_j \leq$	4	6,...,8	6,...,12	6,...,17
	$\delta=7, y(\delta+3-y)-3$	6	13	18	21
Case 2 $t_0/M^2 \leq 8$; $t_0=1$	$\Sigma\alpha_j m_j \leq$	4	10	18	20
Case 6 $t_0/M^2 \leq 8$; $t_0=3,4$	$\Sigma\alpha_j m_j \leq$	4	10	18	20
Case 15 $t_0/M^2 \leq 8$; $t_0=5,6,7$	$\Sigma\alpha_j m_j \leq$	4	10	16,17	16,...,20

Therefore all those surfaces exist.

(1.2.1) Remark: Consider case 4. Denote by (X_i,L_i) the pair such that $L_i.L_i=i$, i=8,...,11. Then: 1) L_8 is very ample and $h^0(L_8)=5$; 2) L_{11} is very ample and $h^0(L_{11})=7$; 3) L_{10} is spanned (see [3, (2.3)]). Note that from 2) and 3) it follows that $h^0(L_{10})=6$. Suppose now that L_9 is very ample. From [15,pg.434] it follows that $h^0(L_9)\neq 5$. Thus $h^0(L_9)=6$ which contradicts the fact that L_{10} is spanned and $h^0(L_{10})=6$. Therefore L_9 has a base point and $h^0(L_9)=5$. Let y denote the image of such base point of L_9 on X_{10} and let x denote the point on X_{10} which we obtain contracting the exceptional line on X_9. It is easy to see that x and y cannot be separated. Therefore L_9 and L_{10} are not very ample. We like to remind that the points satisfy (0.3).

(1.3) Consider now the remaining cases (not contained in (1.1)) i.e. the cases 5,7,8,14,16,17,18, 19. In order to apply (0.9) we have to construct the test divisors E_y. Since $M^2 \geq 9$, we restrict the values of t_0 found in [5] to the one as in the following table

δ	y	cases	t_4	t_3	t_2	$t_1=t_0$
9	1,...,5	5		7	1	0,...,3
		7		6	4	0
		14		7		0,...,5
		17		6	3	0,...,3
		18		5	6	0
10	1,...,6	8		10		0
12	1,...,7	16	8			0,1,2
		19	6	4		0

where t_j denotes the number of points of multiplicity m_j.

First of all we need to compute the multiplicity α_j of the points p_j respect to E_y. This can be done using the condition (iii) of (0.7). (We like to remind that m_j denotes the multiplicity of p_j respect to $D \in |L|$ and α_j that of p_j respect to E_y). The situation is summarized in the following table:

y	$m_j=4$	$m_j=3$	$m_j=2$	$m_j=1$
1,2	$\delta=12$, $\alpha_j=0,1$	$\delta=9,10,12$, $\alpha_j=0,1$	$\delta=9,10,12$, $\alpha_j=0,1$	$\delta=9,10,12$, $\alpha_j=0,1$
3,4	$\delta=12$, $\alpha_j=0,1,2$	$\delta=9,10,12$, $\alpha_j=0,1,2$	$\delta=9,10,12$, $\alpha_j=0,1$	$\delta=9,10,12$, $\alpha_j=0,1$
5	$\delta=12$, $\alpha_j=0,1,2$	$\delta=9,10$, $\alpha_j=1,2$ $\delta=12$, $\alpha_j=0,1,2$	$\delta=9$, $\alpha_j=1$ $\delta=10,12$, $\alpha_j=0,1$	$\delta=9,10,12$, $\alpha_j=0,1$
6	$\delta=12$, $\alpha_j=1,2$	$\delta=10$, $\alpha_j=2$ $\delta=12$, $\alpha_j=1,2$	$\delta=10$, $\alpha_j=1$ $\delta=12$, $\alpha_j=0,1$	$\delta=10$, $\alpha_j=1$ $\delta=12$, $\alpha_j=0,1$
7	$\delta=12$, $\alpha_j=2$	$\delta=12$, $\alpha_j=2$	$\delta=12$, $\alpha_j=1$	$\delta=12$, $\alpha_j=1$

We want now to apply the condition (ii) of (0.7) i.e. $g(E_y) \geq 0$. Looking at the above table we see that the test divisors can pass through the points p_j, j=1,...,s, at least with multiplicity two. If we denote by p (q) the number of possible double (simple) points of E_y then $g(E_y)=-p+(y-1)(y-2)/2 \geq 0$ and $p \leq \mathrm{Min}\{(y-1)(y-2)/2, t_4+t_3\}$. Since $E^2_y>0$ implies $(M-E_y)\cdot E_y \geq 3$ and since $M \not\equiv 3E_y$ in all the cases by (0,9), in order to have L to be very ample, we have to check that:

(i) if $E^2_y<0$ then $(M-E_y)\cdot E_y\geq 3$; (ii) if $E^2_y=0$ then $M\cdot E_y\neq 1,2$.

Since $E^2_y=y^2-4p-q$, it follows that the test divisors E_y such that $E^2_y\leq 0$ are those with the pairs (p,q) satisfying the following condition:

$$(1.3.1) \qquad \begin{cases} (y^2-q)/4\leq p\leq \mathrm{Min}\{(y-1)(y-2)/2, t_4+t_3\} \\ p+q\leq \sum t_j. \end{cases}$$

(1.3.2) Remark: From (0.10) we get that if y=7, since $7=[(\delta+3)/2]$, then $(M-E_7)\cdot E_7\geq 3$ for any E_7. So we have to check that the test divisors E_y satisfy the conditions of (0.9) only for y=1,...,6 and $E^2_y\leq 0$.

y	1	2	3	4	Case 19. In column 1 there are the values of $p/E^2_y\leq 0$
6	9,10	(9.1)		(10,0)	
5	5,6	(6,1)...(6,3) (5,0)...(5,5)	(6,4)		In column 2 there are the values of(p,q)/$(M-E_y).E_y\geq 3$
4	2,3	(3,0)...(3,6) (2,0)...(2,8)	(3,7)		In column 3 there are the values of (p,q)/$(M-E_y).E_y\leq 2$
3	0,1	(1,0)...(1,7) (0,0)...(0,9)	(1,8),(1,9) (0,10)		but by (0.3) those test divisors are not effective. In column 4 there are the values of (p,q)/$(M-E_y).E_y\leq 2$
2	0	(0,0)...(0,5)	(0,6),...(0,10)		but by (0.11) those test divisors are not effective. Thus L is always very ample.
1	0	(0,0)...(0,2)	(0,3)...(0,10)		

y	1	2	3	Case 5
5	5,6	(5,3)...(5,6) (6,2)...(6,5)		In column 1 there are the values of $p/E_y^2 \leq 0$
4	2,3	(3,0)...(3,8) (2,0)...(2,9)		In column 2 there are the values of $(p,q)/(M-E_y).E_y \geq 3$
3	0,1	(1,0)...(1,7) (0,0)...(0,9)	(1,8)...(1,10) (0,10),(0,11)	In column 3 there are the values of $(p,q)/(M-E_y).E_y \leq 2$
2	0	(0,0)...(0,5)	(0,6)...(0,11)	but by (0.3) those test divisors are not effective. Thus L is always very ample.
1	0	(0,0)...(0,2)	(0,3)...(0,11)	

y	1	2	3	Case 7
5	5,6	(6,4) (5,5)		In column 1 there are the values of $p/E_y^2 \leq 0$
4	2,3	(3,0)...(3,7) (2,0)...(2,8)		In column 2 there are the values of $(p,q)/(M-E_y).E_y \geq 3$
3	0,1	(1,0)...(1,7) (0,0)...(0,9)	(0,10) (1,8)...(1,9)	In column 3 there are the values of $(p,q)/(M-E_y).E_y \leq 2$
2	0	(0,0)...(0,5)	(0,6)...(0,10)	but by (0.3) those test divisors are not effective. Thus L is always very ample.
1	0	(0,0)...(0,2)	(0,3)...(0,10)	

y	1	2	3	4
6	9,10			(10,0)
5	5,6		(6,4)	
4	2,3	(3,0)...(3,6) (2,0)...(2,8)	(3,7)	
3	0,1	(1,0)...(1,7) (0,0)...(0,5)	(1,8)(1,9) (0,10)	
2	0	(0,0)...(0,5)	(0,6)...(0,10)	
1	0	(0,0)(0,1)(0,0)	(3,3)...(0,10)	

Case 8

In column 1 there are the values of $p/E_y^2 \leq 0$

In column 2 there are the values of $(p,q)/(M-E_y).E_y \geq 3$

In column 3 there are the values of $(p,q)/(M-E_y).E_y \leq 2$

but by (0.3) those test divisors are not effective.
In column 4 there are the values of $(p,q)/(M-E_y).E_y \leq 2$

but by (0.11) those test divisors are not effective.
Thus L is always very ample.

y	1	2	3
5	5,6	(6,1)...(6,6) (5,2)...(5,7)	
4	2,3	(3,0)...(3,9) (2,0)...(2,10)	
3	0,1	(1,0)...(1,8) (0,0)...(0,10)	(1,9)...(1,11) (0,11),(0,12)
2	0	(0,0)...(0,5)	(0,6)...(0,12)
1	0	(0,0)...(0,2)	(0,3)...(0,12)

Case 14

In column 1 there are the values of $p/E_y^2 \leq 0$

In column 2 there are the values of$(p,q)/(M-E_y).E_y \geq 3$

In column 3 there are the values of $(p,q)/(M-E_y).E_y \leq 2$

but by (0.3) those test divisors are not effective.
Thus L is always very ample.

y	1	2	3	Case 16
5	5,6	(6,0)...(6,4) (5,0)...(5,5)		In column 1 there are the values of $p/E_y^2 \leq 0$ In column 2 there are the values of $(p,q)/(M-E_y).E_y \geq 3$ In column 3 there are the values of $(p,q)/(M-E_y).E_y \leq 2$ but by (0.3) those test divisors are not effective. Thus L is always very ample.
4	2,3	(3,0)...(3,7) (2,0)...(2,8)		
3	0,1	(1,0)...(1,6) (0,0)...(0,9)	(1,7)...(1,9) (0,10)	
2	0	(0,0)...(0,5)	(0,6)...(0,10)	
1	0	(0,0)...(0,2)	(0,3)...(0,10)	

y	1	2	3	Case 17
5	5,6	(6,3)(6,4)(6,5) (5,4)...(5,7)	(6,6)	In column 1 there are the values of $p/E_y^2 \leq 0$ In column 2 there are the values of $(p,q)/(M-E_y).E_y \geq 3$ In column 3 there are the values of $(p,q)/(M-E_y).E_y \leq 2$ but by (0.3) those test divisors are not effective. Thus L is always very ample.
4	2,3	(3,0)...(3,8) (2,0)...(2,10)	(3,9)	
3	0,1	(1,0)...(1,7) (0,0)...(0,9)	(1,8)...(1,11)	
2	0	(0,0)...(0,5)	(0,6)...(0,12)	
1	0	(0,0)(0,1)(0,2)	(0,3)...(0,12)	

y	1	2	3
5	5	(5,6)	
4	2,3	(3,0)...(3,7)	(3,8)
3	0,1	(1,0)...(1,8) (0,0)...(0,9)	(1,9)...(1,10) (0,10),(0,11)
2	0	(0,0)...(0,5)	(0,6)...(0,11)
1	0	(0,0)...(0,2)	(0,3)...(0,11)

Case 18.
In column 1 there are the values of $p/E_y^2 \leq 0$

In column 2 there are the values of$(p,q)/(M-E_y).E_y \geq 3$

In column 3 there are the values of $(p,q)/(M-E_y).E_y \leq 2$

but by (0.3) those test divisors are not effective.
Thus L is always very ample.

§2 Rational surfaces whose minimal model are Fe.

(2.0) Let L be a very ample line bundle on a smooth, connected, projective surface X_s. Suppose $g \leq 7$. The main purpose of this section is to exhibit the existence of some of "the candidate surfaces" $X_s=Fe_s$ described in [5], [7] and precisely those whose minimal model is an Hirzebruch surface Fe. The situation is described in the following table.

Instructions to understand the table:

1) We have included in our table the invariant $\gamma_0=d-c_1{}^2$ so it will be easy to identify the surfaces quoted in the tables II,III,IV in [5] and that in [7]. Since $L \cdot C_0 \geq 1$, we have that $e \neq 3$ in the cases 13,21,26 and $e \neq 4$ in the case 23, (see the table below) which appeared as possible in the recalled papers. All the other invariants of the surfaces can be found in the mentioned tables. We like to remind that t_0 denotes the number of exceptional lines that we have to contract on Fe_s in order to get a minimal reduction. 2) In A we list the numbers of points blowing up which on Fe we get surfaces Fe_s whose existence follows from [15]. 3) In B we list the numbers of points blowing up which on Fe we get surfaces Fe_s whose existence follows assuming that the points satisfy the definitions (0.14) and (0.16). 4) In C we list the numbers of points blowing up which on Fe we get surfaces Fe_s whose existence follows assuming that the points satisfy the definition (0.19). 5) In D we list the numbers of points blowing up which on Fe we get surfaces Fe_s with either $M^2 \leq 8$ or $M^2 \geq 9$ but there exists a Reider divisor (case 23) and so we are not able to say anything about their existence. 6) We write $|L|$ for the sublinear system on Fe corresponding to L. So when we write $|L| = aC_0+bf-\sum_{1 \leq j \leq s} m_j p^{k_j}$, we mean the sublinear system of aC_0+bf which passes trough m_j points of multiplicity k_j.

	g	γ_0	d	$h^0(L)$	$\lvert L \rvert$	e	A	B	C	D
1)	0		2k-e		C_0+kf	$e\leq k-1$	$t_0=0$			
2)	1	0	8,...,3	9,...,4	$2C_0+2f$	e=0	$0\leq t_0\leq 5$			
3)	2	4	12,...,5	12,...,5	$2C_0+(e+3)f-t_0p$	e=0,1,2	$t_0=0$	$1\leq to\leq e+2+\varepsilon(e)$	$e+2+\varepsilon(e)\leq t_0\leq 7$	
4)	3	8	16,...,7	15,...,6	$2C_0+(e+4)f-t_0p$	$0\leq e\leq 3$	$t_0=0$	$1\leq t_0\leq e+3+\varepsilon(e)$	$e+4+\varepsilon(e)\leq t_0\leq 9$	
5)	4	12	20,...,8	18,...,6	$2C_0+(e+5)f-t_0$	$0\leq e\leq 4$	$t_0=0$	$1\leq t_0\leq e+4+\varepsilon(e)$	$e+5+\varepsilon(e)\leq t_0\leq 11$	$t_0=12$
6)	4	10	18,...,8	16,...,6	$3C_0+3f-t_0p$	e=0	$t_0=0$	$1\leq t_0\leq 5$	$6\leq t_0\leq 10$	
7)	5	16	24,...,9	21,...,6	$2C_0+(e+6)f-t_0p$	$0\leq e\leq 5$	$t_0=0$	$1\leq t_0\leq e+5+\varepsilon(e)$	$e+6+\varepsilon(e)\leq t_0\leq 13$	$t_0=14,15$
8)	5	11	12,...,9	9,...,6	$4C_0+(2e+5)f-7p^2-t_0p$	e=0,1,2			$0\leq t_0\leq 3$	
9)	5	13	21,...,9	18,...,6	$3C_0+5f-t_0p$	e=1	$t_0=0$	$1\leq t_0\leq 4$	$5\leq t_0\leq 11$	$t_0=12$
10)	6	20	28,...,9	24,...,6	$2C_0+(e+7)f-t_0p$	$0\leq e\leq 6$	$t_0=0$	$1\leq t_0\leq e+6+\varepsilon(e)$	$e+7+\varepsilon(e)\leq t_0\leq 15$	$16\leq t_0\leq 19$
11)	6	16	24,...,9	20,...,6	$3C_0+4f-t_0p$	e=0				
					$3C_0+7f-t_0p$	e=2	$t_0=0$	$1\leq t_0\leq 6$	$7\leq t_0\leq 12$	$13\leq t_0\leq 15$
12)	6	14	16,...,9	12,...,6	$4C_0+(2e+5)f-6p^2-t_0p$	e=0,1,2			$0\leq t_0\leq 5$	$t_0=6,7$
13)	6	13	12,...,9	8,...,6	$4C_0+(2e+6)f-9p^2-t_0p$	e=0,1,2			$t_0=0,1$	$t_0=2,3$
14)	6	12	10,9	6	$5C_0+5f-10p^2-t_0p$	e=0				$t_0=0,1$
15)	6	11	9	6	$6C_0+(3e+7)f-7p^3-3p^2$	e=0,1,2				$t_0=0$
16)	7	24	32,...,10	27,...,6	$2C_0+(e+8)f-t_0p$	$0\leq e\leq 7$	$t_0=0$	$1\leq t_0\leq e+7+e(e)$	$e+8+e(e)\leq t_0\leq 17$	$18\leq t_0\leq 22$
17)	7	12	10	6	$8C_0+(4e+9)f-7p^4-2p^3-p^2-t_0p$	e=0,1,2				$t_0=0$
18)	7	11	10	6	$12C_0+(6e+13)f-7p^6-2p^5-t_0p$	e=0,1,2				$t_0=0$
19)	7	19	27,...,10	22,...,6	$3C_0+6f-t_0p$	e=1	$t_0=0$	$1\leq t_0\leq 5$	$6\leq t_0\leq 14$	$15\leq t_0\leq 17$
20)	7	17	20,...,10	15,...,6	$4C_0+(2e+5)f-5p^2-t_0p$	e=0,1,2			$0\leq t_0\leq 7$	$8\leq t_0\leq 10$
21)	7	16	16,...,10	11,...,6	$4C_0+(2e+6)f-8p^2-t_0p$	e=0,1,2			$0\leq t_0\leq 3$	$t_0=4,5,6$
22)	7	15	14,...,9	9,...,5	$5C_0+5f-9p^2-t_0p$	e=0			$t_0=0,1,2$	$t_0=3,4,5$
23)	7	15	12,...,9	7,6,5	$4C_0+(2e+7)f-11p^2-t_0p$	$0\leq e\leq 3$				$0\leq t_0\leq 3$
24)	7	14	13,...,10	8,7,6	$6C_0+(3e+7)f-7p^3-2p^2-t_0p$	e=0,1,2			$t_0=0,1$	$t_0=2,3$
25)	7	14	11,10	7,6	$5C_0+8f-11p^2-t_0p$	e=1				$t_0=0,1$
26)	7	13	11,10	7,6	$6C_0+(3e+8)f-9p^3-p^2-t_0p$	e=0,1,2				$t_0=0,1$
27)	7	13	10	6	$6C_0+(3e+7)f-6p^3-5p^2-t_0p$	e=0,1,2				$t_0=0$

How we obtain the results described in the above table.

(2.1) In the cases 14,15,17,18,25,26,27, $M^2\leq 8$ for every value of t_0. Therefore we don't know if those surfaces Fe_s do really exist. Moreover in the cases 16,19,20 and 21, using Castelnuovo's Inequality and [15,pg.434], we see that if d=9, which seemed to be possible in [5] and [7], then $h^0(L)=6$ and we get a contradiction by (0.23).

(2.2) Consider now the cases (not contained in (2.1)) in which Fe_s is a conic bundle i.e.the cases 3,4,5,7,10,16. We can write $L\equiv 2C_0+(g+1+e)f-\sum_{1\leq j\leq t_0} P_j$, $0\leq e\leq g$. It is easy to check that $M^2\geq 9$ if $t_0\leq 2g+3$. In order to apply (0.18) we have to construct the test divisors $E_{x,y}$. From (0.17) it follows that x=1, $\mathrm{Max}\{e,1\}\leq y\leq (2e+2g+5)/4$ and $\alpha_j=0,1$. Moreover $(M-E_{1,y})\cdot E_{1,y}\geq 3$ if $\sum\alpha_j\leq g+2y-e$.

Thus if $t_0\leq e+g+\varepsilon(e)$ where $\varepsilon(e)=\begin{cases} 0 & \text{if } e>0 \\ 2 & \text{if } e=0 \end{cases}$ then L is very ample supposing that the points satisfy (0.14) and (0.16). If $t_0>e+g+\varepsilon(e)$ then L is very ample supposing (0.19).

(2.3) Consider now the cases 6,9,11,19 (not contained in (2.1)) in which Fe_S is a 3-bundle. We can write $L\equiv 3C_0+((g+2+3e)/2)f-\sum_{1\le j\le t0}P_j$. It is easy to check that $M^2\ge 9$ if $t_0\le(5g+21)/4$. In order to apply (0.18) we have to construct the test divisors $E_{x,y}$. From (0.17) it follows that either $x=1$, $\mathrm{Max}\{e,1\}\le y\le (g+5+2e)/4$ or $x=2$, $\mathrm{Max}\{2e,1\}\le y\le(g+5+4e)/4$. In both cases $\alpha_j=0,1$. Moreover $(M-E_{1,y})\cdot E_{1,y}\ge 3$ if $\sum\alpha_j\le(g+6y-3e)/2$ and $(M-E_{2,y})\cdot E_{2,y}\ge 3$ if $\sum\alpha_j\le g+y-2e+5$. Thus:

case	L very ample supposing (0.14),(0.16)	L very ample supposing (0.19)
6	$0\le t_0\le 5$	$6\le t_0\le 10$
9	$0\le t_0\le 4$	$5\le t_0\le 11$
11	$0\le t_0\le 6$	$7\le t_0\le 12$
19	$0\le t_0\le 5$	$6\le t_0\le 14$

(2.4) Consider now the cases (not contained in (2.1)) in which Fe_S is a 4-bundle i.e. the cases 8,12, 13,20,21,23. We can write $L\equiv 4C_0+((g+t_2+3+6e)/3)f-2\sum_{1\le j\le t2}P_j-\sum_{t2+1\le j\le t2+t0}P_j$, where we let t_2 be the number of double points. It is easy to check that $0\le e\le[(g+t_2)/6]$ and that $M^2\ge 9$ if $t_0\le(4g-5t_2+27)/4$. In order to apply (0.18) we have to construct the test divisors $E_{x,y}$. From (0.17) it follows that in all the cases $\alpha_j=0,1$. In case 23 there exists $E_{2,e+3}\equiv 2C_0+(e+3)f-\sum_{1\le j\le 11}P_j$ which is effective and such that $E_{2,e+3}\in \mathcal{D}_1$, $g(E_{2,e+3})=2$, $M\equiv 3E_{2,e+3}$, $E^2_{2,e+3}=1$ and $(M-E_{2,e+3})\cdot E_{2,e+3}=2$. Therefore we are not able to decide if L is very ample or not. In all the remaining cases e=0,1,2. It is easy to see that either $x=1$, $\mathrm{Max}\{e,1\}\le y\le(g+t_2+6+3e)/6$ or $x=2$, $\mathrm{Max}\{2e,1\}\le y\le(g+t_2+6+6e)/6$. Moreover $(M-E_{1,y})\cdot E_{1,y}\ge 3$ if $\sum\alpha_j m_j\le(g+t_2+12y-6e)/3$ and $(M-E_{2,y})\cdot E_{2,y}\ge 3$ if $\sum\alpha_j m_j\le(2g+2t_2+6y-6e+9)/3$. Thus in all these cases L is very ample by (0.19) if $t_0\le(4g-5t_2+27)/4$.

(2.5) Consider now the case (not contained in (2.1)) in which Fe_S is a 5-bundle i.e. case 22. We have $L\equiv 5C_0+5f-2\sum_{1\le j\le 9}P_j-\sum_{10\le j\le 9+t0}P_j$, e=0. It is easy to check that $M^2\ge 9$ if $t_0\le 2$. In order to apply (0.18) we have to construct the test divisors $E_{x,y}$. From (0.17) it follows that $\alpha_j=0,1$; $x\le 3$ and $y\le 3$. We can assume $y\ge x$. Moreover $(M-E_{x,y})\cdot E_{x,y}\ge 3$ if $\sum\alpha_j m_j\le 7x+7y-2xy-3$. In the usual way we see that L is very ample by (0.19).

(2.6) Consider now the case (not contained in (2.1)) in which Fe_S is a 6-bundle i.e. case 24. We have $L\equiv 6C_0+(3e+7)f-3\sum_{1\le j\le 7}P_j-2P_8-2P_9-\sum_{10\le j\le 9+t0}P_j$, e=0,1,2. It is easy to check that $M^2\ge 9$ if

$t_0 \leq 1$. In order to apply (0.18) we have to construct the test divisors $E_{x,y}$. From (0.17) it follows $\alpha_j=0$, 1,2; $x\leq 3$ and $\mathrm{Max}\{xe,1\}\leq y\leq 4+xe/2$. Moreover $(M-E_{x,y})\cdot E_{x,y}\geq 3$ if $\sum\alpha_j(m_j+1-\alpha_j)\leq(8-2x)y-(4-x)xe+9x-3$. In the usual way we see that L is very ample by (0.19).

§3 Surfaces whose minimal model is a geometrically ruled surface .

(3.0) Let L be a very ample line bundle on a smooth, connected, projective surface X_S. Suppose $g \leq 7$. The main purpose of this section is to exhibit the existence of some of "the candidate surfaces" found in [7] and precisely those whose minimal model is a geometrically ruled surface X with irregularity $q\geq 1$. The situation is described in the following table.

Instructions to understand the table:

1) We have included in our table the invariant $\gamma_0=d-c_1^2$ so it will be easy to identify the quoted surfaces in the tables II, III, IV in [5] and that in [7]. All the other invariants of the surfaces can be found in the mentioned tables. We like to remind again that t_0 denotes the number of exceptional lines that we have to contract on X_S in order to get a minimal reduction. 2) In A we list the numbers of points blowing up which on X we get surfaces X_S whose existence follows either from [15] or [8]. 3) In B we list the numbers of points blowing up which on X we get surfaces X_S whose existence follows assuming that the points satisfy the definition (0.14). 4) In D we list the numbers of points blowing up which on X we get either surfaces X_S with $M^2\leq 8$ or surfaces with divisors which, if effective, they are Reider divisors and so we are not able to decide if those surfaces do really exist. 5) We write $|L|$ for the sublinear system on X corresponding to L. So when we write $|L|=aC_0+bf-\sum_{1\leq i\leq s} m_j p^{k_j}$ we mean the sublinear sustem of aC_0+bf which passes trough m_j points of multiplicity k_j.

	g	γ_0	d	$h^0(L)$	q	$\lvert L\rvert$	e	A	B	D
1)	3	8	8	6	1	$2C_0+f$	-1	$t_0=0$		
2)	4	12	12,...,8	9,...,6	1	$2C_0+(e+3)f-t_0p$	-1,0	$t_0=0$	$1\leq t_0\leq -e$	$-e+1\leq t_0\leq 4$
3)	4	9	9	6	1	$3C_0$	-1	$t_0=0$		
4)	5	16	16,...,9	12,...,6	1	$2C_0+(e+4)f-t_0p$	-1,0,1	$t_0=0$	$1\leq t_0\leq -e+1$	$e+2\leq t_0\leq 8$
5)	6	20	20,...,9	15,...,6	1	$2C_0+(e+5)f-t_0p$	-1,...,2	$t_0=0$	$1\leq t_0\leq -e+2$	$-e+3\leq t_0\leq 11$
6)	6	15	15,...,10	10,...,6	1	$3C_0+f-t_0p$	-1	$t_0=0$	$t_0=1$	$2\leq t_0\leq 5$
7)	6	20	12,...,9	6	2	$2C_0+f-t_0p$	-2			$0\leq t_0\leq 3$
8)	7	24	24,...,9	18,...,6	1	$2C_0+(e+6)f-t_0p$	-1,...,3	$t_0=0$	$1\leq t_0\leq -e+3$	$-e+4\leq t_0\leq 15$
9)	7	18	18,...,10	12,...,6	1	$3C_0+3f-t_0p$	0	$t_0=0$		$1\leq t_0\leq 8$
10)	7	16	16,...,10	10,...,6	1	$4C_0-t_0p$	-1	$t_0=0$	$t_0=1$	$2\leq t_0\leq 6$
11)	7	15	15,...,10	9,...,6	1	$5C_0-f-t_0p$	-1	$t_0=0$		$1\leq t_0\leq 5$
12)	7	15	12,11,10	7,6	1	$4C_0+(2e+3)f-3p^2-t_0p$	-1,0			$t_0=0,1,2$
13)	7	24	16,...,10	9,...,6	2	$2C_0+(e+4)f-t_0p$	-2		$t_0=0,1$	$2\leq t_0\leq 6$
						$2C_0+(e+4)f-t_0p$	-1	$t_0=0$		$1\leq t_0\leq 6$

(3.0.1) Remark: If $h^0(L)=5$ then q=1 and:

g	d	γ_0	$\|L\|$	e	D
7	10	15	$5C_0-f-t_0p$	-1	$t_0=5$
7	10	15	$4C_0+(2e+3)f-3p^2-t_0p$	-1,0	$t_0=2$

Proof: Let $h^0(L)=5$. By [7] it follows that the only missing case is g=5, d=8, $\gamma_0=16$ which doesn't exist by [29].

(3.0.2) Remark: We have not listed in the above table the scrolls. In this case g=q, $-q\leq e\leq q-1$, $L\equiv C_0+((q+e)/2)f$.

(3.0.3) Remark: In case 13 using Castelnuovo's Inequality and [15,pg.434], we see that if d=9, which seemed to be possible in [5], then $h^0(L)=6$ and we get a contradiction by (0.23).

(3.1) In case 7, $M^2\leq 8$ for every value of t_0. So we don't know if in this case X_s do really exist.

(3.2) Consider now the cases (not contained in (3.1)) in which X_s is a conic bundle i.e. the cases 2, 4,5,8,13. We can write $L\equiv 2C_0+(g+1+e-2q)f-\sum_{1\leq j\leq t_0} P_j$. An easy computation shows that if $M^2\geq 9$ then $t_0\leq(8g-32q+15)/4$. In order to apply (0.18) we have to construct the test divisors $E_{x,y}$. From (0.15) x=1, $0\leq y\leq(e+g+2-4q)/2$ and $\alpha_j=0,1$. Moreover $(M-E_{1,y})\cdot E_{1,y}\geq 3$ if then $\sum\alpha_j\leq g+2y-e-4q$.It is easy at this point to check that when q=2 and e=-2 then L is always very ample; that when q=2 and e=-1 then L is very ample if $t_0=0$; and that when q=1 then L is very ample if $t_0\leq g-e-4$.

(3.3) Consider now the cases 6,9 (not contained in (3.1)) in which X_s is a 3-bundle. We can write $L\equiv 3C_0+((g-1+3e)/2)f-\sum_{1\leq j\leq t_0} P_j$. The usual computations show that $M^2\geq 9$ if $t_0\leq(5g-14)/4$. In order to apply (0.18) we have to construct the test divisors $E_{x,y}$. From (0.15) either x=1, $0\leq y\leq(2e+g-2)/4$, or x=2, $0\leq y\leq(4e+g-2)/4$. Again $\alpha_j=0,1$. Moreover $(M-E_{1,y})\cdot E_{1,y}\geq 3$ if $\sum\alpha_j\leq(g+6y-3e-7)/2$ and $(M-E_{2,y})\cdot E_{2,y}\geq 3$ if $\sum\alpha_j\leq g+y-e-4$. It is easy at this point to check that L is very ample if $t_0\leq(g-3e-7)/2$.

(3.4) Consider now the cases (not contained in (3.1)) in which X_s is a 4-bundle ie the cases 10,12. Case 10. $M^2\geq 9$ if $t_0\leq 3$. In order to apply (0.18) we have to construct the test divisors $E_{x,y}$. From (0.15) it follows that either x=1, y=0 or x=2 and y=-1. Again $\alpha_j=0,1$. Moreover $(M-E_{1,0})\cdot E_{1,0}\geq 3$ if $\sum\alpha_j\leq 1$ and $(M-E_{2,-1})\cdot E_{2,-1}\geq 3$ if $\sum\alpha_j\leq 2$. It is easy at this point to check that L is very ample if $t_0\leq 1$. Case 12. $M^2\geq 9$ if $t_0=0$. In order to apply (0.18) we must construct the test divisors $E_{x,y}$. From (0.15) it follows that x=1 and $\begin{cases} y=0 & \text{if } e=-1 \\ y=0,1 & \text{if } e=0 \end{cases}$ or x=2 and $\begin{cases} y=0,1 & \text{if } e=0 \\ y=-1,0 & \text{if } e=-1 \end{cases}$.

Again $\alpha_j=0,1$. Moreover $(M-E_{1,x})\cdot E_{1,x}\geq 3$ if $\sum\alpha_j\leq 2y-e$ and $(M-E_{2,x})\cdot E_{2,x}\geq 3$ if $\sum\alpha_j\leq y-e-(3/2)$. Unfortunately in this case there are the following divisors which, if effective, they are Reider divisors and so we are not able to decide if L is very ample or not.

Let e=0 then $E_{1,0}\equiv C_0-P_i$, $i=1,2,3$; $E_{1,0}\equiv C_0-P_i-P_j$, $i\neq j$; $E_{1,0}\equiv C_0-P_1-P_2-P_3$

$E_{2,0}\equiv 2C_0-P_i-P_j$, $i\neq j$; $E_{2,0}\equiv 2C_0-P_1-P_2-P_3$

Let e=-1 then $E_{1,0}\equiv C_0-P_i-P_j$, $i\neq j$; $E_{1,0}\equiv C_0-P_1-P_2-P_3$

$E_{2,0}\equiv 2C_0-P_1-P_2-P_3$

(3.5) Consider now the case (not contained in (3.1)) in which X_s is a 5-bundle i.e. the case 11. $M^2\geq 9$ if $t_0\leq 3$. In order to apply (0.18) we have to construct the test divisors $E_{x,y}$. From (0.15) either x=1, y=0 or x=2, y=-1 or x=3, y=-1. Moreover $\alpha_j=0,1$ and $(M-E_{1,0})\cdot E_{1,0}\geq 3$ if $\sum\alpha_j\leq 1$, $(M-E_{2,-1})\cdot E_{2,-1}\geq 3$ if $\sum\alpha_j\leq 0$ and $(M-E_{3,-1})\cdot E_{3,-1}\geq 3$ if $\sum\alpha_j\leq 2$. As above we see that if $t_0=0$ then L is very ample. If $t_0=1,2,3$ then there are the following divisors which, if effective, they are Reider divisor. In this case we don't know if L is very ample or not.

$t_0=1$ $E_{2,-1}\equiv 2C_0-f-P$

$t_0=2$ $E_{1,0}\equiv C_0-P_1-P_2$; $E_{2,-1}\equiv 2C_0-f-P_1-P_2$; $E_{2,-1}\equiv 2C_0-f-P_i$, $i=1,2$

$t_0=2$ $E_{1,0}\equiv C_0-P_i-P_j$; $i\neq j$; $i,j=1,2,3$ $E_{2,-1}\equiv 2C_0-f-P_i-P_j$; $i\neq j$; $i,j=1,2,3$ $E_{2,-1}\equiv 2C_0-f-P_i$, $i=1,2,3$

$E_{2,-1}\equiv 2C_0-f-P_1-P_2-P_3$ $E_{3,-1}\equiv 3C_0-f-P_1-P_2-P_3$

§4 Surfaces of non negative Kodaira dimension $\kappa(X)$.

(4.0) Let L be a very ample line bundle on a smooth, connected, projective surface X_s. Suppose $g\leq 7$ and $\kappa(X_s)\geq 0$. The main purpose of this section is to exhibit the existence of some of "the candidate surfaces" found in [24] and [6]. By (0.20) we reduce the list to the following:

	g	d	$h^0(L)$	c_1^2	q	p_g	t_0	r	$\kappa(X_s)$	(Y,L_Y)
1)	3	4	4	0	0	1	0	0	0	K3
2)	4	6	5	0	0	1	0	0	0	K3
3)	5	8	6	0	0	1	0	0	0	K3
4)	5	7	5	-1	0	1	1	0	0	K3
5)	6	5	4	5	0	4	0	0	2	Gen.Type
6)	6	7	5	0	0	2	0	0	1	Elliptic
7)	6	8	5	-1	0	1	0	1	0	K3
8)	6	10	5	0	2	1	0	0	0	Abelian
9)	6	10	5	0	1	0	0	0	0	Bielliptic
10)	6	10,9	6,5	0,-1	0	0	0,1	0	0	Enriques
11)	6	10,9	7,6,5	0,-1	0	1	0,1	0	0	K3
12)	7	8	5	0	0	2	0	0	1	Elliptic
13)	7	9	5	0	1	2	0	0	1	Elliptic
14)	7	9	5	0	0	1	0	0	1	Elliptic
15)	7	9	6	0	0	1,2	0	0	1	Elliptic
16)	7	9	6	-1	0	1	0	1	0	K3
17)	7	10	6	0	0	0	0	0	1	Elliptic
18)	7	10	6	-1	0	0	0	1	0	Enriques
19)	7	10,9	6	-1,-2	0	1	0,1	1	0	K3
20)	7	11	7,6	0	1	0	0	0	1	Elliptic
21)	7	11,10	7,6	0,-1	0	0	0,1	0	1	Elliptic
22)	7	12,11,10	7,6	0,-1,-2	0	0	0,1,2	0	0	Enriques
23)	7	12,11	7,6	0,-1	1	0	0,1	0	0	Bielliptic
24)	7	12,...,9	8,7,6	0,...,-3	0	1	0,...,3	0	0	K3
25)	7	12,11	6	0,-1	2	1	0,1	0	0	Abelian
26)	7	9	6	1	0	1	0	0	2	General Type

(4.0.1) Remark: We write (Y,L_Y) for the minimal model of (X_s,L) and we let $(X,\mathcal{L})$ denote the minimal reduction of (X_s,L). X_s is gotten by blowing up at most t_0 points on X and X is gotten by blowing up r points on Y.

(4.0.2) Remark: As it was done in the case $\kappa(X)=-\infty$ we see that by (0.23) the cases 16 and 26 cannot happen. Moreover, in the cases 19 and 24, $d\neq 9$ and in the case 15, $p_g\neq 1$.

(4.2) Study of surfaces birational to a K3 surface. The existence of 1,...,4 is classical. The existence of 7 has been proved in [29]. Consider case 11 i.e. g=6, d=10,9, $h^0(L)=7,6,5$, $c_1^2=0,-1$, $t_0=0,1$, r=0 i.e. X=Y. Let d=10. Then s=0. Therefore $h^0(L)=h^0(K_Y\otimes L)=7$ and $h^1(L)=h^1(K_Y\otimes L)=0$. Thus if d=10 then $h^0(L)=7$ whose existence is classical since it is a K3 surface of degree 2g-2 in $\mathbf{P}^g$. Let d=9. Then $s=t_0=1$. By Clifford Theorem we have $h^0(L)=6,5$. If $h^0(L)=5$ we get a contradiction using [15,pg.434]. So $h^0(L)=6$. Since $\chi(L)=6$ it follows that $h^1(L)=h^2(L)$. Thus by the long cohomology sequence associated to the short exact sequence

(4.2.1) $$O\to \mathcal{O}_{X_1}\to L\to L_C\to 0$$

we get that $h^0(K_{X_1}\otimes L^{-1})=h^2(L)=h^1(L)=0,1$. If $h^1(L)=1$ then $K_{X_1}\otimes L^{-1}$ has a section moreover, in order the surface to exist L, has to be very ample. Thus $L\cdot(K_{X_1}-L)>0$. But $L\cdot(K_{X_1}-L)=-8$. Therefore

$h^0(K_{X_1}\otimes L^{-1})=0$ which implies that $h^1(L)=0$ and (X_1,L) exists by [9] (in fact it is a 2n-1 surface in $\mathbf{P}^n$ with n=5 and $h^1(L)=0$). Consider now case 19 i.e. g=7, d=10, $h^0(L)=6$, $c_1^2=-1$, $t_0=0$, r=1. In this case s=r=1. Exactly as in the above case we see that this surface does not exists by [9]. In fact it is a surface with regular hyperplane section of degree 2n in $\mathbf{P}^n$ where n=5. We can now consider the last possibility i.e. case 24). We have g=7, d=12,11,10, $h^0(L)=8,7,6$, $c_1^2=0,-1,-2$, $t_0=0,1,2$, r=0. X=Y. Let d=12. Then s=0, $h^0(L)=8$, $h^1(L)=0$. So Y is a K3 surface of degree 2g-2 in $\mathbf{P}^g$ whose existence is classical. Let d=11. Then $s=t_0=1$. From Clifford Theorem it follows that $h^0(L)\leq 7$. If $h^0(L)=7$ exactly as in the case {11): g=6, d=9 and $h^0(L)=6$}, we get that (X_1,L) exists. If $h^0(L)=6$ then from (4.2.1) and $\chi(L)=7$ it follows that $h^1(L)=0$ and $h^2(L)=1$. But L has to be very ample thus we get a contradiction since $0< L\cdot(K_{X_1}-L)=-10$. Therefore $h^0(L)\neq 6$. Let d=10. Then $s=t_0=2$. From Clifford Theorem it follows that $h^0(L)\leq 7$. Moreover, since $L\cdot(K_{X_2}-L)=-8$, we have $h^2(L)=0$. Thus $h^0(L)-h^1(L)=\chi(L)=6$. Let $h^0(L)=6$. Then $h^1(L)=h^2(L)=0$ and the existence of the surface follows from [9] in fact it is a 2n surface in $\mathbf{P}^n$ with n=5 and $h^1(L)=0$. We have so obtained that the following surfaces birational to a K3 surfaces exist.

Surfaces Birational to a K3 with $g(L)\leq 7$.

	g	d	$h^0(L)$	c_1^2	(Y,L_Y)
1)	3	4	4	0	$(H_4,\mathcal{O}_{\mathbf{P}^3}(1))$
2)	4	6	5	0	$Y_{(2,3)}$
3)	5	8	6	0	$Y_{(2,2,2)}$
4)	5	7	5	-1	$Y_8\subset\mathbf{P}^5$
7)	6	8	5	-1	$Y_{12}\subset\mathbf{P}^7$
11)	6	9	6	-1	$Y_{10}\subset\mathbf{P}^6$
11)	6	10	7	0	$Y_{10}\subset\mathbf{P}^6$
24)	7	10	6	-2	$Y_{12}\subset\mathbf{P}^7$
24)	7	11	7	-1	$Y_{12}\subset\mathbf{P}^7$
24)	7	12	8	0	$Y_{12}\subset\mathbf{P}^7$

(4.3) Study of Surfaces Birational to an Enriques Surface: Consider case 10 i.e. g=6, d=10,9, $h^0(L)=6,5$, $c_1^2=0,-1$, $t_0=0,1$, r=0. Let d=10. Then s=0. From [15,pg.434] it follows that $h^0(L)=6$. This surface, $V_1(W)$, exists, see [14.pg.749]. Let d=9. Then $s=t_0=1$. From the existence of the above surface it follows that if $h^0(L)=6$ then L has to have a base point. Thus $h^0(L)=5$. In this case from Riemann-Roch Theorem and (4.2.1) it follows that $h^1(L)=h^2(L)$. But $(K_{X_1}-L)\cdot L=-8$ therefore $h^1(L)=h^2(L)=0$. Applying now [9,Corollary (3.2)] it follows that such surface exists. In fact it is a degree 2n+1 surface in $\mathbf{P}^n$ with $h^1(L)=0$. Consider now case 18 i.e. g=7, d=10, $h^0(L)=6$, $c_1^2=-1$, $t_0=0$, r=1. In this case s=r=1. Moreover $(K_{X_1}+L)^2=13$, $h^0(K_{X_1}\otimes L)=7$. Therefore the pair $(X_1, K_{X_1}\otimes L)$ is a 2n+1 surface in $\mathbf{P}^n$ with $h^1(K_{X_1}\otimes L)=0$. By [9,(2.3)], $K_{X_1}\otimes L$ is very ample and (X_1,L) is gotten by blowing up one double point on the minimal Enriques surface of degree 14 in $\mathbf{P}^7$. We don't know if this surface exists. We can now consider the last case i.e. case 22). We have g=7, d=12,11,10, $h^0(L)$= 7,6, c_1^2=0,-1,-2, t_0=0,1,2, r=0. Let d=12. Then s=0, thus $h^1(L)=0$ and $h^0(L)=7$, therefore X is a degree 2n surface in $\mathbf{P}^n$ with regular hyperplane section i.e. $h^1(L)=0$. The existence of such surface follows from [9]. Let d=11. Then $s=t_0=1$. If $h^0(L)=7$ then L has a base point. Thus we can assume $h^0(L)=6$. With the usual computations we see that $\chi(L)=6$, $h^2(L)=0$ and $h^1(L)=0$. Also in this case the existence of the surface follows from [9, Corollary (3.2)]. Let d=10. Then $s=t_0=2$. If $h^0(L)=7$ then n=6 and d=2n-2. Therefore we get, see [9], that X is either ruled or K3 which gives a contradiction. If $h^0(L)=6$ then, from the existence of the above case it follows that L has a base point. We have so obtained that the following surfaces birational to an Enriques Surface exist. (We remind that in case 18 we are not able to show that (X,L) exists).

Surfaces Birational to an Enriques Surface with $g(L)\leq 7$.

	g	d	$h^0(L)$	c_1^2	(Y,L_Y)
10)	6	9	5	-1	$V_1(W)$
10)	6	10	6	0	$V_1(W)$
22)	7	12	7	0	$Y_{12}\subset\mathbf{P}^6$
22)	7	11	6	-1	$Y_{12}\subset\mathbf{P}^6$

(4.4) Remark: To complete our investigation in the case $\kappa(X)=0$ it remains to consider the cases in which X is birational either to an Abelian surface, i.e. the cases 8 (for the existence see [13] or [17]) and 25, or to a Bielliptic surface, i.e. the cases 9 and 23. Unfortunately we are not able to say anything about there existence.

(4.5) Study of Surfaces with $\kappa(X)=1$: In all the cases $h^2(L)=h^0(K_{X_s}-L)=0$ since $(K_{X_s}-L)\cdot L<0$. The existence of 6 and 12 has been proved in [28] and [29]. In case 15 by (4.0.2) we have $p_g=2$. Thus from (0.23) it follows that X is a Castelnuovo surface. For its existence see either [16] or [9]. In case 13 X_s=X=Y, and we get a contradiction. In fact the elliptic fibration $\psi:Y\to\Delta$ is given by K_Y. Then $F\cdot L=K_Y\cdot L=3$, where F is the general fiber of the elliptic fibration i.e. g(F)=1. Let f_i be the reduced component of a fiber of multiplicity m_i. Then $F\equiv m_i f_i$ and $K_Y=f^*\delta+\sum_i(m_i-1)f_i$ where $\delta\in \mathrm{Div}(\Delta)$ and $\deg\delta=2g(\Delta)-2+\chi(\mathcal{O}_Y)$, see[14,pg.572]. With an easy computation we see that $g(\Delta)=0$ and $\chi(\mathcal{O}_Y)=2$. Thus $\deg\delta$=O and $3=K_Y\cdot L=\sum_i(mi-1)f_i\cdot L$. Therefore there exist multiple fibers and this contradicts the fact that the fibers are elliptic plane cubics. Consider now case 14. We have g=7, d=9, $h^0(L)=5$, $h^1(L)=h^2(L)=0$, q=0, $p_g=1$, X_s=X=Y. Unfortunately in this case we don't know if the surface exists. What we can say is that $d_1=(K_Y+L)\cdot(K_Y+L)=4(g-1)-\gamma_0=15$, $g_1=3g-3-\gamma_0+1=10$, $h^0(K_Y\otimes L)=8$. Moreover since $K_Y\otimes L$ is very ample by [40], $h^1(K_Y\otimes L)=0$ and $(Y, K_Y\otimes L)$ is a degree $d_1=2n+1$ in $\mathbf{P}^n$, n=7, by [9] we know that $2K_Y$ gives the elliptic fibration and $P_2=2$. It remains now to investigate the cases in which $p_g=0$. Consider case 17 i.e. g=7, d=10, $h^0(L)=6$, $c_1^2=0$, q=0, $p_g=0$, $t_0=0$, r=0. So $X_s=X=Y$. By Castelnuovo's Criterion $P_2\neq 0$. Let $D\in|2K_Y|$. Then $D\cdot D=0$, $D\cdot L=4$. This is a 2n surface in $\mathbf{P}^n$, n=5 and $h^1(L)=1$. We don't know if this surface exists. Consider now case 20 i.e. g=7, d=11, $c_1^2=0$, $h^0(L)=7,6$, q=1, $p_g=0$. Again $X_s=X=Y$. If $P_2\neq 0$ then $D\cdot L=2$ and $g(D)=1$ where $D\in|2K_Y|$ and we get a contradiction since by degree consideration D has to be a conic but $g(D)=1$. Therefore $P_2=0$. We don't know if this surface exists. Consider now case 21 i.e. g=7, d=11,10, $h^0(L)=7,6$, $c_1^2=0,-1$, q=0, $p_g=0$. Let d=11, then $X_s=X=Y$ and we get a contradiction. In fact by Castelnuovo's Criterion $P_2\neq 0$. Let $D\in|2K_Y|$. Then $D\cdot L=2$ since $K_Y\cdot L=1$. Moreover since $c_1^2=0$ and Y is a minimal model the moving part of D forms a pencil of rational curves which contradicts the fact that Y is an elliptic surface. Let d=10. Then $s=t_0=1$. Moreover $h^0(L)=6$ otherwise X_1 would be either ruled or K3. As in the above case we have that $P_2\neq 0$. So there exists a $D\in|K_{X_1}|$. But $D\cdot L=4$. Then by adjunction we have $2g(D)-2=D^2+D\cdot K_{X_1}=-6$ which implies that $g(D)=-2$. Thus we get a contradiction if D is irreducible. Let D be reducible and let D=M+F where M is the moving part and F is the fixed part of K_{X_1}. If M is non empty then $F\cdot L\geq 2$. To show that $F\cdot L\geq 2$ we argue in the following way. From the adjunction formula we get $P\cdot K_{X_1}=-1$, where with P we denote as usual the exceptional curve on X_1 such that $L\cdot P=1$. But $\deg(2K_{X_1}|_P)=-2$ so $h^0(2K_{X_1}|_P)=0$. Then by the long cohomology sequence associated to the short exact sequence $0\to 2K_{X_1}\otimes[P]^{-1}\to 2K_{X_1}\to 2K_{X_1}|_P\to 0$ it follows that $h^0(2K_{X_1}\otimes[P]^{-1})=h^0(2K_{X_1})$ thus $P\subset F$. But $\deg(2K_{X_1}\otimes[P]^{-1}|_P)=-1$ so $h^0(2K_{X_1}\otimes[P]^{-1}|_P)=0$ and by the long cohomology sequence associated to the short exact sequence $0\to 2K_{X_1}\otimes 2[P]^{-1}\to 2K_{X_1}\otimes[P]^{-1}\to 2K_{X_1}\otimes[P]^{-1}|_P\to 0$ it follows that $h^0(2K_{X_1}\otimes 2[P]^{-1})=h^0(2K_{X_1}\otimes[P]^{-1})$ i.e. $2P\in[2K_{X_1}-P]$ i.e. $P\subset 2F$. Therefore $F\cdot L\geq 2$. But $(M+F)\cdot L=2K_{X_1}\cdot L=4$. So $M\cdot L\leq 2$. Thus the moving part of D forms a pencil of rational curves which contradicts the fact that X_1 is an elliptic surface. Suppose now that M is empty i.e. $2K_{X_1}=F$. Then $P_2=1$ and since $\kappa(X_1)=1$ then $P_3\neq 0$ otherwise it will be an Enriques surface. We don't know if this surface exists. Our results are described in the following tables:

$\kappa(X)=1$, Exist

	g	d	$h^0(L)$	c_1^2	q	p_g	(Y,L_Y)
6)	6	7	5	0	0	2	$Y(2,4)$
12)	7	8	5	0	0	2	
15)	7	9	6	0	0	2	

$\kappa(X)=1$, We don't know if they exist.

	g	d	$h^0(L)$	c_1^2	q	p_g
14)	7	9	5	0	0	1
17)	7	10	6	0	0	0
20)	7	11	7,6	0	1	0
21)	7	10	7,6	-1	0	0

(4.6) Study of Surfaces of General Type:

Since the case 26 cannot happen by (0.23) we have that if $g \leq 7$ then:

General Type

g	d	$h^0(L)$	c_1^2	q	p_g	t_0	r	(Y, L_Y)
6	5	4	5	0	4	0	0	$(H_5, \mathcal{O}_{\mathbf{P}^3}(1))$

References

[1] Beltrametti M., Biancofiore A. and Sommese A.J., Projective N-folds of Log General Type, I. To appear on Trans. Amer.Math.Soc.

[2] Bese E., On the spannedness and very ampleness of certain line bundles on the blow-ups of $\mathbf{P}^2$ and F_r. Math.Ann.262,225-238(1983)

[3] Biancofiore A., On the hyperplane sections of blow-ups of complex projective plane. To appear on Can.J.Math.

[4] Biancofiore A.,On the hyperplane sections of ruled surfaces. Preprint

[5] Biancofiore A. and Livorni E.L., On the iteration of the adjunction process in the study of rational surfaces. Ind. Univ. Math. J. Vol 36, No.1,167-188(1987).

[6] Biancofiore A. and Livorni E.L., Algebraic non-ruled surfaces with sectional genus equal to seven. Ann. Univ. Ferrara. Sez.VII-Sc. Mat. Vol. XXXII,1-14(1986).

[7] Biancofiore A. and Livorni E.L., On the iteration of the adjunction process for surfaces of negative Kodaira dimension. To appear on Manuscripta Mathematica.

[8] Biancofiore A. and Livorni E.L., On the genus of a hyperplane section of a geometrically ruled surface. Annali di Matematica pura ed applicata (IV), Vol.CXLVII,173-185(1987)

[9] Buium A., On surfaces of degree at most 2n+1 in $\mathbf{P}^n$. Proceedings of the week of algebraic geometry, Bucharest 1982. Lect.Notes Math. Vol.1056. Berlin, Heidelberg, New York. Springer 47-60 (1984).

[10] Castelnuovo G., Sulle superficie algebriche le cui sezioni piane sono curve iperellittiche. Memorie Scelte, XII, Zanichelli, Bologna (1939).

[11] Castelnuovo G.,Sulle superficie algebriche le cui sezioni sono curve di genere 3. Memorie Scelte, XIII, Zanichelli, Bologna (1939).

[12] Castelnuovo G. and Enriques F., Sur quelques resultats nouveaux dans la theorie des surfaces algebrique. Note V in [30] below.

[13] Comessatti A., Sulle superficie di Jacobi semplicemente singolari. Mem.Soc.Ital. delle Scienze (deiXL) (3)21, 45-71(1919)

[14] Griffiths P.A. and Harris J., Principles of Algebraic Geometry. A.Wiley-Interscience publication, (1978)

[15] Hartshorne R., Algebraic Geometry. Springer Verlag, New York (1977).

[16] Harris J., A bound on the geometric genus of projective varieties. Ann.Scuola Norm.Sup. Pisa, 35-68(1981)

[17] Horrocks G. and Munford D.,Topology. Pergman Press 12,63-81(1973).

[18] Ionescu P., An enumeration of all smooth projective varieties of degree 5 and 6. I.N.C.R.E.S.T. Preprint Series Math. 74(1981).

[19] Ionescu P., Embedded projective varieties of small invariants. Proc. of the week of Algebraic Geometry, Bucharest (1982). Lect. Notes Math. Vol. 1056 Berlin, Heidelberg, New York. Springer Verlag (1984).

[20] Ionescu P., On varieties whose degree is small with respect to codimension. Math.Ann. 27, 339-348(1985).

[21] Lanteri A., Sulle superfici di grado sette. Ist.Lombardo (Rend.Sc.) A115,171-189(1981).

[22] Lanteri A. and Palleschi M., Sulle superfici di grado piccolo in $\mathbf{P}^4$. Ist.Lombardo (Rend.Sc.) A113, 224-241(1979).

[23] Livorni E.L., Classification of algebraic surfaces with sectional genus less then or equal to six: Rational surfaces. Pac. J. Math. Vol.113, No.1, 93-114(1984).

[24] Livorni E.L., Classification of algebraic non-ruled surfaces with sectional genus less than or equal to six. Nagoya J. Math. Vol.100, 1-9(1985).

[25] Livorni E.L., Classification of algebraic surfaces with sectional genus less than or equal to six. II Ruled surfaces with dim $\Phi_{KX\otimes L}(X)=1$. Can.J.Math.Vol.XXXVII,No.4, 1110-1121(1986).

[26] Livorni E.L., Classification of algebraic surfaces with sectional genus less than or equal to six. II Ruled surfaces with dim $\Phi_{KX\otimes L}(X)=2$. Math. Scand. 58,9-29(1986).

[27] Okonek C., Moduli reflexiver Garben und Flächen von kleinem Grad. in P^4. Math.Z.184 , 549-572(1983).

[28] Okonek C., Über 2-codimensionale Untermannigfaltigkeiten vom Grad 7 in $\mathbf{P}^4$ und $\mathbf{P}^5$. Math.Z. 187, 209-219(1984).

[29] Okonek C., Flächen vom Grad 8 in $\mathbf{P}^4$. Math.Z. 191, 207-223(1986).

[30] Picard E. and Simart G., Théories des Fonctions Algebriques de Deux Variables Indépendantes, Chelsea Pub.Co.,Bronx,New York (1971)

[31] Reider I., Vector bundles of rank 2 and linear systems on algebraic surfaces. Ann.Math. 127,309-316(1988).

[32] Room T.G., The Geometry of Determinantal loci. Cambridge University Press (1938)

[33] Roth L., On surfaces of sectional genus four. Proc. Cambridge Phil.Soc.29, 184-194(1933).

[34] Roth L., On surfaces of sectional genus five. Proc. Cambridge Phil.Soc.30, 123-133 (1934).

[35] Roth L., On the regularity of surfaces I,II,III. Proc. Cambridge Phil.Soc.30, 4-14, 271-286,404-408(1934).

[36] Roth L., On surfaces of sectional genus six. Proc. Cambridge Phil.Soc.32, 355-365(1936).

[37] Roth L., On the projective classification of surfaces. Proc. London Math. Soc.(2) 42, 142-170 (1937).

[38] Serrano F., The adjunction mapping and hyperelliptic divisors on a surface. Preprint.

[39] Sommese A.J.,Hyperplane sections of projective surfaces I. The adjunction mapping. Duke Math. J.,46, 377-401(1979).

[40] Sommese A.J. and Van de Ven A., On the adjunction mapping. Math.Ann.278, 593-603 (1987)

[41] Van de Ven A., On the 2-connectedness of very ample divisors on a surface. Duke Math.J. 46, 403-407(1979).

On the pluriadjoint maps of polarized normal Gorenstein surfaces

Cristina Oliva

Let X be a complex normal Gorenstein projective surface and L an ample Cartier divisor on X. In [So] many results concerned with the n-th adjoint map of (X, L) (i.e. the map associated with $\Gamma(n(K_X + L)))$ for $n >> 0$, are proven. In particular the map is a morphism and an embedding unless (X, L) fits into a precise list of well known pairs.

A recent method by Reider [R] provided a powerful technique in studying the spannedness and the very ampleness of adjoint bundles on a smooth surface. By using this method, the properties of the 2− and 3−adjoint maps of a smooth polarized surface have been investigated in [La].

In this paper, we carry out the same investigation for normal Gorenstein polarized surfaces, by using a recent version of Reider's method elaborated by Sakai in the normal case [Sa]. In particular, following the convention of not distinguishing between a Cartier divisor and its corresponding invertible sheaf, if (X, L) is a polarized normal Gorenstein projective surface, we prove that:

i) $2(K_X + L)$ is spanned unless (X, L) is either $(\mathbf{P}^2, \mathcal{O}_{\mathbf{P}^2}(1))$ or $(\mathbf{P}^2, \mathcal{O}_{\mathbf{P}^2}(2))$ or a quadric or a scroll or a conic bundle with some multiple fibres (see section 0. for terminology);

ii) if $K_X + L$ is ample, then $3(K_X + L)$ is very ample off the locus of non rational singularities of X;

iii) if $K_X + L$ is ample, then $2(K_X + L)$ is very ample off the locus of non rational singularities of X unless either $g(L) = 2$ or X contains an irreducible curve E such that $L \cdot E = 1$ and $0 \leq E^2 \leq 4/(K_X + 2L)^2$.

The exceptions to the ampleness of $K_X + L$ have been investigated in [So], while a detailed classification of pairs (X, L) where $g(L) = 2$ can be found in [Be+S].

I would like to thank Prof. F. Sakai and Prof. A. Lanteri for making available to me a preliminary version of their papers [Sa] and [La] and for many helpful conversations during the Conference on Hyperplane sections held in L'Aquila, of which this paper benefited.

0. Background material

Let X be a complex analytic normal Gorenstein projective surface and K_X a Cartier divisor corresponding to the invertible dualizing sheaf ω_X. Following Sakai's notation, $Irr(X)$ will denote the locus of non rational singularities on X. We say that a Cartier divisor L on X is *nef* (numerically effective) if $L \cdot C \geq 0$ for every curve $C \subset X$ and *spanned* if the line bundle associated to L is spanned by its global sections. We define sectional genus of X the number $g(L) = 1+\frac{1}{2}(L^2+L\cdot K_X)$. A pair (X,L) consisting of an ample Cartier divisor on a normal Gorenstein surface X is called a *scroll* if X is a holomorphic $\mathbf{P}^1$-bundle $p: X \to R$ over a non singular curve R and the restriction L_f to a fibre f of p is $O_f(1)$. (X,L) is called a *quadric* if X is biholomorphic to a possibly singular quadric $Q \subseteq \mathbf{P}^3$ and L is isomorphic to the restriction of $\mathcal{O}_{\mathbf{P}^3}(1)$ to Q. (X,L) is called a *conic bundle* if there is a holomorphic surjection $p: X \to R$ with connected fibres onto a smooth curve R with the property that for some $n > 0$ and some very ample Cartier divisor E on R,

$$n(K_X + L) \sim p^\star E.$$

The above classes will be denoted as follows:

$$\mathcal{B} = \{\text{scrolls}\},$$

$$\mathcal{C} = \{\text{conic bundles}\},$$

$$\mathcal{Q} = \{\text{quadrics}\}.$$

We denote by $\mathcal{C}^*$ the subclass of $\mathcal{C}$ consisting of conic bundles admitting some multiple fibres. A classification of the possible multiple fibres of conic bundles can be found in [Sa2] p.269. We also let $\mathcal{A} = \{(\mathbf{P}^2, \mathcal{O}_{\mathbf{P}^2}(e)); \quad e = 1,2\}$. We recall the following facts:

(0.1) THEOREM ([So]): Let L be an ample Cartier divisor on a normal Gorenstein surface X. Then $K_X + L$ is nef iff $(X,L) \notin \mathcal{A} \cup \mathcal{B} \cup \mathcal{Q}$;

(0.2) THEOREM ([Sa]): Let X be a normal Gorenstein surface and let D be a nef Cartier divisor on X.

a) If $D^2 > 4$, then $K_X + D$ is spanned unless there exists a nonzero effective divisor E such that

$$D \cdot E = 0, \qquad -1 \leq E^2 < 0, \text{or}$$

$$D \cdot E = 1, \qquad 0 \leq E^2 \leq 1/D^2.$$

b) if $D^2 > 8$, then $\mid K_X + D \mid$ separates two distinct points on X and separates tangent vectors everywhere off $Irr(X)$ unless there exists a nonzero effective divisor E satisfying one of the following conditions:

$$D \cdot E = 0, \qquad -2 \leq E^2 \leq 0,$$

$$D \cdot E = 1, \qquad -1 \leq E^2 \leq 1/D^2,$$

$$D \cdot E = 2, \qquad 0 \leq E^2 \leq 4/D^2,$$

$$D \equiv 3E \quad (D \text{ numerically equivalent to } E), \qquad E^2 = 1.$$

1. Spannedness and very ampleness properties of $n(K_X + L)$.

According to [So], there exists a positive integer n such that $n(K_X + L)$ is spanned iff $(X, L) \notin \mathcal{A} \cup \mathcal{B} \cup \mathcal{Q}$.

(1.1) PROPOSITION: $2(K_X + L)$ is spanned unless $(X, L) \in \mathcal{A} \cup \mathcal{B} \cup \mathcal{Q} \cup \mathcal{C}^\star$.

Proof: Write $2(K_X + L)$ as $K_X + M$, where $M = N + L$ and $N = K_X + L$. If $(X, L) \notin \mathcal{A} \cup \mathcal{B} \cup \mathcal{Q}$, then N is nef by (0.1), hence M is ample. Since L is ample, $(K_X + L) \cdot L \geq 0$, with equality iff $(K_X + L) \equiv 0$; in this case $K_X + L$ is also linearly trivial([Sa1]), hence $K_X + L$ itself is spanned. If $(K_X + L) \cdot L > 0$ then $(K_X + L) \cdot L \geq 2$, in fact, let $\pi: \tilde{X} \to X$ be a resolution; since $\pi^\star L \in Div(\tilde{X}, \mathbf{Z})$, by the genus formula and Mumford's Q-valued intersection theory on normal surfaces, we have $g(\pi^\star L) = 1 + \frac{1}{2} L \cdot (K_X + L)$, hence $(K_X + L) \cdot L \geq 2$ and $M^2 \geq 5$. By (0.2.a) $K_X + M$ is spanned on X unless there exists an effective divisor E such that

$$M \cdot E = 1 \qquad 0 \leq E^2 \leq 1/M^2.$$

If such a divisor exists, by the ampleness of L, the condition $M \cdot E = 1$ implies $L \cdot E = 1$ and $(K_X + L) \cdot E = 0$ and, having already studied the case $K_X + L \equiv 0$, it only remains to consider the case $K_X + L \not\equiv 0$. If $K_X + L \not\equiv 0$, we can have $(K_X + L)^2 = 0$, in which case (X, L) is a conic bundle ([So]), or $(K_X + L)^2 > 0$, but this second possibility cannot occur by the Hodge Index Theorem and the relations $M \cdot E = 1$ and $0 \leq E^2 \leq 1/M^2$. If $(X, L) \in \mathcal{C}$, since $(K_X + L) \cdot E = 0$, there are two possibilities:

i) E is a component of a reducible fibre of X, in which case $E^2 < 0$ and it would contradict $0 \leq E^2 \leq 1/M^2$;

ii) E is the reduced component of a multiple fibre.

(1.2) PROPOSITION: Assume that $K_X + L$ is ample. Then $\mid 3(K_X + L) \mid$ has no base points on X, separates distinct points on X and separates tangent vectors off $Irr(X)$. In particular $3(K_X + L)$ is very ample off $Irr(X)$.

Proof: Write $3(K_X + L)$ as $K_X + M$, where $M = 2N + L$ and $N = K_X + L$. Since M is ample and $M^2 = 4N^2 + 4N \cdot L + L^2 \geq 9$, (0.2.b) applies. By the ampleness of L and N, $M \cdot E = 0, 1, 2$ cannot occur. If $M \equiv 3E$ and $E^2 = 1$, then $L^2 = 1$ and $K_X \cdot L = 0$. By the same argument as used in (1.1), it follows that $g(\pi^\star L) = 3/2$, but this gives a contradiction, hence the assertion follows.

Let X be a normal Gorenstein projective surface with $K_X \equiv 0$. Then by replacing L with $L - K_X$ we get from (1.1) and (1.2) the following extension of a result proven in [LP].

(1.3) COROLLARY: Let L be an ample Cartier divisor on a normal Gorenstein surface X with $K_X \equiv 0$. Then $2L$ is spanned on X and $3L$ is very ample off $Irr(X)$.

Let's come back to the general case. What can be said about the very ampleness of $2(K_X+L)$? The following proposition gives a partial answer to this question.

(1.4) PROPOSITION: Assume that $K_X + L$ is ample. Then $2(K_X + L)$ is very ample off $Irr(X)$ unless, either X has sectional genus $g(L) = 2$ or X contains an irreducible curve E such that $L \cdot E = 1$ and $0 \leq E^2 \leq 4/(K_X + 2L)^2$.

Proof: Write $2(K_X + L)$ as $K_X + M$, where $M = N + L$ and $N = K_X + L$. M is ample and $M^2 \geq 2+2(K_X+L)\cdot L$. In view of the genus formula, we have $(K_X+L)\cdot L \geq 2$ with equality iff $g(L) = 2$, in which case a detailed classification can be found in [Be+So]. If $g(L) \neq 2$, then $(K_X + L) \cdot L \geq 4$ and $M^2 \geq 10$. If $K_X + M$ is not very ample off $Irr(X)$, then X contains an effective divisor E such that $M \cdot E = 2$, which implies $(K_X + L) \cdot E = L \cdot E = 1$, hence E is irreducible, and $0 \leq E^2 \leq 4/M^2$.

References

[Be+So] BELTRAMETTI, SOMMESE, On generically polarized Gorenstein surfaces of sectional genus two. Preprint.

[La] LANTERI, Pluriadjoint bundles of polarized surfaces. Preprint.

[LP] LANTERI, PALLESCHI, Adjunction properties of polarized surfaces via Reider's method. Preprint.

[R] REIDER, Vector bundles of rank 2 and linear systems on algebraic surfaces. Preprint.

[Sa] SAKAI, Reider-Serrano's method on normal surfaces. Preprint.

[Sa1] SAKAI, Ample Cartier divisors on normal surfaces. J. reine angew. Math., 366 (1986), 121-128.

[Sa2] SAKAI, Ruled fibrations on normal surfaces. J. Math. Soc. Japan, 40 (1988), 249-269.

[So] SOMMESE, Ample divisors on normal Gorenstein surfaces. Abh. Math. Sem. Univ. Hamburg, 55 (1985), 151-170.

On the adjoint line bundle to an ample and spanned one

Marino Palleschi
Dipartimento di Matematica dell'Università di Milano
Via C. Saldini, 50, 20133 Milano, Italy

Introduction

Let S be a nonsingular complex projective surface and let $\mathbf{L}=\mathcal{O}_S(L)$ be a line bundle on it. In [SV] Sommese and Van de Ven investigated the very ampleness properties of $\mathbf{K_S}\otimes\mathbf{L}$, when $\mathbf{L}$ is very ample, in this way completing a study started in [S] and [VdV] and also undertaken by Serrano [Se]. Results essentially equivalent to those of [SV] were also independently obtained by Ionescu [I].

In this paper we still look at the very ampleness of $\mathbf{K_S}\otimes\mathbf{L}$ – up to contracting (-1)-lines– when $\mathbf{L}$ is simply an ample line bundle spanned by its global sections and under the assumption $L^2\geq 10$ (Theorem (2.1)), this leading to a large number of exceptions.

As an application of this result we prove (Proposition (2.2)) that, if $L^2\geq 10$ and such an $\mathbf{L}$ is 3-connected, then $\mathbf{K_S}\otimes\mathbf{L}$ is very ample. This is a well-known result by Van de Ven [VdV], here recovered in the wider context of ample and spanned line bundles and dropping any assumptions on $H^0(S,L)$.

1. Notation and background.

(1.1) Let S be a complex projective smooth surface and $\mathbf{D}=\mathcal{O}_S(D)$ a line bundle on S. Let

$|\mathbf{D}|$ = the complete linear system defined by $\mathbf{D}$;
$h^i(D)=\dim_{\mathbf{C}}H^i(S,\mathbf{D})$;
CD = the intersection index of $C,D\in \mathrm{Div}(S)$;
D^2 = DD;
$\mathbf{K_S}=\mathcal{O}_S(K_S)$ the canonical bundle;
$g(D) = 1 + (D^2+K_SD)/2$ the arithmetic genus of D.

Divisor D is said **nef** (numerically effective) if $DC\geq 0$ for any curve $C\subset S$.

We shall always assume that S is polarized by an ample line bundle $\mathbf{L}=\mathcal{O}_S(L)$. Let us list a number of pairs (S,L) frequently occurring in the sequel. (S,L) is said to be a **scroll** if S is a $\mathbf{P}^1$-bundle over a smooth curve and $\mathbf{L}_{|f}=\mathcal{O}_{\mathbf{P}^1}(1)$ for every fibre f of S. (S,L) is a **conic bundle** if there is a morphism $p:S\to C$ over a smooth curve C whose general fibre F satisfies $\mathbf{L}_{|F}=\mathcal{O}_{\mathbf{Q}}(1)$, $\mathbf{Q}$ being the smooth conic.

We say that **L** is **spanned** to mean that it is spanned by its global sections. If this is the case, for any irreducible reduced curve $D^* \subset S$ we also have the obvious formula

(1.1.1) $$LD^* = \deg \Phi_{|D^*} \cdot \deg \Phi(D^*),$$

Φ being the morphism defined by **L**.

(1.2) Let $\mathbf{L} = \mathcal{O}_S(L)$ be an ample and spanned line bundle on S. A smooth rational curve $E \subset S$ is said a **(-1)-line** (relative to **L**) if $E^2=-1$ and $EL=1$. Recall that there is only a finite number of (-1)-lines and that they are disjoint unless (S,**L**) is a conic bundle. Apart from this case, there is a birational morphism $r:S \to S'$ onto a smooth surface S' contracting all the (-1)-lines of S to a finite set $F \subset S'$. The line bundle $\mathbf{L}' = \mathcal{O}_{S'}(L') = r_*\mathbf{L}$ is ample and (S',**L'**) is referred to as the **reduction** of (S,**L**). Furthermore we have

(1.2.1) $$L'^2 \geq L^2,$$

equality holding if and only if (S,**L**) is already the reduction of itself.

If D^* stands for the proper transform via r of any divisor D on S', we get

(1.2.2) $$LD^* \leq L'D,$$

(1.2.3) $$D^{*2} \leq D^2$$

and in both cases equality holds if and only if $F \cap \mathrm{Supp}\, D = \phi$.

The main tool used here is Reider's method, which we recall in the following form

(1.3) **Theorem.** ([R], see also [S V]) *Let* L' *be a numerically effective divisor on a smooth projective surface* S'.

(1.3.1) *If* $L'^2 \geq 5$ *and* $\mathbf{K}_{S'} \otimes \mathcal{O}_{S'}(L')$ *is not spanned, then there exists an effective divisor* D *on* S' *satisfying either*

$$L'D=0,\ D^2=-1 \quad \text{or} \quad L'D=1,\ D^2=0\ .$$

(1.3.2) *If* $L'^2 \geq 10$ *and* $\mathbf{K}_{S'} \otimes \mathcal{O}_{S'}(L')$ *is not very ample, then there exists an effective divisor* D *on* S' *satisfying either*

(i) $L'D=0,\ D^2=-1$ *or* -2,

(ii) $L'D=1,\ D^2=-1$ *or* 0 , *or*

(iii) $L'D=2,\ D^2=0$.

Furthermore, in each case there exists an $L' \in |\mathcal{O}_{S'}(L')|$ *such that* $L'=D+R$, R *effective divisor.*

2. The very ampleness of the adjoint line bundle.

Let $\mathbf{L}=\mathcal{O}_S(L)$ be an ample and spanned line bundle on S and consider the reduction (S',L') of (S,L). Let $r:S\to S'$ and $\mathbf{L'}=\mathcal{O}_{S'}(L')$ be as in (1.2). The main result of this paper is the following

(2.1) **Theorem.** *Assume that* (S,L) *is neither a scroll nor a conic bundle. If* $L^2\geq 10$ *and* $\mathbf{K}_{S'}\otimes\mathbf{L'}$ *is not very ample, then*

(2.1.1) S' *is either an elliptic or a hyperelliptic fibration such that* $L'f=2$ *for any fibre* f;

(2.1.2) S' *contains an irreducible reduced curve* D *which doubly covers* $\mathbf{P}^1$, *such that* $D^2=0$ $L'D=2$ *and* $h^0(D)=1$;

(2.1.3) S' *contains two smooth rational curves* D_1, D_2 *such that* $D_iL'=1$, $D_i^2\leq -2$, $D_1D_2\geq 2$, $(D_1+D_2)^2=0$ *and* $h^0(D_1+D_2)=1$.

Proof. First of all $\mathbf{K}_{S'}\otimes\mathbf{L'}$ is ample [LP] and assume that it is not very ample. By (1.2.1) Reider's results can be used in order to get an effective divisor D on S' as in (1.3.2). Let D* be its proper transform via the reduction morphism r and let F be as in (1.2).

Cases (1.3.2) (ii) and (iii) can only occur, L' being ample.

In the former case, (1.2.2) is an equality and so $F\cap \mathrm{Supp}\, D=\phi$, which says that $r_{|D^*}$ is an isomorphism. Moreover, as (1.2.2) reads

(2.1.4) $$LD^*=L'D=1,$$

D* is an irreducible reduced smooth curve. Formula (1.1.1) can thus be applied and in view of (2.1.4) it shows that D* is a smooth $\mathbf{P}^1$ and so is D, as $r_{|D^*}$ is an isomorphism. As a consequence, recalling that $D^2=-1,0$, we get

$$(K_{S'}+L')D=2g(D)-2-D^2+L'D\leq -1,$$

contradicting the ampleness of $\mathbf{K}_{S'}\otimes\mathbf{L'}$.

Now assume that D is as in case (1.3.2) (iii) and so that $LD^*=2$ or 1 by (1.2.2).

<u>1st case</u>. D is irreducible and reduced. If $LD^*=2$, (1.2.2) is an equality, which yields $F\cap \mathrm{Supp}\, D=\phi$ and so D and D* are isomorphic. Furthermore (1.1.1) says that D* is either a smooth $\mathbf{P}^1$ or a double cover of a line and so is D. As before the ampleness of $\mathbf{K}_{S'}\otimes\mathbf{L'}$ rules out the case when D is rational, since $L'D=2$ and $D^2=0$.

If $h^0(D)\geq 2$, since $D^2=0$, $|\mathcal{O}_{S'}(D)|$ is base-point free, its general element is a double cover of $\mathbf{P}^1$ of genus $g(D)\geq 1$ and $|\mathcal{O}_{S'}(D)|$ defines a morphism exhibiting S' as in (2.1.1). If $h^0(D)=1$, we are in case (2.1.2).

Let us now assume that $LD^*=1$. Formula (1.1.1) says that D* is a smooth $\mathbf{P}^1$. Furthermore,

$LD^*=L'D-1$ and so Supp D contains one point $x\in F$ only and 1 is the multiplicity of D at x. As a consequence,

$$D^{*2}=D^2-1=-1$$

and so D^* is a (-1)-line relative to L: a contradiction.

2nd case . D is not irreducible and reduced. If $h^0(D)\geq 2$, the general element $D^{\#}\in|\mathcal{O}_{S'}(D)|$ is irreducible and reduced by Bertini's theorem and equality $D^2=0$. Then reason as in the 1st case on $D^{\#}$ substituting for D and get case (2.1.1).

Now assume that $h^0(D)=1$. As $L'D=2$ and L' is ample, we can only have

$$D=D_1+D_2,\ D_i>0,\ L'D_i=1 \text{ (possibly } D_1=D_2).$$

As D_i is irreducible and reduced, its proper transform D_i^* via r is a smooth $\mathbf{P}^1$ in view of (1.2.2) and (1.1.1) and so is D_i, (1.2.2) being an equality. This gives

$$(K_{S'}+L')D_i= -2-D_i^2+1$$

and so $D_i^2\leq -2$, $K_{S'}\otimes L'$ being ample. As a consequence $D_1\neq D_2$ because the inequality $0=D^2\leq -4+2D_1D_2$ gives $D_1D_2\geq 2$. All this gives case (2.1.3). ////

As explicit examples of pairs (S',L') occurring in class (2.1.1) consider the product of two elliptic curves $S'=C_1xC_2$. Let f be a fibre of the projection on C_1 and B the sum of two fibres of the other projection. By Reider's result $\mathcal{O}_{S'}(B+nf)$ can be shown to be ample and spanned for n>>0. Choose L'=B+nf.

As an almost immediate consequence, we can recover the following well-known result by Van de Ven [VdV], concerned with the very ampleness of the line bundle adjoint to a 3-connected one. Note that this property can now be obtained in the wider context of ample and spanned line bundles **L** without any assumptions on $h^0(\mathbf{L})$.

To this end recall that a line bundle **L** is said to be **3-connected** if for every $L\in|\mathbf{L}|$ and every splitting $L=L_1+L_2$, L_i effective, the inequality $L_1L_2\geq 3$ holds.

(2.2) **Proposition.** (see [VdV] pp.405-406). *Let* $\mathbf{L}=\mathcal{O}_S(L)$ *be an ample and spanned line bundle on* S *and assume that* $L^2\geq 10$. *If* **L** *is* 3-*connected, then* $\mathbf{K}_S\otimes\mathbf{L}$ *is very ample.*

Proof. First of all (S,**L**) cannot be a scroll. Otherwise, if f were a general fibre, as $h^0(\mathbf{L})\geq 3$, there would be an $L\in|\mathbf{L}|$ passing through two points of f at least. As Lf=1, we should have L=f+R, with R effective divisor. If this were the case, we should get

$$fR = (L-f)f = 1,$$

contradicting the 3-connectedness of L.

An analogous argument rules out the conic bundle case at least when $h^0(L)\geq 4$.

Let now (S,L) be a conic bundle, f a general fibre and assume that $h^0(L)=3$. The morphism Φ_L associated with $\Gamma(L)$ exhibits S as a cover of $\mathbf{P}^2$. Furthermore,

$$2 = Lf = \deg \Phi_{L\,|f} \cdot \deg \Phi_L(f)$$

and so $\Phi_L(f)$ is either a conic C or a line. In the first case let $\Phi_L{}^* C = f+R$, R effective divisor. First of all, by the projection formula we get

$$2 = L\,\Phi_L{}^*C = fL + Rl,$$

which yields R=0 as L is ample and so $\Phi_L{}^*C = f$. On the other hand $\Phi_L{}^* C \in |\Phi_L{}^*\mathcal{O}_{\mathbf{P}^2}(2)| = |\mathbf{L}^{\otimes 2}|$; a contradiction, as the above formula gives $2=2L^2$.

Let now $\Phi_L(f)$ be a line l. Let $\Phi_L{}^* l = f+R$ and note that $\Phi_L{}^* l \in |\mathbf{L}|$. This contradicts the 3-connectedness of **L** again, as fR= f(L-f)=fL=2.

As a second fact (S,L) is the minimal reduction of itself. Otherwise if S contained a (-1)-line E (with respect to **L**), as $h^0(\mathbf{L})\geq 3$ and LE=1, there would exist an $L\in|\mathbf{L}|$ such that L=E+R, R effective divisor. This would again contradict the 3-connectedness assumption as

$$ER = (L-E)E = 1-E^2 = 2.$$

By contradiction asssume that $\mathbf{K_S}\otimes\mathbf{L}$ is not very ample. Theorem (2.1) shows that we are either in case (2.1.1), (2.1.2), or in case (2.1.3). In each case S=S' contains a curve C such that

(2.2.1) $$LC = 2\,,\ C^2 = 0\,,$$

(C=f, D, D_1+D_2 respectively). Furthermore, Reider's method shows that there exists an $L\in|\mathbf{L}|$ such that L=C+R, R effective divisor. Therefore we get again a contradiction to the 3-connectdedness assumption as

$$RC = (L-C)C = LC = 2,$$

by (2.2.1). ////

To finish with, we can now show that the very ampleness of $\mathbf{K_S}\otimes\mathbf{L}$ is almost equivalent to **L** being 3-connected.

(2.3) **Proposition.** *Let* $\mathbf{L} = \mathcal{O}_S(L)$ *be an ample and spanned line bundle on* S. *Assume both that* (S,**L**) *is neither a scroll nor a conic bundle and that its reduction* (S', **L'**) *is neither as* *in* (2.1.2) *nor as in* (2.1.3). *If* $L^2 \geq 10$, *then the following facts are equivalent:*

(2.3.1) $\mathbf{K}_{S'} \otimes \mathbf{L}'$ *is very ample* ;

(2.3.2) (S',L') *is not as in* (2.1.1) ;

(2.3.3) **L'** *is* 3-*connected*.

Proof. Proposition (2.2) shows that (2.3.3) implies (2.3.1), whereas (2.3.2)→(2.3.3) follows from [BL] Th.B. It only remains to prove that (2.3.1)→(2.3.2). By contradiction assume that (S', L') is as in (2.1.1) and let f be a general fibre of the elliptic or hyperelliptic fibration. Let $\mathbf{K}_f$ stand for the canonical bundle of f and let $\mathbf{L}_f$ denote the restriction of **L'** to f. The tensor product of a section $\eta \in \Gamma(\mathbf{K}_f)$ with a section $\tau \in \Gamma(\mathbf{L}_f)$ induces a finite to one map

$$\psi : |\mathbf{K}_f| \times |\mathbf{L}_f| \to |\mathbf{K}_f \otimes \mathbf{L}_f|.$$

On the other hand, $h^0(\mathbf{K}_f) = g$, where g is the genus of f, and $h^0(\mathbf{K}_f \otimes \mathbf{L}_f) = g+1$ by the Riemann-Roch theorem, recalling that Lf=2. Moreover |**L'**| cuts out a $g^1{}_2$ on f and so $h^0(\mathbf{L}_f) \geq 2$. All this shows that ψ is onto (and that $h^0(\mathbf{L}_f) = 2$). As a consequence every section $\sigma \in \Gamma(\mathbf{K}_f \otimes \mathbf{L}_f)$ can be written in the form

$$\sigma = \eta \otimes \tau.$$

Let $x \in f$ be a point such that $\sigma(x)=0$ and let $y \in f$ be the point such that $x+y \in |\mathbf{L}_f| = g^1{}_2$. If $\tau(x)=0$, we also have $\tau(y)=0$ and so $\sigma(y)=0$. Otherwise, it can only be $\eta(x)=0$. Now recall that $|\mathbf{K}_f|$ is multiple of the $g^1{}_2$ and so $\eta(x)=0$ yields $\eta(y)=0$ as well. Since $\mathbf{K}_{X|f} = \mathbf{K}_f$, this shows that $\Gamma(\mathbf{K}_X \otimes \mathbf{L})$ does not separate points on f. ////

REFERENCES

[BL] M. Beltrametti, A. Lanteri, *On the 2 and the 3-connectedness of ample divisors on a surface.* Manuscripta Math., 58 (1987), 109-128; *Erratum to "On the 2 and the 3-connectdedness of ample divisors on a Surface"*. Manuscripta Math., 59 (1987), 130.

[I] P. Ionescu, *Ample and very ample divisors on a surface*. Revue Roumaine Math. Pur. Appl. 33 (1988), 349-358.

[LP] A. Lanteri-M. Palleschi, *About the adjunction process for polarized algebraic surfaces* J. reine angew. Math, 352 (1984), 15-23.

[R] I. Reider, *Vector bundles of rank 2 and linear systems on algebraic surfaces*. Ann. of Math., 127 (1988).

[Se] F. Serrano, *The adjunction mapping and hyperelliptic divisors on a surface*. J. reine angew. Math., 381 (1987), 90-109.

[S] A.J. Sommese, *Hyperplane sections of projective surfaces I - The adjunction mapping*. Duke Math. J., 46 (1979), 377-401.

[SV] A.J.Sommese-A. Van de Ven, *On the adjunction mapping*. Math. Ann., 278 (1987), 593-603.

[VdV] A. Van de Ven, *On the 2-connectedness of very ample divisors on a surface*. Duke Math. J., 46 (1979), 403-407.

Quadrics through a canonical surface

Miles Reid, University of Warwick

No profit grows
where is no pleasure ta'en.
The Taming of the Shrew I.1

(0.0) Motivation. I want to consider the geography of surfaces X of general type, especially the region $K^2 < 4\chi$ (usually with $p_g >> 0$); my aim is to extend results known if $K^2 < 3\chi$.

Conjecture. For $g = 2, 3,..$ there exist rational numbers a_g and b_g with

$$a_2 < a_3 < .. \quad \text{and} \quad \lim a_g = 4 \text{ as } g \longrightarrow \infty,$$

such that for every surface X of general type,

$$K^2 \leq a_g\chi - b_g \implies X \text{ has a pencil of curves of genus } \leq g.$$

As in [Reid3, Xiao], this statement would imply that the algebraic fundamental group $\pi_1 X$ of a surface X with $K^2 < 4\chi$ is either finite or a finite extension of π_1 of a curve; and that the Albanese morphism of X maps to a curve.

(0.1) The classical results in this direction are due to Max Noether, Castelnuovo, Horikawa and Xiao Gang: every surface of general type satisfies $K^2 \geq 2p_g - 4$; if $p_g \geq 4$ and $K^2 < 3p_g - 7$ then the 1-canonical map $\varphi_K : X \dashrightarrow \varphi(X) \subset \mathbb{P}^{p_g-1}$ is 2-to-1 onto a surface ruled by lines or conics. The conjecture for $g = 2, 3$ with $a_2 = 2^2/_3$ and $a_3 = 3$ is closely related to this, and was proved by Horikawa. If a surface X of general type is hyperelliptic, in the sense that it is birational to a 2-to-1 cover of a ruled surface, then Xiao has proved that

$$K^2 < \frac{4s}{s+1}\chi - \frac{9s^2+8s}{2(s+1)} \implies$$

X has a pencil of hyperelliptic curves of genus $\leq s$;

see for example [Horikawa1-2, Xiao, Reid4].

(0.2) There is thus no loss of generality in assuming that X is a surface for which the 1-canonical map $\varphi_K : X \dashrightarrow \varphi(X) \subset \mathbb{P}^{p_g-1}$ is birational. The following conjecture of mine is more than 10 years old [Reid2-3].

Conjecture. Suppose that X is a surface for which the 1-canonical map $\varphi_K: X \dashrightarrow Y \subset \mathbb{P}^{p_g-1}$ is birational. Suppose that

$$K^2 < 4p_g - 12.$$

Then $Y \subset W \subset \mathbb{P}^{p_g-1}$, where W is a variety of dimension ≥ 3 contained in the intersection of all quadrics through Y.

Remark. It is easy to see that if W in the conclusion of Conjecture 0.2 exists then necessarily $\dim W = 3$ and $\deg W \leq K^2 - 2p_g + 4$; the relation with Conjecture 0.0 is that the general hyperplane section $S = W \cap \mathbb{P}^{p_g-2}$ of the 3-fold W is a surface for which K_S is not quasi-effective, so that the methods of [Xiao, Reid4] are applicable to give a pencil of rational curves of bounded degree on S. In fact it follows that $K_W + H$ is not quasieffective on W itself, so that it may be possible to generalise the method of [Reid4] to prove that suitable inequalities imply that W has a pencil of surfaces of small degree cutting out the required pencil on X. If W is smooth then results of Sommese [Sommese] are already available: W is a $\mathbb{P}^1$-bundle over a smooth surface, a quadric bundle over a curve, or a (possibly blown-up) Fano 3-fold of index ≥ 2.

(0.3) Contents of the paper. Unfortunately, I am still not able to make any substantial advance towards proving Conjecture 0.2, although the paper does contain several numerological considerations each of which points to the inequality $K^2 < 4p_g - 12$ as relevant (see especially (1.5, iii), (2.4), (3.2), (3.5, (7)) and (3.11)). §1 tries to explain the background, and §2 contains an *ad hoc* proof of the first case of the key Conjecture 1.5. This material is about 100 years old (and is for the most part covered in [Castelnuovo, Fano, Babbage, Harris, Ciliberto]), but I enjoyed rediscovering it and writing it out, and I believe that it may serve as a useful introduction to Conjecture 0.2.

§3 discusses a hopeful approach to the problem that uses vector bundles, but that doesn't work yet, and raises several open questions. I should say that Conjecture 1.5 has occupied me on and off for more than 10 years, and the more I fail to prove it, the more I am convinced of its truth; my main aim in publishing this tentative material is to recruit the interest of people with more competence in these matter or more time than me; a secondary aim is to get the numerology of (1.5) and (3.2) written down in one place so that I don't have to do it all over again from scratch every time I come back to the problem.

(0.4) Acknowledgements. Much of §§1–2 of this paper is a reworking of a letter to Tyurin (circulated in [Reid2]); this is closely related to ideas of Tyurin in [Tyurin, Chap. II, §5], and I have benefited from many discussions with him since 1972. I am very grateful to Ciro Ciliberto for pointing out the references [Fano, Ciliberto]. The idea of a set being uniform w.r.t. forms of degree k occuring in (1.2) and (2.6) is due to Harris and Eisenbud, and fills a painful gap in my argument; their detailed work-out of my letter [Eisenbud] has been useful throughout §2, although I dispute that the proof of (2.2) in (2.7–9) is 'exceedingly ugly'. The material of §3 was worked out during a research visit to Pisa in December 1987, and I would like to thank the Italian CNR for financial support, and Fabrizio Catanese for stimulating discussions, and for wonderful hospitality and *punctuality*.

§1. Counting quadrics

(1.1) Terminology and its abuse. Let $X \subset \mathbb{P}^N = \mathbb{P}$ be a subscheme with ideal sheaf I_X; I assume throughout that X spans $\mathbb{P}$. Then $H^0(\mathbb{P}, I_X \cdot \mathcal{O}_{\mathbb{P}}(2))$ is the space of quadrics through X, and its dimension is the *number* of quadrics through X; I usually write

$$h^0(\mathbb{P}, I_X \cdot \mathcal{O}_{\mathbb{P}}(2)) = \binom{N+2}{2} - f,$$

and say that X *imposes* f *conditions on quadrics*. Equivalently $H^0(I_X \cdot \mathcal{O}_{\mathbb{P}}(2))$ is the kernel of the restriction map $\rho\colon H^0(\mathbb{P}^N, \mathcal{O}_{\mathbb{P}}(2)) \to H^0(X, \mathcal{O}_X(2))$, and $f = f_X = \operatorname{rank} \rho$.

I will say that an irreducible variety $X \subset \mathbb{P}^N$ is *generically an intersection of quadrics* to mean that X is one component of the intersection of all quadrics through X; the opposite possibility is that the intersection of all quadrics through X contains a component of dimension at least $\dim X + 1$ containing X.

(1.2) Reduction to the zero-dimensional case. Question: how to give lower bounds for f_X, or equivalently, how to find an upper bound for the number of quadrics through X? The difficulty here is that ρ is bilinear in $H^0(\mathbb{P}^N, \mathcal{O}_{\mathbb{P}}(1))$, whereas the obvious methods are usually linear.

First, passing to any hyperplane section $X' \subset \mathbb{P}^{N-1} = \mathbb{P}'$, the number of quadrics can only increase:

$$H^0(I_{X}\cdot\mathcal{O}_{\mathbb{P}}(2)) \hookrightarrow H^0(I_{X''}\cdot\mathcal{O}_{\mathbb{P}'}(2));$$

because the kernel $H^0(I_X\cdot\mathcal{O}_{\mathbb{P}}(1))$ consists of hyperplanes through X. Now assume that X is irreducible, and the hyperplane sections are chosen generically; then the process can be continued down to a reduced finite set of points $\Sigma \subset \mathbb{P}^n$ spanning $\mathbb{P}^n$, where $n = \operatorname{codim}(X \subset \mathbb{P})$. A traditional argument shows that the finite set Σ consists of points that are *linearly general,* or *in general position with respect to hyperplanes* (see for example [4 authors, pp.107–113] in characteristic 0; or the argument of [Andreotti, §6, p.815] gives the same result for char p large compared to the degree). Moreover, Σ is *uniform* (or in *uniform position*) with respect to hypersurfaces of any degree $k \geq 1$: that is, all subsets of Σ with the same number of elements impose the same number of conditions on forms of degree k; this follows from the Lefschetz–Harris principle that if Σ is a generic 0-dimensional section of the irreducible variety X, it is irreducible and its Galois group over its field of definition is the full symmetric group on Σ.

(1.3) Proposition. Let $\Sigma \subset \mathbb{P}^n$ be a finite set of d points spanning $\mathbb{P}^n$; suppose that Σ is linearly in general position.

(i) If $d \leq 2n+1$ then Σ imposes d conditions on quadrics.

(ii) If $d \geq 2n+3$ and Σ imposes $\leq 2n+1$ conditions on quadrics then Σ is contained in a rational normal curve $C \subset \mathbb{P}^n$.

Suppose in addition that Σ is uniform with respect to quadrics.

(iii) If $d \geq 2n+5$ and Σ imposes $\leq 2n+2$ conditions on quadrics then Σ is contained in a curve $C \subset \mathbb{P}^n$ of degree $\leq n+1$ (necessarily with $p_aC \leq 1$).

Proof of (i). This is rather trivial: I add $2n+1-d$ general points of $\mathbb{P}^n$ to Σ to give a set Σ' of $2n+1$ points not contained in any pair of hyperplanes. Then by construction, the (projectivised) space of quadrics through Σ' is disjoint from the $2n$-dimensional space of quadrics of rank 2; this implies that $H^0(I_{\Sigma'}\cdot\mathcal{O}(2))$ has codimension $\geq 2n+1$ in all quadrics, so that Σ' imposes independent conditions.

(ii) is a 100-year old result of Castelnuovo (see [Castelnuovo, Babbage, §6]) and (iii) (due to Fano [Fano, Ciliberto]) is obtained by working harder with the same ideas. I discuss the proofs later in §2.

(1.4) Harmless though it may appear, Proposition 1.3, (i), together with the argument of (1.2), has a large number of consequences. It is essentially equivalent to the free pencil trick of Castelnuovo and Mumford (see [Mumford, §2, Segre] and (1.7)).

Corollaries. (i) Let $W \subset \mathbb{P}^N$ be an irreducible variety spanning $\mathbb{P}^N$ of dimension w, and set $n = N - w$; then W imposes

$$\geq \binom{N+2}{2} - \binom{n+2}{2} + \min(\deg W, 2n+1)$$

conditions on quadrics.

(ii) ('Clifford plus') Let C be a smooth curve and $D = g^r{}_d$ a divisor (of degree d with $h^0(D) = r+1$), such that the rational map φ_D is a birational embedding; then

$$h^0(2D) \geq \min(r + \deg \varphi_D(C), 3r);$$

if D is a special divisor then $\deg \varphi_D(C) \geq 2r$ by Clifford's theorem so that $h^0(2D) \geq 3r$; if moreover 2D is special then $d \geq h^0(2D) - 1 \geq 3r - 1$.

(iii) Let X be a surface of general type for which the 1-canonical map φ_K is a birational embedding; then

$$K^2 \geq 3p_g + q - 7.$$

Proof. (i) follows directly from (1.3, i) and (1.2).

(ii) Throwing out the fixed part of the linear system, I assume that $|D|$ is free, so $\varphi_D(C) = \Gamma \subset \mathbb{P}^r$ is an irreducible curve of degree d. Now $H^0(C, 2D) \supset \operatorname{Im} S^2H^0(C, D)$, which can be identified with the image of $\rho_\Gamma : H^0(\mathbb{P}^r, \mathcal{O}(2)) \to H^0(\Gamma, \mathcal{O}_\Gamma(2))$; by (1.3, i), rank $\rho_\Gamma = f_\Gamma$ is at least

$$r + 1 + \min(d, 2r - 1);$$

Clifford's theorem for D gives $d \geq 2r$, which proves that $h^0(2D) \geq 3r$. Now 2D is again effective and special, so Clifford's theorem for 2D gives

$$2d = \deg 2D \geq 2(h^0(2D) - 1),$$

as required.

(iii) Suppose that $\varphi_K: X \dashrightarrow \varphi(X) \subset \mathbb{P}^{p_g-1}$ is birational. Then quadrics of $\mathbb{P}^{p_g-1}$ cut out on $\varphi(X)$ a space that can be identified with a subspace of $H^0(X, \mathcal{O}(2K))$, so that using (1.2), it follows from (ii) that

$$1 - q + p_g + K^2 = P_2 \geq p_g + 3(p_g - 2). \quad \text{Q.E.D.}$$

(1.5) The main problem. Proposition 1.3 raises the following hope:

Conjecture. Let $\Sigma \subset \mathbb{P}^n$ be a finite set spanning $\mathbb{P}^n$, linearly general and uniform with respect to quadrics; suppose that Σ consists of d points and imposes $f = f_\Sigma$ conditions on quadrics. Then for $p = 0,.. n-2$,

$$d \geq 2n+2p+1 \text{ and } f \leq 2n+p \implies$$

the intersection of all quadrics through Σ contains a curve C containing Σ.

Remarks. (i) I don't know that C is irreducible, although that is certainly the case if Σ is a general hyperplane section of a curve. The statement is equivalent to saying that Σ is contained in a curve C of degree $\leq n+p$. Also, (1.4, i) guarantees that no variety of dimension ≥ 2 spanning $\mathbb{P}^n$ can be contained in so many quadrics.

(ii) The bound $d \geq 2n+2p+1$ is best possible: let $C \subset \mathbb{P}^n$ be a curve of genus p embedded by a complete linear system of degree $n+p$. In view of $p \leq n-2$, it follows that $\mathcal{O}_C(1)$ has degree $\geq 2p+2$ and is necessarily nonspecial, and C is an intersection of quadrics imposing exactly $2n+p+1$ conditions on quadrics. Therefore, if $\Sigma = C \cap Q$ is the intersection of C with a quadric then Σ consists of $2n+2p$ points, and imposes $2n+p$ conditions on quadrics.

(iii) The bound $p \leq n-2$ is also best possible, since the intersection $\Sigma = C \cap Q$ of a quadric with a canonical curve $C \subset \mathbb{P}^n$ of genus $n+1$ has degree $4n$ and imposes $3n-1$ conditions on quadrics. This example suggests that $d \geq 4n$ and $f \leq 3n-1$ imply either Σ is not generically an intersection of quadrics, or equalities throughout and Σ is ideal-theoretically an intersection of quadrics.

(iv) The conjecture is an easy exercise if Σ is contained in a linearly normal curve $C \subset \mathbb{P}^n$ of degree $\leq 2n-1$; in §2 I will prove that it also holds if Σ is contained in a surface $F \subset \mathbb{P}^n$ of degree $n-1$, a result presumably known to Fano and Castelnuovo.

(vi) It might be interesting to formulate an analogous conjecture in the style of Mark Green for the higher syzygies:

Σ of small degree and an intersection of quadrics $\implies$

higher syzygies aren't generated in lowest degree?

(1.6) Conditional results. Conjecture 1.5 implies the following.

(i) Let C be a curve and $D = g^r{}_d$ a divisor with $2D$ special such that the rational map $\varphi_D\colon C \to \Gamma \subset \mathbb{P}^r$ is a birational embedding; suppose that the image $\varphi_D(C) = \Gamma$ is generically an intersection of quadrics in the sense of (1.1); then

$$d \geq h^0(2D) - 1 \quad \text{and} \quad h^0(2D) \geq 4r - 5.$$

(ii) Let X be a surface of general type for which the 1-canonical image $\varphi_K(X) \subset \mathbb{P}^{p_g-1}$ is generically an intersection of quadrics; then X satisfies

$$K^2 \geq 4p_g + q - 12.$$

Proof. (i) This is exactly the same as (1.4, ii). Without loss of generality I can assume $|D|$ is free, and set

$$d = \deg D \quad \text{and} \quad p = \left[\frac{d-1}{2}\right] - n,$$

where $n = r - 1$ as in (1.2). Then since $\deg \varphi_D(C) \geq 2n + 2p + 1$, Conjecture 1.5 would imply that

$$h^0(\mathcal{O}_C(2D)) \geq 3n + 1 + \min(p, n-2).$$

However, Clifford's theorem for the special divisor 2D gives

$$d \geq h^0(\mathcal{O}_C(2D)) - 1,$$

and it follows from these inequalities that $p \geq n - 2$, and thus $h^0(2D) \geq 4n - 1 = 4r - 5$.

For (ii), note that (1.2) and (i) give

$$1 - q + p_g + K^2 = P_2 \geq p_g + 4(p_g - 2) - 5. \qquad \text{Q.E.D.}$$

Remark. The cases $p = 1$ and $p = 2$ of (1.5) are contained in (1.3), proved in §2, so that it follows for example that for a surface X with $K^2 = 3p_g - 6$ and birational φ_K, the image $\varphi_K(X) \subset \mathbb{P}^{p_g-1}$ is contained in a 3-fold W of degree $\leq p_g - 2$. When $p_g \geq 12$ the only possibilities for W are a rational normal 3-fold scroll or a double cone over an elliptic curve, or linear projections of these.

(1.7) Relation with the free pencil trick. Let $C \subset \mathbb{P}^{n+1}$ be a curve. The free pencil trick is a classical method of giving a lower bound on the rank of the restriction map $\rho_C \colon H^0(\mathcal{O}_{\mathbb{P}}(2)) \to H^0(\mathcal{O}_C(2))$: fix $n+2$ general points $P_0, \ldots P_{n+1} \in C$, and choose the coordinates $x_0, \ldots x_{n+1}$ so that $x_i(P_j) = \delta_{ij}$. Then x_0, x_1 span the pencil of hyperplanes through $\Pi = \langle P_2, \ldots P_{n+1} \rangle = \mathbb{P}^{n-1}$, which cuts out residually on C the linear system $|H - A|$, where H is the hyperplane section and $A = P_2 + \ldots P_{n+1}$; then the exact sequence of sheaves

$$0 \to \mathcal{O}_C(A) \xrightarrow{x_0,\, x_1} \mathcal{O}_C(1) \oplus \mathcal{O}_C(1) \xrightarrow{\substack{-x_1 \\ x_0}} \mathcal{O}_C(2H - A) \to 0$$

shows that the $2n + 4$ quadratic monomials

$$\{x_0 x_i, x_1 x_i\}_{i=0,\dots n+1} \in H^0(\mathcal{O}_C(2))$$

are linearly independent except for relations of the form $s(x_0x_1 - x_1x_0)$ with $s \in H^0(\mathcal{O}_C(A))$. For example, if $g(C) \geq n$ then $h^0(\mathcal{O}_C(A)) = 1$, and this implies that the $3n + 3$ monomials

$$\{x_0^2, x_0x_1, x_1^2, x_0x_i, x_1x_i, x_i^2\}_{i = 2,\dots n+1}$$

are linearly independent, so $h^0(\mathcal{O}_C(2H)) \geq 3(n + 1)$.

This is the same result as (1.4, ii), and the proof is just a mutation of the proof in (1.2) and (1.3, i): it boils down to saying that after taking a general hyperplane section Σ: $(x_0 = 0)$ $\subset C$, the only quadrics of rank 2 containing Σ with vertex in the linear subspace $(x_1 = 0)$ are of the form $x_1 \cdot \lambda$ where λ is a linear form vanishing at $\Sigma - A$.

(1.8) It's important to understand the weakness of this argument: the bilinear problem of estimating the rank of ρ_C is reduced to a linear one, but at the expense of considering only the $(2n+3)$-dimensional subspace of forms involving x_0 and x_1.

§2. Rational normal scrolls and the proof of (1.3)

The remaining assertions (ii) and (iii) of (1.3) will follow from the two following theorems.

(2.1) Theorem. Let $\Sigma \subset \mathbb{P}^n$ be a set of $d \geq 2n + 2p + 1$ points, linearly general and uniform with respect to quadrics. Suppose that $p \leq n - 2$ and that Σ imposes $f \leq 2n + p$ conditions on quadrics. Then Σ is contained in a p-dimensional rational normal scroll F (possibly a cone; if $p = 1$, F is a rational normal curve).

(2.2) Theorem. Suppose that $\Sigma \subset F \subset \mathbb{P}^n$ is contained in a rational normal surface scroll of degree $n - 1$. Then Conjecture 1.5 holds for Σ.

(2.3) Plan of proof of (2.1). Write

$$\Pi = \langle P_1,.. P_{n-1}\rangle = \mathbb{P}^{n-2}$$

for the codimension 2 space spanned by $n-1$ elements of Σ. A dimension-count shows that $\Pi \cup \Sigma$ is contained in at least $n-p$ linearly independent quadrics, and the intersection of these will consist of Π together with the required scroll F. This is a classic construction for rational normal scrolls, and the only possible way it can degenerate would contradict the fact that Σ is linearly general. The tricky part is to show that the initial points $P_1,.. P_{n-1}$ are also contained in the residual intersection F, and this is where the uniform assumption on Σ is used.

(2.4) Dimension count. Since quadrics of $\Pi = \mathbb{P}^{n-2}$ vanishing at the $n-1$ given points $\{P_1,.. P_{n-1}\}$ form a vector space of dimension $\binom{n-1}{2}$, it follows that Π imposes at most this number of conditions on quadrics of $\mathbb{P}^n$ through Σ. Therefore the vector space of quadrics through Π and Σ has dimension

$$\geq \binom{n+2}{2} - (2n+p) - \binom{n-1}{2} = n-p,$$

as required.

(2.5) Classic construction for scrolls. Consider the blow-up $\sigma\colon F_0 \longrightarrow \mathbb{P}^n$ of $\mathbb{P}^n$ in Π; with its projection $\pi\colon F_0 \longrightarrow \mathbb{P}^1$, this is the n-dimensional scroll $F_0 = \mathbb{P}(\mathcal{E}_0)$, where $\mathcal{E}_0$ is the rank n vector bundle over $\mathbb{P}^1$

$$\mathcal{E}_0 = \mathcal{O}(1) \oplus \mathcal{O}^{\oplus(n-1)}.$$

F_0 contains a negative divisor $B_0 = \sigma^{-1}\Pi \cong \mathbb{P}^1 \times \mathbb{P}^{n-2}$, and I write A_0 for the fibre of π. The morphism $\sigma\colon F_0 \longrightarrow \mathbb{P}^n$ is given by the complete linear system $|A_0 + B_0|$. Let $Q_1,.. Q_{n-p}$ be the linearly independent quadrics of $\mathbb{P}^n$ through Π provided by (2.4). Then since $Q_i \supset \Pi$, it follows that each $\sigma^*Q_i = B_0 + Q_i'$, where $Q_i' \sim 2A_0 + B_0$. Now for $k = 1,.. n-p$, set

$$F_k = \bigcap_{j=1}^{k} Q_j' \subset F_0.$$

By induction on k, suppose F_k is irreducible and is a $\mathbb{P}^{n-k-1}$-bundle over $\mathbb{P}^1$, having degree $k+1$ under the morphism σ to $\mathbb{P}^n$ defined by the divisor $A_k + B_k$ (where I write A_k and B_k for the restriction of A_0 and B_0 to F_k). Then $F_{k+1} \subset F_k$ is in the divisor class $2A_k + B_k$, and so has degree $k+2$ under σ, and has a unique component that is a $\mathbb{P}^{n-k-1}$-bundle over $\mathbb{P}^1$. Thus if reducible, F_{k+1} could be written as a sum of A_k with a divisor in $|A_k + B_k|$; then

$\sigma(F_{k+1}) \subset \sigma(F_k) \subset \mathbb{P}^n$ would be contained in the union of two hyperplane sections, of which the one containing $\sigma(A_k)$ can be chosen to pass through Π. Since Σ is certainly contained in $\sigma(F_{k+1}) \cup \Pi$, this contradicts the assumption that Σ is linearly general.

Therefore $F = \sigma(F_{n-p}) \subset \mathbb{P}^n$ is a p-dimensional scroll as required.

Remark. If irreducible, $B_k \subset F_k$ is itself a scroll mapping birationally to $\sigma(B_k) \subset \Pi$ if $k \geq 1$. It can certainly happen that B_k is reducible, but in any case $\sigma(B_k) \subset \Pi$ has codimension $k - 1$ and degree k.

(2.6) Key technical point: $\Sigma \subset F$. The points of Σ other than $\{P_1,.. P_{n-1}\}$ belong to each quadric Q_i and not to Π, and hence are in the residual component F of $\bigcap Q_i = \Pi \cup F$.

I now prove that the points P_i for $i = 1,.. n-1$ also belong to F, following an argument kindly supplied by Eisenbud and Harris. For this, let $P_n, P_{n+1} \in \Sigma$ be two more points, and choose the coordinates x_n, x_{n+1} so that x_n vanishes on $\langle \Pi, P_{n+1} \rangle$ and x_{n+1} on $\langle \Pi, P_n \rangle$; then of course Π: $(x_n = x_{n+1} = 0)$.

Lemma. After possibly reordering $\{P_1,.. P_{n-1}\}$, the equations of F can be written in the determinantal form

$$\operatorname{rk} \begin{bmatrix} x_n & \lambda_1 & \lambda_2 & .. & \lambda_{n-p} \\ x_{n+1} & \mu_1 & \mu_2 & .. & \mu_{n-p} \end{bmatrix} \leq 1, \qquad (*)$$

where the λ_i are linear forms such that $\lambda_i(P_j) = \delta_{ij}$ for $i, j = 1,.. n-p$. Here the $n - p$ quadrics through $\Sigma \cup \Pi$ of (2.4) are given by

$$Q_i = \det \begin{bmatrix} x_n & \lambda_i \\ x_{n+1} & \mu_i \end{bmatrix} \quad \text{for } i = 1,.. n-p. \qquad (**)$$

First of all, the lemma implies $\Sigma \subset F$, and thus proves Theorem 2.1. In fact, the remaining quadrics of the determinantal are

$$Q_{ij} = \det \begin{bmatrix} \lambda_i & \lambda_j \\ \mu_i & \mu_j \end{bmatrix} \quad \text{for } i, j = 1,.. n-p \text{ with } i \neq j.$$

Now Q_{ij} vanishes on F, hence on all the points of Σ except for the P_i, and the ever-vigilant reader will be able to see from the form of Q_{ij} that it also vanishes at P_k for $k = 1,.. n-p$, $k \neq i, j$. Thus in total it vanishes at

$$\geq 2n+2p+1-(n-1)+(n-p-2) = 2n+p$$

points of Σ. Hence by the uniform assumption on Σ, Q_{ij} vanishes at all points of Σ.

Proof of the lemma. Every quadric through Π is of the form $x_n\mu - x_{n+1}\lambda$, so the equations of the $n-p$ quadrics Q_i can certainly be put in the form (∗∗) for some linear forms λ_i and μ_i. It follows that F is given by equations (∗). Now I claim that the λ_i restrict to $n-p$ linearly independent forms on Π. For otherwise some nonzero linear combination of the Q_i would be of the form

$$Q = \det \begin{bmatrix} x_n & \lambda \\ x_{n+1} & \mu \end{bmatrix}$$

with λ a linear combination of x_n and x_{n+1}; but $Q(P_n) = 0$ implies that λ would have to be a multiple of x_n, since $x_n(P_n) = 0$ and $x_{n+1}(P_n) \neq 0$. This is absurd, since Σ is not contained in a quadric of rank 2.

Suitable linear combinations of the λ_i achieve the statement of the lemma. Q.E.D.

(2.7) Plan of proof of (2.2). The proof of (2.2) considers the linear system L with assigned base points cut out on F by quadrics of $\mathbb{P}^n$ through Σ (the case of F a cone causes no problem, just resolve the vertex). Firstly, since $h^0(F, \mathcal{O}(2)) = 3n$ and Σ imposes $\leq 2n+p$ conditions on quadrics, it follows that

$$h^0(F, I_\Sigma\cdot\mathcal{O}(2)) \geq n-p, \quad \text{that is,} \quad \dim L \geq n-p-1;$$

write $L = M + D$ with M mobile and D the fixed part. Clearly since F is a scroll, $\mathcal{O}(2)$ has degree 2 relative to the projection $\pi\colon F \to \mathbb{P}^1$, and there are 3 possibilities for the decomposition $L = M + D$:

Case 1. M has degree 2 over $\mathbb{P}^1$, and D is a union of $\beta \geq 0$ fibres of π.

Case 2. M and D both have degree 1 over $\mathbb{P}^1$.

Case 3. M is a union of $\geq n-p-1$ fibres of π and the base locus D has degree 2 over $\mathbb{P}^1$.

In Case 3, there is nothing to prove: the base locus $D \supset \Sigma$ and is the intersection of all quadrics through Σ; clearly $\deg D \leq n+p$. It is therefore enough to prove that Case 1 leads to a contradiction; and that in Case 2, $\Sigma \subset D$ and $\deg D \leq n+p$.

(2.8) Idealised proof of (2.2). I first illustrate the proof by deriving a contradiction from the assumptions that L has base locus exactly the reduced set Σ, and that its general element Γ is irreducible and nonsingular; (*a priori* the main case?). Write L_Γ for the free linear system cut out on Γ by L after subtracting the base locus Σ.

Then by (2.7),

$$\dim L_\Gamma \geq n - p - 2.$$

On the other hand, $(2H)^2 = 4\deg F = 4(n-1)$ (where $H = \mathcal{O}_F(1)$), so that

$$\deg L_\Gamma = 4(n-1) - \deg \Sigma \leq 2n - 2p - 5.$$

But it's easy to see from properties of scrolls that

$$H^0(\mathbb{P}, \mathcal{O}(2)) \longrightarrow H^0(\Gamma, \mathcal{O}_\Gamma(2H))$$

is surjective; therefore Σ does not impose independent conditions on $H^0(\Gamma, \mathcal{O}_\Gamma(2H))$, and so $H^1(\Gamma, \mathcal{O}_\Gamma(2H-\Sigma)) \neq 0$ and the linear system L_Γ is special. Hence the two most recently acquired inequalities contradict Clifford's theorem.

(2.9) Case 1 is impossible. Since the fibre A of $\pi\colon F \longrightarrow \mathbb{P}^1$ satisfies $LA = 2$, $A^2 = 0$, I get

$$M^2 = 4(n-1) - 4\beta.$$

Suppose that the points of Σ distribute themselves as

$$2n + 2p + 1 \leq \deg \Sigma = d = a + b,$$

with a points on the fixed part D and b base points of M outside D. Clearly $a \leq 2\beta$.

Step 1. The general element of M is irreducible.

For if M is composed of a pencil, $M = kE$ with $k \geq 2$ and E has b base points, so

$$M^2 = kE^2 \geq 4E^2 \geq 4b \geq 4(2n + 2p + 1 - 2\beta),$$

giving $\beta \geq n + 2p + 2$, which is absurd.

Step 2. Write Γ for the normalisation of the general element of M; consider the free linear system M_Γ cut out on Γ by M (after subtracting off the base locus). Then

$$\deg M_\Gamma \leq M^2 - b \leq 4(n-1) - 4\beta - (2n + 2p + 1 - 2\beta)$$
$$\leq 2n - 2p - 2\beta - 5;$$

and as before,

$$\dim M_\Gamma \geq n - p - 2,$$

which contradicts Clifford's theorem if M_Γ is special.

Step 3. Therefore M_Γ is nonspecial, so I get a bound on the genus of g from RR, which will lead to a contradiction.

I intend to use the classical language, writing $\sum m_i P_i$ for the actual base locus of M, including infinitely near points; the reader who is unduly distressed by this can perform the easy exercise of translating the following argument in terms of successive blow-ups of the base locus. First,

$$\deg M_\Gamma = M^2 - \sum m_i^2 = 4n - 4 - 4\beta - \sum m_i^2$$

and

$$g(\Gamma) = n - 2 - \beta - \sum \binom{m_i}{2}.$$

Here $n - 2 - \beta$ is the genus of the general element of $|2H - \beta A|$ on F (exercise using the adjunction formula).

Also, since M_Γ is a nonspecial linear system, RR gives

$$\begin{aligned} n - p - 1 \leq h^0(\Gamma, M_\Gamma) &= 1 - g(\Gamma) + \deg M_\Gamma \\ &= 3n - 3 - 3\beta - \sum \binom{m_i + 1}{2}. \end{aligned}$$

However, M has at least

$$b \geq 2n + 2p + 1 - 2\beta$$

assigned base points, and hence

$$2n + 2p + 1 - 2\beta \leq b \leq \sum \binom{m_i + 1}{2} \leq 2n + p - 2 - 3\beta,$$

which is absurd.

(2.10) Proof in Case 2. Write $M \sim H - \beta A$ and $D \sim H + \beta A$ where $H = \mathcal{O}_F(1)$, A is the fibre of π, and $\beta \in \mathbb{Z}$. Let $\Gamma \in M$ be a general element; notice that Γ projects isomorphically to $\mathbb{P}^1$ under π, so that I am spared the cases of M being composed of a pencil or the linear system M_Γ special.

As before, suppose that the points of Σ distribute themselves as

$$2n + 2p + 1 \le \deg \Sigma = d = a + b,$$

with a points on the fixed part D and b base points of M outside D. Let M_Γ be the free linear system cut out by M on Γ (after subtracting the base points). Now

$$\deg M_\Gamma \le M^2 - b = n - 1 - 2\beta - b,$$

and

$$h^0(M_\Gamma) \ge h^0(F, I_\Sigma \cdot \mathcal{O}(2)) - 1 \ge n - p - 1.$$

Therefore

$$n - p - 1 \le n - 2\beta - b, \quad \text{that is} \quad b + 2\beta \le p + 1,$$

and hence

$$a = d - b \ge 2n + p + 2\beta.$$

However, since $M \sim H - \beta A$ moves in an irreducible linear system on F it follows that $h^0(F, \mathcal{O}_F(M)) = \chi(\mathcal{O}_F(M)) = n + 1 - 2\beta$, and it is obvious that $D \sim 2H - M$ imposes

$$h^0(F, \mathcal{O}(2)) - h^0(F, \mathcal{O}_F(M)) = 3n - n - 1 + 2\beta = 2n - 1 + 2\beta$$

conditions on quadrics. Therefore the a points of $\Sigma \cap D$ impose dependent conditions on quadrics, and by the uniform assumption, this implies that $\Sigma \subset D$. The remaining assertion, that $\deg D \le n + p$ follows from the fact that $2\beta \le p + 1$. Q.E.D.

§3. A vector bundle approach

(3.1) This section starts off with some numerology. The following material on quadrics of small rank or containing large linear subspaces is well-known and will be used throughout.

Proposition. The (projective) space of quadrics of rank $\le k$ in $\mathbb{P}^n$ is a symmetric determinantal $S_{k,n}$ in the space S_n of quadrics of $\mathbb{P}^n$, and has dimension

$$k(n - k + 1) + \binom{k+1}{2} - 1;$$

it has a resolution that is a projective bundle over the Grassmanian $Gr(n - k + 1, n + 1)$ of (n–k)-planes in $\mathbb{P}^n$, with fibre S^2E, where E is the tautological rank k quotient bundle on Gr (the point of Gr is the vertex $\Lambda \subset \mathbb{P}^n$, and the fibre consists of quadrics on a complementary $\mathbb{P}^{k-1}$, the base of the cone).

If $\Lambda \subset \mathbb{P}^n$ is a fixed (n-k)-plane and Q any quadric with Sing Q = Λ then Q is a nonsingular point of $S_{k,n}$, and the (projective) space of quadrics of $\mathbb{P}^n$ containing Λ is the tangent space $T_{S_{k,n},Q} \subset S_n$ to $S_{k,n}$ at Q; in particular, it has the above dimension. □

(3.2) Numerology. Conjecture 1.5 deals with a set $\Sigma \subset \mathbb{P}^n$ of d points imposing f conditions on quadrics, such that d is 'sufficiently large', and $f = 2n + p$ is in the range

$$2n + 1 \le f \le 3n - 2.$$

This range of values has the following peculiarities:

(a) If $\Pi = <P_1,.. P_{n-1}> = \mathbb{P}^{n-2}$ is an (n-2)-plane spanned by points of Σ, then $h^0(I_{\Pi \cup \Sigma} \cdot \mathcal{O}(2)) \ge 3n - f \ge 2$; that is, the Castelnuovo argument of Theorem 2.1 can at least start working, and Σ is contained in a scroll of dimension $f - 2n \le n - 2$.

(b) If $\Lambda = \mathbb{P}^{n-3}$ is a general (n-3)-plane of $\mathbb{P}^n$ then again

$$h^0(I_{\Pi \cup \Sigma} \cdot \mathcal{O}(2)) \ge 3n - f \ge 2;$$

(c) The (projective) dimension of the space of quadrics of rank ≤ 3 through Σ is at least 1.

A curve $C \subset \mathbb{P}^{n+1}$ imposing $f_C \le 4n$ conditions on quadrics is contained in:

(a) a positive-dimensional linear system of quadrics through an (n-2)-plane $\Pi = <P_1,.. P_{n-1}> = \mathbb{P}^{n-2}$ spanned by n - 1 (general) points of C;

(b) a positive-dimensional linear system of quadrics through a general (n-3)-plane $\Lambda = \mathbb{P}^{n-3} \subset \mathbb{P}^{n+1}$;

(c) a positive-dimensional family of quadrics of rank 4.

Likewise, let X be a surface of general type with $K^2 < 4p_g - 12$ and for which the 1-canonical map $\varphi_K : X \longrightarrow \varphi(X) \subset \mathbb{P}^{p_g-1}$ is birational; set $n = p_g - 3$. Then the canonical image $\varphi(X)$ is contained in

(a) a positive-dimensional linear system of quadrics through an (n-2)-plane $\Pi = <P_1,.. P_{n-1}> = \mathbb{P}^{n-2} \subset \mathbb{P}^{p_g-1}$ spanned by n - 1 points of X;

(b) a positive-dimensional linear system of quadrics through any (n-3)-plane $\Lambda = \mathbb{P}^{n-3} \subset \mathbb{P}^{p_g-1}$;

(c) a positive-dimensional family of quadrics of rank 5.

(3.3) Since the proof of Proposition 1.3, (i) just used the fact that the set Σ cannot be contained in too many quadrics of rank 2, it seems reasonable to ask if quadrics of small rank couldn't be persuaded to provide some entertainment.

(3.4) Suppose that $X \subset \mathbb{P}^{p_g-1}$ is a nonsingular surface with $K_X = \mathcal{O}_X(1)$ and $K^2 \le 4p_g - 12$; write $n = p_g - 3$, so that $K^2 = \deg X \le 4n$. Then X is contained in a quadric of rank 5,

$$X \subset Q_5 \subset \mathbb{P}^{p_g-1}.$$

I assume in addition that X does not meet the vertex of Q_5.

Remark. These assumptions look harmless, since I could presumably blow up and subtract off the base locus, etc. In fact, I would be happy to make progress under even stronger assumptions, like X is projectively normal, and its homogeneous ideal is generated by quadrics; unfortunately, these don't seem to be too much help. I am ignoring the case that X is contained in quadrics of rank ≤ 4 not forced by dimension count; if this happens I get at once a decomposition of K_X.

(3.5) The motivation for the vector bundle construction of (3.7–11) is as follows: let $C = X \cap \mathbb{P}^{n+1}$ be a general hyperplane section of X through Sing Q_5, so that C is a nonsingular curve with $K_C = \mathcal{O}_C(2)$, and is contained in $Q_4 = Q_5 \cap \mathbb{P}^{n+1}$, a quadric of rank 4. The two rulings of Q_4 by generators $\mathbb{P}^{n-1}$ cut out pencils E and E' on C with

$$\mathcal{O}_C(1) = E + E'. \tag{1}$$

The pencils E are free, but not necessarily complete linear systems. I abuse the notation by writing E for a divisor in the pencil E; a general E is made up of distinct points, and I write $\langle E \rangle$ for their span. Note that $E + E'$ is a hyperplane section of C, so that for any choice of E, E' in their pencils,

$$\langle E, E' \rangle = \mathbb{P}^n \subset \mathbb{P}^{n+1}. \tag{2}$$

In view of $\mathcal{O}_C(1) = E + E'$, obviously

$$\dim |E| \ge n - \dim \langle E' \rangle. \tag{3}$$

Now, the two families of generators of Q_4 are interchanged by monodromy as the hyperplane $\mathbb{P}^{n+1}$ moves, so that

$$\deg E = \deg E' = \tfrac{1}{2}\deg X = \tfrac{1}{2}K^2, \tag{4}$$

$$\dim \langle E \rangle = \dim \langle E' \rangle \quad \text{and} \quad h^0(E) = h^0(E'); \tag{5}$$

similarly, points of a general E impose independent conditions on forms of degree k if and only if the same holds for E', and so on.

(3.6) Wake up! The rest of §3 aims to prove that E is very special: it moves in a big linear system on C, it spans $\langle E\rangle = \mathbb{P}^\nu$ with ν small, and it imposes few conditions on quadrics. The main result of §3 is (3.11): roughly speaking, the point set $E \subset \mathbb{P}^\nu$ itself satisfies the assumptions of Conjecture 1.5. Though the statement is tentative, it is very striking; since E has half the degree of Σ and most of its good properties, this looks like an opening for an attack by induction.

Let me now try to explain what's so special about E. Duality in RR asserts that E imposes $\deg E - \dim |E|$ conditions on $H^0(\mathcal{O}_C(K_C)) = H^0(\mathcal{O}_C(2))$; so *a fortiori*, E imposes

$$f = f_E \le \deg E - \dim |E| \tag{6}$$

conditions on quadrics of $\mathbb{P}^{n+1}$. Since

$$d = \deg E = \tfrac{1}{2}K^2 \le 2n, \tag{7}$$

there is tension with (1.3, i): E is cut out on C by a pencil of $\mathbb{P}^{n-1} \subset Q_4$, and $E \subset \mathbb{P}^{n-1}$ is a set of $d \le 2n$ points imposing dependent conditions on quadrics. If $K^2 < 4p_g - 12$ then $d \le 2n - 1$, and this contradicts Proposition 1.3, (i) if the points of E are linearly general.

On the other hand, E should satisfy some weak uniform position property, coming from the fact that C is irreducible, which should then imply that $\langle E\rangle = \mathbb{P}^\nu$ with $\nu < n - 1$. By (3) and (5), this should imply that $|E|$ is more than a pencil, so that in turn, by (6), its points impose even fewer conditions on quadrics, ..

Making this argument work in terms of $C \subset \mathbb{P}^{n+1}$ seems to be hard; for example, the first thing to prove is the following:

Conjecture. If $E \subset \mathbb{P}^\nu$ is a set of points defined and irreducible over a field K, and imposing dependent conditions on quadrics, then

$$\deg E \ge 2\nu + 2.$$

This is a version of (1.3) in which the condition of linear generality is replaced by the weaker uniform condition: if E has a subset $\{P_1,.. P_k\}$ of k points spanning a ℓ-plane $\mathbb{P}^\ell$, then every point $P \in E$ is contained in such a subset. This and other uniform properties of E come from the fact that C is a fairly general hyperplane section of X, and E general in a pencil on C; so it is appropriate to work directly on X.

(3.7) Construction and properties of $\mathcal{E}$. I use the notation of the start of (3.5). First of all, how does one get pleasure or profit out of a quadric of rank 5? I view the nonsingular quadric $Q \subset \mathbb{P}^4$ as a hyperplane section of the Klein quadric $Gr = Gr(2, 4) = Q_6 \subset \mathbb{P}^5$. Let $\mathcal{E}_0$ be one of the two tautological quotient bundles on Gr, and

$$V = H^0(\mathrm{Gr}, \mathcal{E}_0), \quad \dim V = 4 \tag{8}$$

so that $\mathrm{Gr} = \mathrm{Gr}(2, V)$; then

$$\mathcal{O}_{\mathrm{Gr}}(1) = \textstyle\bigwedge^2 \mathcal{E}_0, \quad H^0(\mathcal{O}_{\mathrm{Gr}}(1)) = \textstyle\bigwedge^2 V \tag{9}$$

and a nonsingular hyperplane section of Gr corresponds to the isotropic spaces of a nondegenerate skew bilinear form $\psi\colon V \times V \longrightarrow k$.

Now under the assumptions of (3.5), $X \subset Q_5$ and $X \cap \mathrm{Sing}\, Q_5 = \emptyset$. This means that X has a projection morphism to the nonsingular $Q \subset \mathbb{P}^4$, so that writing $\mathcal{E}$ for the restriction to X gives the following.

Proposition. $X \subset Q_5$ and $X \cap \mathrm{Sing}\, Q_5 = \emptyset$ gives rise to a vector bundle $\mathcal{E}$ of rank 2 on X such that

$$\textstyle\bigwedge^2 \mathcal{E} \cong K_X, \quad \text{that is,} \quad K_X \otimes \mathcal{E}^* \cong \mathcal{E}, \tag{10}$$

and V a 4-dimensional space of sections spanning $\mathcal{E}$:

$$0 \to \mathcal{F} \longrightarrow V \otimes \mathcal{O} \longrightarrow \mathcal{E} \to 0 \tag{11}$$

V has a nondegenerate skew bilinear form $\psi\colon V \times V \longrightarrow k$ such that at every point, the fibre of $\mathcal{F}$ is isotropic for ψ, inducing an isomorphism

$$\psi\colon \mathcal{F} \xrightarrow{\cong} \mathcal{E}^* = \mathcal{E}(-K_X). \tag{12}$$

(3.8) Lemma. (i) Any general section $s \in H^0(\mathcal{E})$ defines a short exact sequence

$$0 \to \mathcal{O} \longrightarrow \mathcal{E} \longrightarrow I_E \cdot \mathcal{O}_X(K_X) \to 0, \tag{13}$$

where $E \subset X$ is a reduced set of points, so that

$$E = c_2(\mathcal{E}) \quad \text{and} \quad c_1(\mathcal{E}) = K_X. \tag{14}$$

(ii)

$$2E = K^2 \quad \text{as 0-cycles of } X \tag{15}$$

(modulo rational equivalence).

(iii) If $s \in V \subset H^0(\mathcal{E})$ is a general element then E is just a reinterpretation of the construction of (3.5), (1).

Proof. The construction on X is the pull back of a tautological construction on Q. The choice of a tautological bundle $\mathcal{E}_0$ on Gr(2, 4) determines one family of generators of Gr = Gr(2, 4) = $Q_6 \subset \mathbb{P}^5$. A general element $s \in V = H^0(\mathrm{Gr}, \mathcal{E}_0)$ defines an exact sequence over Gr

$$0 \to \mathcal{O}_{\mathrm{Gr}} \longrightarrow \mathcal{E}_0 \longrightarrow I_\pi \cdot \mathcal{O}_{\mathrm{Gr}}(1) \to 0, \tag{16}$$

where $\pi = c_2(\mathcal{E}_0)$ is a generator of Gr in the given family. If π_1 and π_2 are generators of Gr in the two families then $\pi_1 + \pi_2$ is a codimension 2 linear section of Gr; on the other hand, each of them restricts down to a generator of Q, so that on Q, the class of a codimension 2 linear section is twice a generator. (ii) follows from this. The generators of Q move around in a free system, so that the nonsingularity in (i) follows from the separability of $X \longrightarrow Q$. (iii) is an exercise for the reader.

Problem. Assume that the projection $X \longrightarrow Q \subset \mathbb{P}^4$ is a birational embedding; (presumably this is the main case?). If K is the field of definition of the generic section $s \in H^0(\mathcal{E})$, then is it true that the Galois group Gal(K(E)/K) is the full symmetric group on E? This would imply that E is uniform, and would be an analogue for vector bundles of the Lefschetz–Harris principle for very ample divisors (compare (1.2)).

(3.9) Lower bound for $H^0(\mathcal{E})$. If follows immediately from (13), or from RR that

$$\chi(\mathcal{E}) = 2\chi(\mathcal{O}_X) - \deg E. \tag{17}$$

Assuming $K^2 < 4p_g - 12$ and $q = 0$, this gives

$$\chi(\mathcal{E}) = 2p_g + 2 - \tfrac{1}{2}K^2 > 8. \tag{18}$$

Now Serre duality together with (10) gives $h^2(\mathcal{E}) = h^0(\mathcal{E}^*(K_X)) = h^0(\mathcal{E})$, so

$$h^0(\mathcal{E}) = h^2(\mathcal{E}) = p_g + 1 - \tfrac{1}{4}K^2 + \tfrac{1}{2}h^1(\mathcal{E}) \geq 5. \tag{19}$$

On the other hand, the number of hyperplanes of $\mathbb{P}^{p_g-1}$ containing E is $h^0(I_E \cdot \mathcal{O}_X(K))$, and from (13), this is

$$h^0(I_E \cdot \mathcal{O}_X(K)) = h^0(\mathcal{E}) - 2 \tag{20}$$

so that (19) implies

$$\dim \langle E \rangle = p_g - h^0(\mathcal{E}) \leq \tfrac{1}{4}K^2 - 1 \leq n - 2. \tag{21}$$

This shows that E is not linearly general.

(3.10) E is not quadratically general. (11) and (12) give an exact sequence

$$0 \to V \to H^0(\mathcal{E}) \to H^1(\mathcal{E}^*) \to 0. \tag{22}$$

Hence

$$h^1(\mathcal{E}^*) = h^1(\mathcal{E}(K)) = h^0(\mathcal{E}) - 4 > 0. \tag{23}$$

Now tensoring (13) with K_X gives

$$0 \to \mathcal{O} \to \mathcal{E} \to I_E \cdot \mathcal{O}_X(K_X) \to 0, \tag{24}$$

leading to the cohomology exact sequence

$$\begin{aligned} H^1(K_X) = 0 \to H^1(\mathcal{E}(K_X)) \to H^1(I_E \cdot \mathcal{O}_X(2K)) \to \\ \to H^2(\mathcal{O}_X(K_X)) = k \to 0 \end{aligned} \tag{25}$$

Therefore

$$h^1(I_E \cdot \mathcal{O}_X(2K)) = h^1(\mathcal{E}(K_X)) + 1 = h^0(\mathcal{E}) - 3. \tag{26}$$

This proves that E imposes

$$f_E = \deg E - h^0(\mathcal{E}) + 3 \tag{27}$$

conditions on quadrics.

(3.11) Curious conclusion. E satisfies

$$\langle E\rangle = \mathbb{P}^\nu, \quad \deg E \geq 2\nu + 2\pi + 1, \quad f_E = 2\nu + \pi,$$

where $1 \leq \pi \leq \nu$. That is, the numerical assumptions of Conjecture 1.5 hold for E (except possibly for the cases $\pi = \nu - 1, \nu$).

Proof. Set $\langle E\rangle = \mathbb{P}^\nu$; then by (21), (15) and (27),

$$\nu = n + 3 - h^0(\mathcal{E}); \tag{28}$$

$$\deg E = \tfrac{1}{2}K^2; \tag{29}$$

$$f_E = \deg E - h^0(\mathcal{E}) + 3; \tag{30}$$

Now set

$$\pi = \deg E - n - \nu; \tag{31}$$

$$= f_E - 2\nu; \tag{32}$$

adding $2n - \frac{1}{2}K^2$ to both sides of (31) gives

$$n - \nu = h^0(\mathcal{E}) - 3 \tag{33}$$

$$= \pi + 2n - \tfrac{1}{2}K^2. \tag{34}$$

Therefore

$$\deg E = \tfrac{1}{2}K^2 = n + \nu + \pi = 2\nu + \pi + h^0(\mathcal{E}) - 3$$

$$= 2\nu + 2\pi + (2n - \tfrac{1}{2}K^2) \geq 2\nu + 2\pi + 1 \tag{35}$$

and

$$f_E = 2\nu + \pi. \tag{36}$$

The final inequality $\pi \leq \nu$ comes easily from $\pi = \deg E - n - \nu$ and $\deg E \leq 2n - 1$, since (1, 2) imply that two copies of $\langle E \rangle = \mathbb{P}^\nu$ span $\mathbb{P}^n$, so that $2\nu + 1 \geq n$.

Remark. The exceptional cases $\pi = \nu-1, \nu$ of (3.11) only occur if $\deg E$ is close to $2n - 1$ and ν close to $(n-1)/2$; by (19–26), this corresponds to $h^0(\mathcal{E})$, $h^1(\mathcal{E})$, $h^1(I_E \cdot \mathcal{O}_X(K))$ and $h^1(I_E \cdot \mathcal{O}_X(2K))$ close to their maximum. It's quite likely that there are cleverer bounds on these groups than the trivial one using (1, 2) I have used.

(3.12) Final remarks. My feeling is that there is a vague analogy between Conjectures 0.2 and 1.5 and the subject of special linear systems on curves on K3 surfaces, another area of research that has lain dormant for around 10 years, and has recently been opened up again by Lazarsfeld's ideas using vector bundles [Reid1, Lazarsfeld, Green and Lazarsfeld].

In fact, an important part of the original motivation for [Reid1] was the idea (which goes back to Petri [11]) of trying to capture a special linear system on a curve C in terms of a variety of small degree $C \subset V \subset \mathbb{P}^{g-1}$ through the canonical curve; then if C is a hyperplane section of a K3 surface $X \subset \mathbb{P}^g$, the problem is to extend V to a variety $W \subset \mathbb{P}^g$ containing X, and intersections of quadrics are very relevant to this.

References

[Andreotti] A. Andreotti, On a theorem of Torelli, Amer. J. Math **80** (1958), 801–828.

[4 authors] E. Arbarello, M. Cornalba, P.A. Griffiths and J. Harris, Geometry of algebraic curves, vol. I, Springer, 1985.

[Babbage] D.W. Babbage, A note on the quadrics through a canonical curve, J. London Math Soc. **14** (1939), 310–315.

[Castelnuovo] G. Castelnuovo, Ricerche di geometria sulle curve algebriche, Atti R. Accad. Sci. Torino **24** (1889), 196–223.

[Ciliberto] C. Ciliberto, Hilbert functions of finite sets of points and the genus of a curve in a projective space, in Space curves (Rocca di Papa, 1985), F. Ghione, C. Peskine and E. Sernesi (eds.), LNM **1266** (1987), pp. 24–73.

[Eisenbud] D. Eisenbud, letter and private notes, c. 1979.

[Fano] G. Fano, Sopra le curve di dato ordine e dei massimi generi in uno spazio qualunque, Mem. Accad. Sci. Torino **44** (1894), 335–382.

[Harris] J. Harris, Curves in projective space, Séminaire de Math Sup. **85**, Presses Univ. Montréal, 1982.

[Horikawa1] E. Horikawa, Algebraic surfaces of general type with small c_1^2, II, Invent. Math **37** (1976), 121–155.

[Horikawa2] E. Horikawa, Algebraic surfaces of general type with small c_1^2, V, J. Fac. Sci. Univ. Tokyo, Sect IA Math **28** (1981), 745–755.

[Green and Lazarsfeld] M. Green and R. Lazarsfeld, Special divisors on curves on a K3 surface, Invent. math **89** (1987), 357–370.

[Lazarsfeld] R. Lazarsfeld, Brill–Noether–Petri without degenerations, J. diff. geom. **23** (1986), 299–307.

[Mumford] D. Mumford, Varieties defined by quadratic equations, in Questions on algebraic varieties (CIME conference proceedings, Varenna, 1969), Cremonese, Roma, 1970, pp.29–100.

[Petri] K. Petri, Über Spezialkurven I, Math. Ann. **93** (1925), 182–209.

[Reid1] M. Reid, Special linear systems on curves lying on a K3 surface, J. London math soc. **13** (1976), 454–458.

[Reid2] M. Reid, Surfaces with $p_g = 0, K^2 = 2$, unpublished manuscript and letters, 1977.

[Reid3] M. Reid, π_1 for surfaces with small K^2, in LNM **732** (1979), 534–544.

[Reid4] M. Reid, Surface of small degree, Math Ann. **275** (1986), 71–80.

[Sommese] A.J. Sommese, On the birational theory of hyperplane sections of projective threefolds, Notre Dame preprint, c. 1981.

[Tyurin] A.N. Tyurin, The geometry of the Poincaré theta-divisor of a Prym variety, Izv. Akad. Nauk SSSR, **39** (1975), 1003–1043 and **42** (1978), 468 = Math USSR Izvestiya **9** (1975), 951–986 and **12** (1978), No. 2.

[Segre] B. Segre, Su certe varietà algebriche intersezioni di quadriche od a sezioni curvilinee normali, Ann. Mat. Pura App. (4) **84** (1970), 125–155.

[Xiao] Xiao Gang, Hyperelliptic surfaces of general type with $K^2 < 4\chi$, Manuscripta Math **57** (1987), 125–148.

Miles Reid, Math Inst., Univ. of Warwick, Coventry CV4 7AL, England

Electronic mail: miles @ UK.AC.Warwick.Maths

Infinitesimal view of extending a hyperplane section - deformation theory and computer algebra

Miles Reid[1], University of Warwick

§0. *Alla marcia*

(0.1) The extension problem. Given a variety $C \subset \mathbb{P}^{n-1}$, I want to study extensions of C as a hyperplane section of a variety in $\mathbb{P}^n$:

$$\begin{matrix} C & \subset & \mathbb{P}^{n-1} \\ \cap & & \cap \\ X & \subset & \mathbb{P}^n \end{matrix} \qquad \text{with } C = \mathbb{P}^{n-1} \cap X;$$

that is, $C\colon (x_0 = 0) \subset X$, where x_0 is the new coordinate in $\mathbb{P}^n$. I will always take the intersection in the sense of homogeneous coordinate rings, which is a somewhat stronger condition than saying that C is the ideal-theoretical intersection $C = \mathbb{P}^{n-1} \cap X$.

(0.2) Some cases of varieties not admitting any extension were known to the ancients: for example, the Segre embedding of $\mathbb{P}^1 \times \mathbb{P}^2$ in $\mathbb{P}^5$ has no extensions other than cones because all varieties of degree 3 are classified ([Scorza1-2, XXX], compare [Swinnerton-Dyer]); and systematic obstructions of a topological nature to the existence of X were discovered from around 1976 by Sommese and others (see [Sommese1], [Fujita1], [Bădescu], [L'vovskii1-2]). More recent work of Sommese points to the conclusion that very few projective varieties C can be hyperplane sections; for example, Sommese [Sommese2-3] gives a detailed classification of the cases for which K_C is not ample when $C = \mathbb{P}^{n-1} \cap X$ is a smooth hypersection of a smooth 3-fold X; this amounts to numerical obstructions to the existence of a smooth extension of C in terms of the Mori cone of C.

(0.3) The infinitesimal view. Here I'm interested in harder cases, for example the famous problem of which smooth curves C of genus g lie on a K3 surface $C \subset X$; the *infinitesimal*

---

[1] Codice Fiscale: RDE MSN 48A30 Z114K

view of this problem is to study the schemes $C \subset 2C \subset 3C \subset ..$ which would be the Cartier divisors $kC : (x_0{}^k = 0) \subset X$ if X existed. Here each step is a linear problem in the solution to the previous one. For example, assuming that C is smooth, the first step is the vector space

$$\mathbb{H}^{(1)} = \{2C \subset \mathbb{P}^n \text{ extending } C\} = H^0(N_{\mathbb{P}^{n-1}|C}(-1))$$

or dividing out by coordinate changes,

$$\mathbb{T}^1_{(-1)} = \{2C \text{ extending } C\} = \operatorname{coker}\{H^0(T_{\mathbb{P}^{n-1}}(-1)) \to H^0(N_{\mathbb{P}^{n-1}|C}(-1))\}.$$

Singularity theorists know this as the graded piece of degree (-1) of the deformation space $\mathbb{T}^1$ of the cone over C. However, the extension from $(k-1)C$ up to kC is only an affine linear problem (there being no trivial or cone extension of $2C$); in particular 1st order deformations may be obstructed.

(0.4) This paper aims to sketch some general theory surrounding the infinitesimal view, and to make the link with deformation theory as practised by singularity theorists. My main interest is to study concrete examples, where the extension–deformation theory can be reduced to explicit polynomial calculation, giving results on moduli spaces of surfaces; for this reason, I have not taken too much trouble to work in intrinsic terms. It could be said that the authors of the intrinsic theory have not exactly gone out of their way to make their methods and results accessible.

The indirect influence on the material of §1 of Grothendieck and Illusie's theory of the cotangent complex [Grothendieck, Illusie2] will be clear to the experts (despite my sarcasm concerning their presentation); §1 can be seen as an attempt to spell out a worthwhile special case of their theory in concrete terms (compare also [Artin]), and I have groped around for years for the translation given in (1.15, 1.18, 1.21) of the enigma [Illusie1, (1.5–7)]. Thus even a hazy understanding of the Grothendieck ideology can be an incisive weapon, which I fear may not pass on to the next generation.

(0.5) Already considerations of 1st and 2nd order deformations lead even in reasonably simple cases to calculations that are too heavy to be moved by hand. An eventual aim of this work is to set up an algorithmic procedure to determine the irreducibility or otherwise of the moduli space of Godeaux surfaces with torsion $\mathbb{Z}/2$ or $\{0\}$, suitable for programming into computer algebra (although this paper falls short of accomplishing this); see §2 for this motivation and §6 for a 'pseudocode' description of a computer algebra algorithm that in principle calculates moduli spaces of deformations.

(0.6) Acknowledgements. The ideas and calculations appearing here have been the subject of many discussions over several years with Duncan Dicks, and I must apologise to him for the overlap between some sections of this paper and his thesis [Dicks]. I have derived similar (if less obviously related) benefit from the work of Margarida Mendes-Lopes [Mendes-Lopes]. I am very grateful to David Epstein for encouragement.

I would like to thank Fabrizio Catanese for persuading me to go the extremely enjoyable conference at L'Aquila, and Prof. Laura Livorni and the conference organisers for their hospitality. This conference and the British SERC Math Committee have provided me, in entirely different ways, with a strong challenge to express myself at length on this subject.

Contents

Chapter I. General theory

§1. The Hilbert scheme of extensions

This *overture in the French style* is mainly formalism, and the reader should skip through it rapidly, perhaps taking in the main theme Definition 1.7 and its development in Theorem 1.15; Pinkham's example in §2 gives a quick and reasonably representative impression of what's going on.

(1.1) Let $C, \mathcal{O}_C(1)$ be a polarised projective k–scheme (usually a variety), and

$$S = R(C, \mathcal{O}_C(1)) = \bigoplus_{i\geq 0} H^0(C, \mathcal{O}_C(i))$$

the corresponding graded ring. Suppose given a ring $\overline{R} \subset S$ of *finite colength*, that is, such that $S/\overline{R}$ is a finite–dimensional vector space. Often $\overline{R} = S$, but I do not assume this: for example, if $C \subset \mathbb{P}^{n-1}$ is a smooth curve that is not projectively normal, its homogeneous coordinate ring $\overline{R} = k[x_1,.. x_n]/I_C$ is of finite colength in $R(C, \mathcal{O}_C(1))$ (the normalisation of $\overline{R}$).

Throughout, a graded ring R is a graded k–algebra

$$R = \bigoplus_{i\geq 0} R_i$$

graded in positive degrees, with $R_0 = k$.

Main Problem. Given a graded ring $\overline{R}$ and $a_0 \in \mathbb{Z}$, $a_0 > 0$. Describe the set of pairs $x_0 \in R$, where R is a graded ring and $x_0 \in R_{a_0}$ a non–zerodivisor, homogeneous of degree a_0, such that

$$\overline{R} = R/(x_0).$$

Notice that since x_0 is a non–zerodivisor, the ideal $(x_0) = x_0R \cong R$. If R is given, then I write

$$R^{(k)} = R/(x_0{}^{k+1}),$$

and call $R^{(k)}$ the k*th order infinitesimal neighbourhood* of $\overline{R} = R^{(0)}$ in R.

This notation and terminology will be generalised in (1.8).

(1.2) The hyperplane section principle. Let R be a graded ring and $x_0 \in R$ a homogeneous non-zerodivisor of degree $\deg x_0 = a_0 > 0$; set $\overline{R} = R/(x_0)$. The hyperplane section principle says that quite generally, the generators, relations and syzygies of R reduce mod x_0 to those of $\overline{R} = R/(x_0)$, and in particular, occur in the same degrees. In more detail:

Proposition

(i) **Generators.** Quite generally, let $R = \oplus R_i$ be a graded ring, and $\overline{R} = R/(x_0)$, where $x_0 \in R_{a_0}$. Suppose that $\overline{R}$ is generated by homogeneous elements $x_1,.. x_n$ of degree $\deg x_i = a_i$; then R is generated by $x_0, x_1,.. x_n$. That is,

$$\overline{R} = k[x_1,.. x_n]/\overline{I} \Rightarrow R = k[x_0,.. x_n]/I,$$

where $\overline{I} \subset k[x_1,.. x_n]$ and $I \subset k[x_0,.. x_n]$ are the ideals of relations holding in $\overline{R}$ and R. (See (1.3, (3)) for the several abuses of notation involved in the x_i.)

(ii) **Relations.** Keep the notation and level of generality of (i). Suppose that $f(x_1,.. x_n) \in \overline{I}$ is a homogeneous relation of degree d holding in $\overline{R}$; then there is a homogeneous relation $F(x_0,.. x_n) \in I$ of degree d holding in R such that $F(0, x_1,.. x_n) \equiv f(x_1,.. x_n)$.

Let $f_1,.. f_m \in \overline{I}$ be a set of homogeneous relations holding in $\overline{R}$ that generates $\overline{I}$, and for each i, let $F_i(x_0,.. x_n) \in I$ be a homogeneous relation in R such that $F_i(0, x_1,.. x_n) \equiv f_i(x_1,.. x_n)$. Now assume that x_0 is a non-zerodivisor. Then $F_1,.. F_m$ generate I; that is,

$$\overline{I} = (f_1,.. f_n) \Rightarrow I = (F_1,.. F_n) \text{ with } F_i \mapsto f_i.$$

(iii) **Syzygies.** Quite generally, let $F_1,.. F_m \in k[x_0,.. x_n]$ be homogeneous elements, and consider the ideal $I = (F_1,.. F_m)$ and the quotient graded ring $R = k[x_0,.. x_n]/I$. For each i, write $f_i = F_i(0, x_1,.. x_n) \in k[x_1,.. x_n]$, and set $\overline{I} = (f_1,.. f_m)$ and

$$\overline{R} = R/(x_0) = k[x_1,.. x_n]/\overline{I}.$$

Then the following 3 conditions are equivalent:

(a) $x_0 \in R$ is a non-zerodivisor in R;

(b) $(x_0) \cap I = x_0 I \subset k[x_0,.. x_n]$;

(c) for every syzygy

$$\sigma\colon \sum_i \ell_i f_i \equiv 0 \in k[x_1,.. x_n]$$

between the f_i there is a syzygy

$$\Sigma\colon \sum_i L_i F_i \equiv 0 \in k[x_0,.. x_n]$$

between the F_i with $L_i(0, x_1,.. x_n) \equiv \ell_i(x_1,.. x_n)$.

(1.3) Remarks. (1) This is standard Cohen–Macaulay formalism, see for example [Mumford1] or [Saint-Donat, (6.6) and (7.9)]; everything works just as well if the non-zerodivisor x_0 is replaced by a regular sequence $(\xi_1,.. \xi_k)$.

(2) Recall the general philosophy of commutative algebra that 'graded is a particular case of local'. The assumption that R is graded and $a_0 > 0$ is used in every step of the argument to reduce the degree and make possible proofs by induction.

In the more general deformation situation $x_0 \in R$ or $x_0 \in H^0(\mathcal{O}_X)$, one must either assume that R or $\mathcal{O}_X$ is (x_0)-adically complete (for example (R, m) is a complete local ring and $x_0 \in m$); or honestly face the convergence problem of analytic approximation of formal structures. This is the real substance of Kodaira and Spencer's achievement in the global analytical context, and, in the algebraic setup, is one of the main themes of [Artin].

By (ii), R is determined by finitely many polynomials of given degree, so it depends *a priori* on a finite-dimensional parameter space. Morally speaking, rather than graded and degree < 0, the right hypothesis for the material of this section (and for the algorithmic routines of §6) should be that $\mathbb{T}^1$ and $\mathbb{T}^2$ are finite-dimensional.

(3) Abuse of notation. There are two separate abuses of notation in writing x_i: (a) the same x_i is used for the variables in the polynomial ring $k[x_1,.. x_n]$ and for the ring element $x_i = \operatorname{im} x_i \in R = k[x_1,.. x_n]/I$; there is no real ambiguity here, since I usually write $=$ for equality in R and $\equiv$ for identity of polynomials. (b) I identify the variables in the two polynomial rings $k[x_0,.. x_n]$ and $k[x_1,.. x_n]$; this means that there is a chosen lifting $k[x_1,.. x_n] \hookrightarrow k[x_0,.. x_n]$ of the quotient map $k[x_0,.. x_n] \to k[x_1,.. x_n] = k[x_0,.. x_n]/(x_0)$. Notice that from a highbrow point of view, I always work in a given trivial extension of a (smooth) ambient space (with a given retraction or 'face operator'), thus sidestepping the unspeakable if more intrinsic theory of the cotangent complex [Grothendieck, Illusie1, Illusie2, Lichtenbaum and Schlessinger].

(4) Higher syzygies for $\overline{R}$ extend to R in a similar way; in fact (1.2, ii–iii) can be lumped together as a more general statement on modules.

(5) The notation of (1.2) will be used throughout §1. I'll write $d_i = \deg f_i$ and $s_j =$

$\deg \sigma_j$.

(1.4) Proof of (1.2, i). Easy: mod x_0, every homogeneous $g \in R$ can be written as a polynomial in $x_1,.. x_n$, so that

$$g = g_0(x_1,.. x_n) + x_0 g',$$

where $g' \in R$ is of degree $\deg g - a_0 < \deg g$, and induction.

(1.5) Proof of (1.2, ii). It's traditional at this point to draw the commutative diagram

$$\begin{array}{ccccccccc} & & & & 0 & & 0 & & \\ & & & & \downarrow & & \downarrow & & \\ & & (x_0)\cap I & \longrightarrow & I & \longrightarrow & \bar{I} & & \\ & & \downarrow & & \downarrow & & \downarrow & & \\ 0 & \to & (x_0) & \longrightarrow & k[x_0,.. x_n] & \longrightarrow & k[x_1,.. x_n] & \to & 0 \\ & & \downarrow & & \downarrow & & \downarrow & & \\ 0 & \to & x_0R & \longrightarrow & R & \longrightarrow & \bar{R} & \to & 0 \\ & & \downarrow & & \downarrow & & \downarrow & & \\ & & 0 & & 0 & & 0 & & \end{array}$$

with exact rows and columns. Now $I \longrightarrow \bar{I}$ is surjective by the Snake Lemma. Take any $f \in \bar{I}$ homogeneous of degree d and $F \in I$ with $F \mapsto f$. Then $f = F - x_0 g$ (this uses the lift $f \in k[x_1,.. x_n] \subset k[x_0,.. x_n]$). If I take only the homogeneous piece of F and g of degree d then $f = F - x_0 g$ still holds, so $F \mapsto f$.

Now suppose $\{F_1,.. F_n\}$ are chosen to map to a generating set $\{f_1,.. f_n\}$ of $\bar{I}$, and let $G \in I$ be any homogeneous element. Then since $G \mapsto g \in \bar{I} = (f_1,.. f_n)$, I can write

$$g = \sum \ell_i f_i$$

with homogeneous $\ell_i \in k[x_1,.. x_n]$, so

$$H = G - \sum \ell_i F_i \in (x_0) \cap I.$$

Claim. If x_0 is a non-zerodivisor of R then $(x_0) \cap I = x_0 I$. Because

$$H = x_0 H' \in I \;\Rightarrow\; x_0 H' = 0 \text{ in } R \;\Rightarrow\; H' = 0 \text{ in } R \;\Rightarrow\; H' \in I.$$

Thus $G = \sum \ell_i F_i + x_0 G'$ with $G' \in I$, so I'm home by induction.

(1.6) Proof of (1.2, iii). (a) $\Longrightarrow$ (b) has just been proved, and $\Longleftarrow$ is just as elementary. I prove (b) $\Rightarrow$ (c). Write $F_i \equiv f_i + x_0 g_i$, and suppose the syzygy of $\overline{R}$ is $\sigma: \sum \ell_i f_i \equiv 0$. Then

$$I \ni \sum \ell_i F_i \equiv x_0 \sum \ell_i g_i \in (x_0),$$

so that (b) implies that $\sum \ell_i g_i \in I$, and so $\sum \ell_i g_i \equiv \sum m_i F_i$. Then

$$\sum L_i F_i \equiv 0, \quad \text{where} \quad L_i = (\ell_i - x_0 m_i).$$

Conversely, assume (c) and let $g \in k[x_0, \ldots x_n]$ be such that $x_0 g \equiv \sum \ell_i F_i$. Then $\sum \ell_i f_i \equiv 0$ so that by (c) there exist $L_i \mapsto \ell_i$ with $\sum L_i F_i \equiv 0$, and

$$x_0 g \equiv x_0 \sum (\ell_i - L_i) F_i,$$

so cancelling x_0 gives $g \in I$. Q.E.D.

(1.7) The Hilbert scheme of extensions of $\overline{R}$. This solves Problem 1.1: the set of rings R, $x_0 \in R_{a_0}$ such that $\overline{R} = R/(x_0)$ can be given as the set of polynomials F_i extending the relations f_i of $\overline{R}$ such that the syzygies σ_j extend to Σ_j.

To discuss this in more detail, fix once and for all the ring $\overline{R}$, its generators $x_1, \ldots x_n$, relations f_i and syzygies σ_j.

I also fix the polynomial ring $k[x_0, \ldots x_n]$ overlying R and discuss the set of extension rings R together with the data $\{F_i, \Sigma_j\}$ of relations and syzygies as in (1.2). Then

$$\left\{\forall R,\ \{F_i, \Sigma_j\} \mid R/(x_0) \cong \overline{R}\right\} \xleftrightarrow{\text{bij}} BH = \left\{ \begin{matrix} F_i = f_i + x_0 g_i \\ \Sigma_j\colon L_{ij} = \ell_{ij} + x_0 m_{ij} \end{matrix} \;\middle|\; \sum L_{ij} F_i \equiv 0 \right\}.$$

The set on the right-hand side has a natural structure of an affine scheme $BH = BH(\overline{R}, a_0)$, the *big Hilbert scheme of extensions* of $\overline{R}$. For the $g_i \in k[x_0, \ldots x_n]$ and $m_{ij} \in k[x_0, \ldots x_n]$ are finitely many polynomials of given degrees, so their coefficients are finite in number, and can be taken as coordinates in an affine space; the conditions $\sum L_{ij} F_i \equiv 0$ are then a finite set of polynomial relations on these coefficients.

Remark. The (small) Hilbert scheme

$$\mathbb{H}(\overline{R}, a_0) = \left\{\forall R, x_0 \mid R/(x_0) \cong \overline{R}\right\}$$

is part of primeval creation, so can't be redefined: it parametrises ideals $I \subset k[x_0,.. x_n]$ such that $(x_0) \cap I = x_0 I$ and $I/x_0 I = \bar{I}$, and is a locally closed subscheme of the Grassmannian of $I_{\le d} \subset k[x_0,.. x_n]_{\le d}$ (for some large d), as usual in the philosophy of [Mumford2].

Throwing away the extra data $\{F_i, L_{ij}\}$ corresponds to dividing out BH by an equivalence relation. This is rather harmless, and mainly a matter of notation: (1) the equivalence relation is of the form

$$F_i \sim F_i + x_0(\sum m_j F_j), \qquad \text{and similarly for the } \Sigma_j,$$

(because I'm only concerned with the ideal I generated by the F_i, and have fixed $f_i \equiv F_i \bmod x_0$), and is therefore given by a nilpotent group action; (2) by methods of Macaulay and Gröbner, any vector space associated with $k[x_0,.. x_n]$ can be given a preferred ordered monomial basis, in terms of which, for a fixed k-valued point $R \in \mathbb{H}(\bar{R}, a_0)$, there is a 'first choice' for the extra data $\{F_i, L_{ij}\}$; I will refer to $\{F_i, L_{ij}\}$ as the *coordinates* of the corresponding $R \in \mathbb{H}(\bar{R}, a_0)$.

(1.8) Definition. Suppose given a graded ring $\bar{R}$ and a degree a_0. A ring $R^{(k)}$ together with a homogeneous element $x_0 \in R_{a_0}$ such that $\bar{R} = R^{(k)}/(x_0)$ is a *kth order infinitesimal extension* of $\bar{R}$ if $x_0^{k+1} = 0$ and $R^{(k)}$ is flat over the subring $k[x_0]/(x_0^{k+1})$ generated by x_0.

Remark. Write $\mu_i = \mu_{x_0^i}: R^{(k)} \to R^{(k)}$ for the map given by multiplication by x_0^i; flatness is of course equivalent to saying that

$$\operatorname{im} \mu_{k+1-i} = \ker \mu_i \cong R^{(i-1)} = R^{(k)}/(x_0^i)$$

for each $0 < i \le k$. This is the nearest a nilpotent element of order k+1 comes to being a non-zerodivisor: x_0 kills only multiples of x_0^k.

The material of (1.2–7) can be easily rewritten in this context: to give $R^{(k)}$ is the same thing as to give relations $F_i \in k[x_0,.. x_n]/(x_0^{k+1})$ extending f_i and such that the syzygies extend mod x_0^{k+1}.

Thus the *big Hilbert scheme* $BH^{(k)} = BH^{(k)}(\bar{R}, a_0)$ *of kth order infinitesimal extensions of* $\bar{R}$ is given as in (1.7) as the affine scheme parametrising power series

$$F_i = f_i + x_0 f'_i + x_0^2 f''_i +.. x_0^k f^{(k)}_i$$

and

$$L_{ij} = \ell_{ij} + x_0\ell'_{ij} + .. x_0{}^k\ell^{(k)}{}_{ij}$$

for which the syzygies are satisfied up to kth order:

$$\sum L_{ij}F_i \equiv 0 \bmod x_0{}^{k+1}.$$

Similarly, the (small) Hilbert scheme $\mathbb{H}^{(k)} = \mathbb{H}^{(k)}(\overline{R}, a_0)$ parametrises ideals

$$I^{(k)} \subset k[x_0,.. x_n]/(x_0{}^{k+1}) \text{ with } (x_0) \cap I^{(k)} = x_0 I^{(k)} \text{ and } I^{(k)}/x_0 I^{(k)} = \overline{I}.$$

(1.9) There is no convergence problem. In the previous sections I have defined a Hilbert scheme $\mathbb{H}(\overline{R})$ parametrising extension of $\overline{R}$, and kth order Hilbert schemes $\mathbb{H}^{(k)}(\overline{R})$ parametrising kth order infinitesimal extensions $R^{(k)}$, with morphisms $\varphi_k\colon \mathbb{H}^{(k)} \to \mathbb{H}^{(k-1)}$ truncating $R^{(k)}$ into $R^{(k-1)} = R^{(k)}/(x_0{}^k)$. Now it is fortunate and obvious that $\mathbb{H}(\overline{R}) = \mathbb{H}^{(k)}(\overline{R})$ for sufficiently large k; this follows for reasons already described in (1.3, (2)). More precisely, the syzygies $\sum \ell_{ij}f_i \equiv 0$ are identities between homogeneous polynomials of given degrees, and F_i, L_{ij} have the same degrees as f_i and ℓ_{ij}; therefore, as soon as

$$(k+1)\cdot a_0 = \deg x_0{}^{k+1} > \max(\deg \sum \ell_{ij}f_i),$$

I get the implication

$$\sum L_{ij}F_i \equiv 0 \bmod x_0{}^{k+1} \implies \sum L_{ij}F_i \equiv 0.$$

Notice that, for reasons mentioned in (1.3, (2)), the Hilbert schemes of (1.7–8) are finite-dimensional with no restrictions on singularities. This contrasts with deformation theory, where for example a cone over a singular curve already has infinite-dimensional versal deformation; of course, I'm only trading in the negatively graded portion of the deformation theory.

Actually, most of what I do in this paper will work with minor modifications for deformations in degree 0 (that is, considering algebras R or $R^{(k)}$ over $k[[\lambda]]$ or $k[\lambda]/(\lambda^{k+1})$ that are flat deformations of $\overline{R}$ and are graded with $\deg \lambda = 0$). Then the convergence problem is nontrivial, but well understood: the Hilbert scheme of $\overline{R}$ is a (bounded) projective k-scheme H, and a formal deformation R over $k[[\lambda]]$ is a formal curve in H, so can be analytically or algebraically approximated.

In the remainder of this section I give a concrete description of the tower of schemes

$$\mathbb{H} \longrightarrow .. \ \mathbb{H}^{(k)} \longrightarrow \mathbb{H}^{(k-1)} \longrightarrow .. \ \mathbb{H}^{(1)} \longrightarrow \mathbb{H}^{(0)} = \text{pt}.$$

in terms of $\overline{R}$ and the $\overline{R}$-modules $\overline{I}/\overline{I}^2$ (the 'conormal sheaf' to $\overline{R}$); see Theorem 1.15. It is worth treating the 1st order case separately, although logically it is covered by the higher order statement (1.15).

(1.10) 1st order theory

Theorem. $\mathbb{H}^{(1)}(\overline{R}, a_0) = \mathrm{Hom}_{\overline{R}}(\overline{I}/\overline{I}^2, \overline{R})_{-a_0}$.

Notes. (a) The subscript refers of course to the degree $-a_0$ piece of the graded module Hom, consisting of $\overline{R}$-linear maps of degree $-a_0$.

(b) If it helps at all, think of the right-hand side as the group of splittings of

$$0 \to x_0\overline{R} \longrightarrow \overline{I}/\overline{I}^2 \oplus x_0\overline{R} \longrightarrow \overline{I}/\overline{I}^2 \to 0.$$

where the middle term is $(\overline{I}, x_0)/(\overline{I}, x_0)^2 \subset k[x_0,.. x_n]/(\overline{I}, x_0)^2$ split by the lifting (1.3, (3)).

(c) For the relation with $\mathbb{T}^1$ see (1.22) below.

Coordinate-free proof of the theorem. Write $A = k[x_1,.. x_n]$ for brevity. Starting from $\overline{I} \subset A$, look for vector subspaces

$$I^{(1)} \subset A \oplus x_0A$$

satisfying

(a) $I^{(1)} + x_0A = \overline{I} \oplus x_0A$;

(b) $x_0A \cap I^{(1)} = x_0I^{(1)} = 0 \oplus x_0\overline{I}$;

(c) $I^{(1)}$ is an ideal of $A \oplus x_0A$.

Subgroups $I^{(1)}$ enjoying (a) and (b) are in bijection with maps $\varphi\colon \overline{I} \longrightarrow \overline{R}$ by a standard graph argument: if $I^{(1)}$ is given, then by (a), for all $f \in \overline{I}$ there exists f' such that

$f + x_0 f' \in I^{(1)}$, and by (b), f' is uniquely determined $\bmod \bar{I}$, so $f \mapsto f' \bmod \bar{I}$ defines $\varphi\colon \bar{I} \to \bar{R}$. The inverse construction is obvious.

Finally, it's an easy calculation to see that $I^{(1)}$ is an ideal if and only if φ is A–linear. Graded ideals $I^{(1)}$ correspond to graded maps φ of degree $-a_0$.

(1.11) Coordinate proof of (1.10). It is useful to have a coordinate proof, both for algorithmic applications, and to make the link with the projective resolution of $\bar{I}/\bar{I}^2$ in (1.13–17). By construction,

$$BH^{(1)}(\bar{R}, a_0) = \left\{(f_i+x_0f'_i), (\ell_{ij}+x_0\ell'_{ij}) \mid \forall j,\ \sum (\ell_{ij}+x_0\ell'_{ij})(f_i+x_0f'_i) \equiv 0 \bmod x_0^2\right\},$$

since $\sum \ell_{ij}f_i \equiv 0$ is given, strip off this term and divide through by x_0 (thus reducing degrees by a_0); then the defining equation is

$$\sum \ell_{ij}f'_i + \sum \ell'_{ij}f_i \equiv 0 \quad \text{in} \quad A_{s_j - a_0},$$

where $s_j = \deg \Sigma_j$. Since the second summand is an arbitrary element of $\bar{I}$, the condition on $\{f'_i\}$ is

$$\sum \ell_{ij}f'_i = 0 \quad \text{in} \quad \bar{R}_{s_j - a_0} \quad \text{(for all } j).$$

Now $\bar{I}$ is the A–module generated by $\{f_i\}$ with relations $\sum \ell_{ij}f_i = 0$, so that $f_i \mapsto f'_i$ defines an A–linear map $\varphi\colon \bar{I} \to \bar{R}$, hence an element $\varphi \in \mathrm{Hom}_{\bar{R}}(\bar{I}/\bar{I}^2, \bar{R})_{-a_0}$.

Clearly, φ only depends on $f'_i \in \bar{R}$; the equivalence relation $\sim$ on $\{(f'_i, \ell'_{ij})\} \in BH^{(1)}(\bar{R}, a_0)$ defining $\mathbb{H}^{(1)}(\bar{R}, a_0)$ ignores the ℓ'_{ij} completely, and takes account only of the classes $f'_i \in \bar{R}$ of the $f'_i \bmod \bar{I}$ (as described in (1.7)). It is now easy to see that there is a bijection between the 3 sets:

(1) $BH^{(1)}(\bar{R}, a_0)/\sim$;

(2) ideals $I^{(1)} \subset k[x_0, x_1, \dots x_n]/(x_0^2)$ generated by $\{f_i + x_0 f'_i\}$;

(3) $\mathrm{Hom}_{\bar{R}}(\bar{I}/\bar{I}^2, \bar{R})_{-a_0}$. Q.E.D.

(1.12) Preparations for the main theorem. Higher order deformation theory studies extensions of an individual ring $R^{(k-1)}$ to $R^{(k)}$, and on the level of the Hilbert schemes, the morphisms $\varphi_k\colon \mathbb{H}^{(k)} \to \mathbb{H}^{(k-1)}$ defined by the forgetful maps $R^{(k)} \mapsto R^{(k-1)}$.

I first indicate briefly why the higher order problem differs from the 1st. To extend the ring $R^{(1)}$ to $R^{(2)}$ one takes account of new terms in x_0^2; however, x_0 is already involved in $R^{(1)}$, so that x_0^2 is not just a coordinate in a transverse extension of the ambient space. A picture in the simplest case may help:

$k[x]/x^3$ $\qquad\qquad$ $k[x, y]/(x, y)^2$.

(1.13) The extension problem in coordinates and the obstruction ψ. The forgetful maps $R^{(k)} \mapsto R^{(k-1)}$ define morphisms φ_k: $BH^{(k)} \longrightarrow BH^{(k-1)}$ that take the kth order power series

$$f_i + x_0 f'_i + x_0^2 f''_i + .. \, x_0^k f^{(k)}_i \quad \text{and} \quad \ell_{ij} + x_0 \ell'_{ij} + .. \, x_0^k \ell^{(k)}_{ij} \tag{1}$$

into their (k−1)st order truncations

$$F_i = f_i + x_0 f'_i + .. \, x_0^{k-1} f^{(k-1)}_i \quad \text{and} \quad L_{ij} = \ell_{ij} + x_0 \ell'_{ij} + .. \, x_0^{k-1} \ell^{(k-1)}_{ij}. \tag{2}$$

Now suppose given (k−1)st order power series F_i and L_{ij} satisfying

$$\sum L_{ij} F_i \equiv 0 \mod x_0^k; \tag{3}$$

extending these to kth order power series satisfying the same $\bmod\, x_0^{k+1}$ is clearly equivalent to fixing up the new kth order terms $f^{(k)}_i$ and $\ell^{(k)}_{ij}$ to satisfy

$$\sum \ell_{ij} f^{(k)}_i + \sum_{a=1}^{k-1} \sum \ell^{(a)}_{ij} f^{(k-a)}_i + \sum \ell^{(k)}_{ij} f_i \equiv 0 \tag{4}$$

(this has involved using (3) to kill terms not divisible by x_0^k, then dividing through by x_0^k, thus lowering degrees by ka_0); that is,

$$\sum \ell_{ij} f^{(k)}_i + \sum \ell^{(k)}_{ij} f_i \equiv - \sum_{a=1}^{k-1} \sum \ell^{(a)}_{ij} f^{(k-a)}_i. \tag{5}$$

This is a set of inhomogeneous linear equations in the new unknowns $f^{(k)}_i$ and $\ell^{(k)}_{ij}$. As in (1.11), the second term on the left-hand side is just an arbitrary element of $\bar{I}_{s_j - ka_0}$, so working $\bmod\, \bar{I}$ allows me to forget it:

$$\sum \ell_{ij} f^{(k)}_i = - \sum_{a=1}^{k-1} \sum \ell^{(a)}_{ij} f^{(k-a)}_i = \psi_j \in \bar{R}_{s_j - ka_0}. \tag{6}$$

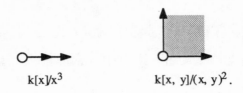

(5) are of course the equations of $BH^{(k)}$ as a scheme over $BH^{(k-1)}$. This is all one needs from the point of view of practical computation: the right-hand side of (6) is a given vector $\psi = \{\psi_j\} \in \oplus_j \bar{R}_{s_j - ka_0}$ (depending in a bilinear way on the given (k−1)st order power series F_i and L_{ij}). Maybe the left-hand side fails to hit ψ at all, so there's an obstruction to extending $R^{(k-1)}$. But if it does hit ψ, the ambiguity in the choice of $f^{(k)}{}_i$ is the vector space $\left\{(f^{(k)}{}_i \mid \sum \ell_{ij} f^{(k)}{}_i = 0 \in \bar{R}_{s_j - ka_0}\right\}$; this space depends only on $\bar{R}$ (not on $R^{(k-1)}$), and is the same space as in (1.11) with a_0 replaced by ka_0. This shows that $BH^{(k)} \to BH^{(k-1)}$ is an affine fibre bundle over its image.

Definition. Write

$$\psi\colon BH^{(k-1)}(\bar{R}, a_0) \to \oplus \bar{R}(s_j)_{-ka_0}$$

for the polynomial map defined by $\psi_j = -\sum_{a=1}^{k-1} \sum_i \ell^{(a)}{}_{ij} f^{(k-a)}{}_i \bmod \bar{I}$ (see (6)).

(1.14) Notation. For brevity, write $A = k[x_1, \dots x_n]$, and let

$$\cdots \xrightarrow{(m_{jn})} \oplus A(-s_j) \xrightarrow{(\ell_{ij})} \oplus A(-d_i) \xrightarrow{(f_i)} \bar{I} \to 0$$

be the projective resolution of $\bar{I}$ as A-module. (Recall $d_i = \deg f_i$ and $s_j = \deg \sigma_j$; the twistings $A(-d_i)$ etc. are a traditional device to make the homomorphisms graded of degree 0.)

Then $\mathrm{Ext}^i_{\bar{R}}(\bar{I}/\bar{I}^2, \bar{R})$ is the homology of the dual complex

$$0 \to \oplus \bar{R}(d_i) \xrightarrow{\delta_0} \oplus \bar{R}(s_j) \xrightarrow{\delta_1} \cdots$$

(the *conormal complex* of $\bar{R} = A/\bar{I}$), and in particular there is an exact sequence

$$0 \to \mathrm{Hom}_{\bar{R}}(\bar{I}/\bar{I}^2, \bar{R}) \to \oplus \bar{R}(d_i) \xrightarrow{\delta_0} \ker \delta_1 \xrightarrow{\pi} \mathrm{Ext}^1_{\bar{R}}(\bar{I}/\bar{I}^2, \bar{R}) \to 0.$$

(1.15) Main theorem. (i) $\psi\colon BH^{(k-1)} \to \oplus \bar{R}(s_j)_{-ka_0}$ factors through a morphism of schemes $\Psi\colon \mathbb{H}^{(k-1)}(\bar{R}, a_0) \to \oplus \bar{R}(s_j)_{-ka_0}$ (the target is a finite-dimensional vector space viewed as an affine variety).

(ii) $\Psi\colon \mathbb{H}^{(k-1)}(\overline{R}, a_0) \longrightarrow \ker \delta_1 \subset \oplus \overline{R}(s_j)_{-ka_0}$.

Therefore, the middle square in the diagram

$$\begin{array}{ccccccccc} & & \mathbb{H}^{(k)}(\overline{R}, a_0) & \xrightarrow{\varphi_k} & \mathbb{H}^{(k-1)}(\overline{R}, a_0) & & & & \\ & & \downarrow & & \downarrow \Psi & & & & \\ 0 \to \mathrm{Hom}_{\overline{R}}(\overline{I}/\overline{I}^2, \overline{R})_{-ka_0} & \to & \oplus \overline{R}(d_i)_{-ka_0} & \xrightarrow{\delta_0} & (\ker \delta_1)_{-ka_0} & \xrightarrow{\pi} & \mathrm{Ext}^1_{\overline{R}}(\overline{I}/\overline{I}^2, \overline{R})_{-ka_0} & \to 0 \end{array}$$

is Cartesian.

In other words, the morphism $\varphi_k\colon \mathbb{H}^{(k)}(\overline{R}, a_0) \longrightarrow \mathbb{H}^{(k-1)}(\overline{R}, a_0)$ has the following structure: its image $\mathrm{im}\,\varphi_k$ is the scheme-theoretic fibre over 0 of the morphism $\pi\circ\Psi\colon \mathbb{H}^{(k-1)}(\overline{R}, a_0) \longrightarrow \mathrm{Ext}^1_{\overline{R}}(\overline{I}/\overline{I}^2, \overline{R})_{-ka_0}$; so $\pi\circ\Psi(R^{(k-1)})$ is the *obstruction to extending* $R^{(k-1)}$ *to* k*th order*. And $\varphi_k\colon \mathbb{H}^{(k)}(\overline{R}, a_0) \longrightarrow \mathrm{im}\,\varphi_k \subset \mathbb{H}^{(k-1)}(\overline{R}, a_0)$ is a fibre bundle (in the Zariski topology), with fibre an affine space over $\mathrm{Hom}_{\overline{R}}(\overline{I}/\overline{I}^2, \overline{R})$.

Given (i) and (ii), the Cartesian square just restates equations (6): a point $R^{(k)} \in \mathbb{H}^{(k)}(\overline{R}, a_0)$ over $R^{(k-1)}$ is $\{f^{(k)}{}_i \in \overline{R}(d_i - ka_0)\}$ which maps to $\Psi(R^{(k-1)})$ on taking $\sum \ell_{ij} f^{(k)}{}_i$, that is, under δ_0.

This can only exist if $\Psi(R^{(k-1)})$ is a boundary, so maps to 0 in Ext^1. Thus the image of $\mathbb{H}^{(k)}(\overline{R}, a_0) \longrightarrow \mathbb{H}^{(k-1)}(\overline{R}, a_0)$ is the fibre over 0 of the composite morphism $\pi\circ\Psi\colon \mathbb{H}^{(k-1)}(\overline{R}, a_0) \longrightarrow \ker \delta_1 \longrightarrow \mathrm{Ext}^1_{\overline{R}}(\overline{I}/\overline{I}^2, \overline{R})$ given by the homology class of ψ; its fibres (when nonempty) are affine spaces under

$$\left\{f^{(k)}{}_i \in \overline{R}(d_i - ka_0) \mid \sum \ell_{ij} f^{(k)}{}_i \equiv 0\right\} = \ker \delta_0 = \mathrm{Hom}^1_{\overline{R}}(\overline{I}/\overline{I}^2, \overline{R})_{-ka_0}.$$

So the only points to prove are (i) and (ii). I first give set-theoretic proofs, working with a $(k-1)$st order infinitesimal extension ring $R^{(k-1)} \in \mathbb{H}^{(k-1)}(\overline{R}, a_0)$ defined over the base field k (sorry about the notation clash), and choosing a lift to $\{F_i, L_{ij}\} \in BH^{(k-1)}(\overline{R}, a_0)$. In fact the same proof works scheme-theoretically; the necessary technicalities are not hard, but are left to (1.20) to allow the reader to skip them.

(1.16) Proof of (i). This is similar to the proof of (1.11). Suppose that (F_i, L_{ij}) and $('F_i, 'L_{ij})$ are two choices of coordinates for $R^{(k-1)}$, with $\left\{F_i, 'F_i \in I^{(k-1)}{}_{d_i}\right\}$ and

$$\sum L_{ij}F_i \text{ and } \sum ('L_{ij})('F_i) \equiv 0 \mod x_0^k.$$

Then the difference

$$\sum L_{ij}F_i - \sum ('L_{ij})('F_i) \in x_0^k \cap I^{(k-1)} = x_0^k\bar{I}.$$

Since ψ_j is just minus the coefficient of x_0^k in $\sum L_{ij} F_i$ reduced mod $\bar{I}$, this proves that it depends only on $I^{(k-1)}$, or $R^{(k-1)} \in \mathbb{H}^{(k-1)}(\bar{R}, a_0)$, so Ψ is well defined.

(1.17) Proof of (ii). I introduce notation for the 2nd syzygies involved in δ_1. These form the next term in the projective resolution of $\bar{I}$ in (1.14), and are identities of the form

$$\mu\colon \sum_j m_j \ell_{ij} \equiv 0$$

holding between the 1st syzygies $\sum_i \ell_{ij} f_i \equiv 0$. Suppose that $\mu_n = (m_{jn})$ is a generating set. Then as mentioned in (1.3, (4)), these identities lift to identities

$$\sum_j M_{jn}L_{ij} = \sum_j (m_{jn} + x_0 m'_{jn} + .. \, x_0^{k-1} m^{(k-1)}{}_{jn})(\ell_{ij} + .. \, x_0^{k-1}\ell^{(k-1)}{}_{ij}) \equiv 0 \mod x_0^k$$

between the 1st syzygies for $R^{(k-1)}$. Now hold on tight, please: write Δ_n for the x_0^k term in

$$\sum_j \left(M_{jn} \sum_i L_{ij}F_i\right) = \sum_i \left(\sum_j M_{jn}L_{ij}F_i\right).$$

I'm going to calculate Δ_n on the two sides. Working on the left, since $\sum_i L_{ij} F_i \equiv -\psi_j x_0^k \mod x_0^{k+1}$ and $M_{jn} \equiv m_{jn} \mod x_0$, the coefficient of x_0^k is $\Delta_n = -\sum_j m_{jn}\psi_j$. On the right, since $\sum_j M_{jn}L_{ij} \equiv 0 \mod x_0^k$ I can pick out the leading term and write

$$\sum_j M_{jn}L_{ij} \equiv \theta_{ni}x_0^k \mod x_0^{k+1};$$

then since $F_i \equiv f_i \mod x_0$, it follows that $\Delta_n = \sum_i \theta_{ni}f_i \in \bar{I}$. Now I've won: δ_1 is the map given on $\oplus \bar{R}(s_j)_{-ka_0}$ by the matrix (m_{jn}), and I've just proved that each coefficient

$$\Delta_n = -\sum_j m_{jn}\psi_j = \sum_i \theta_{ni}f_i = 0 \in \bar{R}. \quad \text{Q.E.D.}$$

(1.18) Normal structure to $R^{(k-1)}$ and obstruction calculus. It is interesting to give an alternative coordinate-free treatment of what Theorem 1.15 means for a given ring $R^{(k-1)}$ (that is, a given k-valued point of $\mathbb{H}^{(k-1)}$).

Same Result. kth order extensions $R^{(k)}$ of a given (k−1)st order extension ring $R^{(k-1)}$ are in bijection with splittings of a certain 'normal' sequence

$$0 \to x_0^k\bar{R} \longrightarrow N \longrightarrow \bar{I}/\bar{I}^2 \to 0 \qquad (*)$$

of graded $\bar{R}$-modules deduced from $R^{(k-1)}$.

Compare with (1.10, Note b). This gives the obstruction $\psi(R^{(k-1)}) \in \mathrm{Ext}^1_{\bar{R}}(\bar{I}/\bar{I}^2, \bar{R})_{-ka_0}$ to the existence of an extension, and if it vanishes, the set of extensions is an affine space under $\mathrm{Hom}_{\bar{R}}(\bar{I}/\bar{I}^2, \bar{R})_{-ka_0}$.

(1.19) Proof. Step 1 (notation and set-up). I start from the ideal

$$J = I^{(k-1)} \subset A[x_0]/(x_0^k) = A \oplus x_0A \oplus .. x_0^{k-1}A$$

defining $R^{(k-1)}$ (so satisfying $J \cap x_0^{k-1}A = x_0^{k-1}J = x_0^{k-1}\bar{I}$). Introduce subgroups

$$L = (\bar{I} + (x_0))\cdot_{new}J \subset M = J \oplus x_0^kA \subset A \oplus x_0A \oplus .. x_0^kA.$$

Here M is the inverse of J under $A[x_0]/(x_0^{k+1}) \longrightarrow A[x_0]/(x_0^k)$, hence an ideal. For L, note that the old multiplication by x_0 in $A[x_0]/(x_0^k)$ does not correspond to multiplication in $A[x_0]/(x_0^{k+1})$; also, $J = J \oplus 0 \subset A[x_0]/(x_0^{k+1})$ is not an ideal (I'm paying here for the abuse of notation (1.3, (3))). So I'm defining L (*a priori* only a subgroup) by

$$L = \bar{I}\cdot J + x_0\cdot_{new}J;$$

the first summand is $\bar{I}\cdot J \oplus 0$, and in the second, the operation $x_0\cdot_{new}$ just shifts down the terms

$$f + x_0f' + .. x_0^{k-1}f^{(k-1)} \in J \mapsto x_0f + x_0^2f' + .. x_0^kf^{(k)}$$

without killing anything off.

Step 2 (definition of $\bar{R}$-module N). (i) L is an ideal of $A[x_0]/(x_0^{k+1})$; (ii) multiplication by x_0 and $\bar{I}$ both take M into L. Therefore the quotient $N = M/L$ is an $\bar{R}$-module.

Proof. (i) Elements of A multiply each summand of L to itself, and x_0 multiplies the first summand into the second, so OK. (ii) Similarly, x_0 and $\bar{I}$ both multiply the first summand $J \oplus 0$ of M to L; the second summand x_0^kA gets killed by x_0 and multiplied into $x_0\cdot_{new}J$ by $\bar{I}$, because $\bar{I} = J \bmod x_0$.

Step 3 (exact sequence (*)). (i) $x_0^kA \cap L = x_0^k\,\bar{I}$; (ii) $x_0^kA + L = (\bar{I}J + x_0J) \oplus x_0^kA \subset J \oplus x_0^kA$. Therefore the module N defined in Step 2 fits into the exact sequence

$$0 \to x_0^k\bar{R} \longrightarrow N \longrightarrow \bar{I}/\bar{I}^2 \to 0 \qquad (*)$$

of graded $\bar{R}$-modules, induced by the split sequence $x_0^kA \hookrightarrow M \twoheadrightarrow J$ in the definition of M.

Proof. (i) Since $x_0^kA \cap \bar{I}J = \emptyset$, this is clear:

$$x_0^kA \cap x_0\cdot_{new}J = x_0(x_0^{k-1}A \cap J) = x_0^k\bar{I}.$$

(ii) Adding x_0^kA kills the difference between $\cdot_{new}$ and $\cdot_{old}$, so

$$L + x_0^kA = \bar{I}J + x_0\cdot_{new}J + x_0^kA = (\bar{I}J + x_0J) \oplus x_0^kA.$$

The terms of the exact sequence follow because $x_0^kA/x_0^k\bar{I} = x_0^k\bar{R}$ and

$$\{J \oplus x_0^kA\}/\{(\bar{I}J + x_0J) \oplus x_0^kA\} = J/(\bar{I}J + x_0J) = \bar{I}/\bar{I}^2.$$

Step 4 (graph argument as in (1.10)). Starting from

$$J = I^{(k-1)} \subset A[x_0]/(x_0^k) = A \oplus x_0A \oplus .. \, x_0^{k-1}A,$$

look for subgroups

$$J' = I^{(k)} \subset M = J \oplus x_0^kA \subset A[x_0]/(x_0^{k+1}) = A \oplus x_0A \oplus .. \, x_0^kA,$$

satisfying

(a) $J' + x_0^kA = J \oplus x_0^kA$;

(b) $x_0^kA \cap J' = x_0^k\bar{I}$;

(c) J' is an ideal of $A \oplus x_0A \oplus .. \, x_0^kA$.

As in (1.10), subgroups J' for which (a) and (b) hold correspond bijectively with maps $\varphi\colon J \longrightarrow \bar{R}$. For if J' is given, then by (a), for all $F = f + x_0f' + .. \, x_0^{k-1}f^{(k-1)} \in J$ there exists $f^{(k)}$ such that $F + x_0^kf^{(k)} \in J'$, and by (b), $f^{(k)}$ is uniquely determined $\bmod \bar{I}$, so $F \mapsto f^{(k)} \bmod \bar{I}$ defines $\varphi\colon J \longrightarrow \bar{R}$. The map φ is not quite what I want to express that J' is an ideal. Try instead the following:

$$\Phi\colon J \longrightarrow (J \oplus x_0^kA)/L = N \quad \text{by} \quad F \mapsto \Phi(F) = F + x_0^kf^{(k)} \bmod L.$$

After passing to quotients as in Step 3, any map of this form (with first component F) obviously splits $(*)$ as a sequence of vector spaces; also, by Step 3, (i), knowing $f^{(k)} \bmod \bar{I}$ is equivalent to $\Phi(F) = F + x_0{}^k f^{(k)} \bmod L$, so that maps Φ also correspond bijectively with subspaces J' satisfying (a) and (b).

Step 5. J' is an ideal if and only if Φ is $A[x_0]/(x_0{}^k)$-linear.

It is trivial to check that Φ is A-linear if and only if $AJ' \subset J'$, so the point is to deal with the multiplications (old and new) by x_0. It's easy to check in turn that the following 3 conditions are equivalent:

(I) J' is closed under multiplication by x_0;

(II) $f + x_0 f' + x_0{}^2 f'' +.. x_0{}^k f^{(k)} \in J' \Rightarrow x_0 f + x_0{}^2 f' +.. x_0{}^k f^{(k-1)} \in J'$;

(III) $F = f + x_0 f' +.. x_0{}^{k-1} f^{(k-1)} \in J \Rightarrow \Phi(x_0 F) = 0$.

Since $\Phi\colon J \longrightarrow (M \oplus x_0{}^k A)/L =$ is $A[x_0]/(x\,{}_0{}^k)$-linear if and only if the induced splitting $J/((\bar{I}J + x_0 J)) = \bar{I}/\bar{I}^2 \longrightarrow N$ is $\bar{R}$-linear, this completes the proof of (1.13). Q.E.D.

Technical appendix

(1.20) Scheme-theoretic proof of (1.15). Required to prove that all the maps in

$$\begin{array}{ccc} BH^{(k-1)} & & \\ \downarrow & \searrow & \\ \mathbb{H}^{(k-1)} & & \\ \downarrow & & \\ \ker \delta_1 & \subset & \oplus \bar{R}(s_j)_{-ka_0} \end{array}$$

are morphisms of schemes. The proof given in (1.16–17) is inadequate because it deals only with k-valued points of $BH^{(k-1)}$ and $\mathbb{H}^{(k-1)}$.

Thus the point is just to use the proper definition of the Hilbert scheme $\mathbb{H}^{(k-1)}$ as the functor of ideals extending $\bar{I}$. More precisely, this is the functor on k-algebras S

$$S \mapsto \left\{ S\text{-submodules } J^{(k-1)} \subset S[x_0,.. x_n]/(x_0{}^k) \text{ s.t. (a–c) hold} \right\},$$

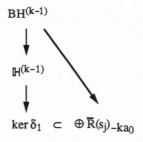

where (a) $J^{(k-1)} + (x_0) = \bar{I}\cdot S[x_1,.. x_n] \oplus (x_0)$; (b) $(x_0) \cap J^{(k-1)} = x_0 J^{(k-1)}$; and (c) $J^{(k-1)}$ is an ideal, that is $x_i J^{(k-1)} \subset J^{(k-1)}$ for each i. Since these conditions are locally closed, $\mathbb{H}^{(k-1)}$ is a representable functor, naturally represented by a subscheme of the Grassmannian of linear subspaces of $\left(k[x_0,.. x_n]/(x_0{}^k)\right)_{\le d}$ for some large d.

There is a universal sheaf of ideals $\underline{J}^{(k-1)} \subset \mathcal{O}_{\mathbb{H}^{(k-1)}}[x_0,.. x_n]/(x_0{}^k)$ over $\mathbb{H}^{(k-1)}$ defined by tautological Grassmannian considerations, and the proof of (1.15) just consists of applying the arguments of (1.16–17) to the stalks of $\underline{J}^{(k-1)}$. That is, let $P \in \mathbb{H}^{(k-1)}$ be a scheme-theoretic point, $S = \mathcal{O}_{\mathbb{H}^{(k-1)},P}$ its local ring and $J^{(k-1)} \subset S[x_0,.. x_n]/(x_0{}^k)$ the stalk of $\underline{J}^{(k-1)}$; then, taking F_i to be elements of $S[x_0,.. x_n]/(x_0{}^k)$ generating $J^{(k-1)}$ and similarly for L_{ij}, M_{jn} etc., the arguments of (1.16–17) go through without modification.

(1.21) The obstruction as a cup product. The fact that the obstruction to extending $R^{(k-1)}$ is bilinear in the coordinates F_i and L_{ij} (see (1.13)) has a high brow interpretations (see [Illusie1, (1.7, iii)]). Consider the diagram

$$\begin{array}{ccc} R^{(k-1)} & \longleftarrow & ? \\ \uparrow & & \uparrow \\ k[x_0]/(x_0{}^k) & \longleftarrow & k[x_0]/(x_0{}^{k+1}) \\ \uparrow & & \\ k & & \end{array}$$

The $\{L_{ij}\}$ can be interpreted as a normal or 'Kodaira-Spencer' class of $R^{(k-1)}$ over $k[x_0]/(x_0{}^k)$. The kernel of $k[x_0]/(x_0{}^{k+1}) \longrightarrow k[x_0]/(x_0{}^k)$ is the 1-dimensional vector space $V = \langle x_0{}^k \rangle$; since multiplication by $x_0{}^k$ is nonzero, this extension has a nonzero class, which pulls back to $R^{(k-1)}$ as multiplication by the F_i. The obstruction can be thought of as a cup product of these two classes.

(1.22) Coordinate changes, $\mathbb{T}^1$ and moduli space of extensions. The Hilbert schemes $\mathbb{H}(\overline{R}, a_0)$ and $\mathbb{H}^{(k)}(\overline{R}, a_0)$ contain every extension of $\overline{R}$, but in a redundant way. One can define the *moduli space* of extensions, and compute it as $\mathbb{H}$ or $\mathbb{H}^{(k)}$ divided out by an equivalence relation given by coordinate changes of the form

$$x_h \mapsto x_h + \alpha_h x_0 \quad \text{with} \quad \deg \alpha_h = a_h - a_0.$$

Thus the answer to the 1st order extension problem for smooth $C \subset \mathbb{P}^{n-1}$ is

$$\mathbb{T}^1{}_{(-1)} = \{2C \text{ extending } C\} = \operatorname{coker}\left\{H^0(T_{\mathbb{P}^{n-1}}(-1)) \longrightarrow H^0(N_{\mathbb{P}^{n-1}|C}(-1))\right\}.$$

here $H^0(N_{\mathbb{P}^{n-1}|C}(-1)) = \mathbb{H}^{(1)}$ is the set of subschemes $2C \subset \mathbb{P}^n$ extending C in a fixed coordinate system $(x_0, x_1, .. x_n)$. Passing to the cokernel consists exactly of dividing out by $H^0(T_{\mathbb{P}^{n-1}}(-1))$, corresponding to coordinate changes $x_h \mapsto x_h + \alpha_h x_0$ fixing $\mathbb{P}^{n-1}$.

Quite generally, for 1st order extensions, these coordinate changes have the effect

$$f'_i \mapsto f'_i + \sum_h \alpha_h \frac{\partial f_i}{\partial x_h},$$

where $\deg \alpha_h = a_h - a_0$; thus in the deformation theory of a hypersurface singularity $V(f)$, the f' can be chosen arbitrarily, so that the Hilbert scheme $\mathbb{H}^{(1)}$ is just one graded piece of a polynomial ring, and $\mathbb{T}^1$ is one graded piece of the quotient by the Jacobian ideal $(\partial f/\partial x_h)$. The Jacobian matrix $(\partial f_i/\partial x_h)$ plays a similar role for complete intersection singularities.

Dividing out by the coordinate changes is not essential for most theoretical purposes: the redundant information contained in the Hilbert scheme does not affect questions such as the connectedness, irreducibility or unirationality of the set of extensions, or obstructions, or forcing into determinantal form. That is why this section has mainly concentrated on Hilbert schemes.

However, in practical (hand or machine) calculations it's often essential to use coordinate changes of this form to cut down the large number of free parameters I have to carry around. Thus in calculations I usually work with $\mathbb{T}^1$ moduli schemes of deformations. See (2.4), (5.13) and (6.3, Step 1) for practical illustrations.

In intrinsic terms, $\mathbb{T}^1(\overline{R}) = \mathrm{Ext}^1_{\overline{R}}(\Omega^1_{\overline{R}}, \overline{R})$; the reader versed in these matters will know that in the derived category there is no essential difference between working with this group or with $\mathrm{Hom}_{\overline{R}}(\overline{I}/\overline{I}^2, \overline{R})$.

(1.23) Abstract versus projective extensions. There is a corresponding 'abstract' extension problem, in which C is not thought of in an ambient projective space, but an ample normal bundle $\mathcal{O}(1)$ is fixed; for example, assuming that C is nonsingular, the set parametrising the abstract extensions $C \subset 2C$ such that $\ker\{\mathcal{O}_{2C} \to \mathcal{O}_C\} \cong \mathcal{O}_C(-1)$, is the whole of $H^1(T_C(-1))$, by analogy with Kodaira–Spencer deformation theory. In this context the graded piece $\mathbb{T}^1_{-1}$ is recovered as the set of abstract extensions $C \subset 2C$, together with a lift $\tilde{x}_i \in H^0(\mathcal{O}_{2C}(1))$ of each $x_i \in H^0(\mathcal{O}_C(1))$. Thus $\mathbb{T}^1_{-1}$ maps to $H^1(T_C(-1))$, with image consisting of abstract extensions that satisfy an infinitesimal analogue of linear normality.

(1.24) The unobstructed case. The ideal case of the extension problem is when every $R^{(k)}$ has an automatic extension to higher order; then

$$\mathbb{H}(\overline{R}, a_0) = \bigoplus_{k>0} \mathbb{H}^{(1)}(\overline{R}, ka_0).$$

The trivial case of this is a hypersurface: if $\overline{R}$ is $k[x_1,.. x_n]/(f_d)$ then its extensions are given by

$$F_d(x_0,.. x_n) = f_d + x_0 f_{d-1} +.. x_0^k f_{d-k} +..$$

where the bits $x_0^k f_{d-k}$ added in at the kth order are chosen freely from a vector space. A less trivial case is provided by the (as yet informal) notion of a *flexible format* (see (5.6)): let me say (vaguely) that a *format* Φ is a way of writing down relations for a ring depending on parameters, (the *entries* of Φ). The example to bear in mind is a generic determinantal. Make the following assumptions:

(i) Flexibility. $\overline{R}$ is given by relations in format Φ, and the format is *flexible*, in the sense that varying freely the entries of Φ (in a small neighbourhood) leads to flat deformations of $\overline{R}$; in determinantal cases this may come about because the syzygies are all consequences of the format (see for example (5.5) and (5.8)).

(ii) Completeness. Suppose that, in addition, every 1st order extension $R^{(1)} \in \mathbb{H}^{(1)}(\overline{R}, ka_0)$ of degree $-ka_0$ (for each $k > 0$) can be written in the same format Φ; this may have to be checked by explicit 1st order calculations (see for example the proof of Theorem 5.11 in (5.13–14)).

Then making this choice automatically lifts $R^{(1)}$, and so all the affine fibre bundles in (1.15) can be trivialised. Thus by induction, for each k, the relations for $R^{(k-1)}$ can be written in the format Φ, and by flexibility, they define an extension to all higher orders. The set of extensions $R^{(k)}$ over a fixed $R^{(k-1)}$ is then just the vector space $\mathbb{H}^{(1)}(\overline{R}, ka_0)$, and the completeness assumption is that every element of this can be obtained by varying the kth order terms of the entries of Φ. It follows in turn that $R^{(k-1)}$ can also be written in the format Φ, and so

$$\mathbb{H}(\overline{R}, a_0) = \bigoplus_{k>0} \mathbb{H}^{(1)}(\overline{R}, ka_0).$$

(1.25) Relation with graded versal deformations. Suppose that $C, \mathcal{O}_C(1)$ is a polarised variety; let $P \in S = \operatorname{Spec} R(C, \mathcal{O}_C(1))$ be the affine cone over C. Since R is a graded ring, S enjoys an action of the multiplicative group $\mathbf{G}_m$. Assume for simplicity that C is nonsingular, so that $P \in S$ is an isolated singularity. Then by Schlessinger's theorem [Artin, Schlessinger] (plus Artin algebraic approximation or its analytic equivalent), there exists a versal deformation

$$\begin{array}{ccc} S & \subset & \Sigma \\ \downarrow & & \downarrow \\ 0 & \in & V; \end{array}$$

V is a finite-dimensional formal scheme (or germ of a local analytic space) with a given deformation $\Sigma \longrightarrow V$ such that every infinitesimal (or local analytic) deformation of $P \in S$ over a parameter scheme T is obtained as the pullback of Σ by a morphism $\varphi_\Sigma\colon T \longrightarrow V$, with the 1st derivative of φ_Σ uniquely determined by Σ.

Pinkham [Pinkham, (2.3)] proved that V and the versal family $\Sigma \longrightarrow V$ also have $\mathbf{G}_m$ actions; this incidentally allows me to be a little vague about the category in which V is defined: by the 'graded is local' principle described in (1.3, (2)), the scheme V is determined by its formal completion at the cone point $0 \in V$. The $\mathbf{G}_m$ action on V corresponds to a grading of the tangent space $\mathbb{T}^1 = T_0V = \sum_i \mathbb{T}^1_i$. An extension of C corresponds to a graded 1-parameter deformation of $P \in S$, hence to a formal (or analytic) curve $(\operatorname{Spec} k[[x_0]]) \longrightarrow (0 \in V)$ with tangent vector in $\mathbb{T}^1_{-1}$, 2nd derivative in $\mathbb{T}^1_{-2}$, etc.

In the unobstructed case, V is local analytically isomorphic to an open ball in $\mathbb{T}^1$, so that the G_m-action on V can be linearised, giving a decomposition of V similar to that in (1.23), and a graded analytic curve in V can be constructed with arbitrary derivatives of each order.

§2. Examples, comments, propaganda

In this lightweight *scherzo*, a transparent example is followed by a brief description of the motivation and some speculative future applications of the infinitesimal view; the material is taken mainly from a recent (so far unsuccessful) research grant application.

(2.1) Pinkham's example ([Pinkham, (8.6)]). Consider the normal rational curve $C_4 \subset \mathbb{P}^4$ given parametrically by $(s^4, s^3t, s^2t^2, st^3, t^4)$. The homogeneous graded ring of C_4 is of the form

$$k[s^4, s^3t, s^2t^2, st^3, t^4] = k[x_1,.. x_5]/I$$

where I is generated by the 6 relations

$$\text{I: rk}\begin{bmatrix} x_1 & x_2 & x_3 & x_4 \\ x_2 & x_3 & x_4 & x_5 \end{bmatrix} \leq 1.$$

Geometrically, if $S \subset \mathbb{P}^5$ is a surface extending C and not a cone, then S is either a quartic scroll or the Veronese surface.

(2.2) 1st order deformation calculation. To find all 1st order extensions of C_4, I must (1) write out the 6 relations coming from the determinantal explicitly, then (2) write out all possible 1st order variation of the relations by adding arbitrary multiples of x_0, and then (3) take account of the syzygies mod x_0^2. This is tedious but wholly mechanical.

$$\begin{array}{r|r|r|llcll}
x_3 & & x_4 & R_{12}: & x_1x_3 & = & x_2^2 & + \\
-x_2 & & & R_{13}: & x_1x_4 & = & x_2x_3 & + \\
x_1 & x_5 & & R_{23}: & x_2x_4 & = & x_3^2 & + \\
 & -x_4 & x_1 & R_{24}: & x_2x_5 & = & x_3x_4 & + \\
 & x_3 & & R_{34}: & x_3x_5 & = & x_4^2 & + \\
 & & -x_2 & R_{14}: & x_1x_5 & = & x_2x_4 & +
\end{array}
\begin{bmatrix}
a_{11}x_1 + a_{12}x_2 + a_{13}x_3 + a_{14}x_4 + a_{15}x_5 \\
a_{12}x_1 + a_{22}x_2 + a_{23}x_3 + a_{24}x_4 + a_{25}x_5 \\
a_{31}x_1 + a_{32}x_2 + a_{33}x_3 + a_{34}x_4 + a_{35}x_5 \\
a_{41}x_1 + a_{42}x_2 + a_{43}x_3 + a_{44}x_4 + a_{45}x_5 \\
a_{51}x_1 + a_{52}x_2 + a_{53}x_3 + a_{54}x_4 + a_{55}x_5 \\
a_{61}x_1 + a_{62}x_2 + a_{63}x_3 + a_{64}x_4 + a_{65}x_5
\end{bmatrix} x_0$$

(2.3) The only syzygies I require are the 3 written out in columns on the left. I work out the effects of the first syzygy slowly before dumping the whole calculation before the reader.

Thus the first syzygy gives

$$\begin{aligned}
0 &= x_3R_{12} - x_2R_{13} + x_1R_{23} \\
&= x_3(-x_1x_3 + x_2^2) - x_2(-x_1x_4 + x_2x_3) + x_1(-x_2x_4 + x_3^2) \\
&\quad + [a_{11}x_1x_3 + a_{12}x_2x_3 + a_{13}x_3^2 + a_{14}x_3x_4 + a_{15}x_3x_5 \\
&\quad - a_{21}x_1x_2 - a_{22}x_2^2 - a_{23}x_2x_3 - a_{24}x_2x_4 - a_{25}x_2x_5 \\
&\quad + a_{31}x_1^2 + a_{32}x_1x_2 + a_{33}x_1x_3 + a_{34}x_1x_4 + a_{35}x_1x_5]x_0.
\end{aligned}$$

The terms on the first line all cancel out (the syzygy for $\overline{R}$), and cancelling x_0 and using the relations in $\overline{R}$ gives the following equality in $\overline{R}$:

$$\begin{aligned}0 = \ & a_{31}x_1^2 + (-a_{21} + a_{32})x_1x_2 + (a_{11} - a_{22} + a_{33})x_1x_3 \\ & + (a_{12} - a_{23} + a_{34})x_2x_3 + (a_{13} - a_{24} + a_{35})x_2x_4 \\ & + (a_{14} - a_{25})x_3x_4 + a_{15}x_3x_5.\end{aligned}$$

Since the 9 monomials

$$x_1^2, x_1x_2, x_1x_3, x_2x_3, x_2x_4, x_3x_4, x_3x_5, x_4x_5, x_5^2$$

form a basis of $\overline{R}_2 = H^0(\mathbb{P}^1, \mathcal{O}(8))$, I can equate coefficients; thus the 3 syzygies imply

$$\begin{array}{lllll}
\text{I.} & a_{31} = 0, & a_{21} = a_{32}, & a_{11} - a_{22} + a_{33} = 0, & \\
& & & a_{12} - a_{23} + a_{34} = 0, & \\
& a_{15} = 0, & a_{14} = a_{25}, & a_{13} - a_{24} + a_{35} = 0; & \\
\text{II.} & a_{35} = 0, & a_{34} = a_{45}, & a_{33} - a_{44} + a_{55} = 0, & \\
& & & a_{32} - a_{43} + a_{54} = 0, & \\
& a_{51} = 0, & a_{41} = a_{52}, & a_{31} - a_{42} + a_{53} = 0; & \\
\text{III.} & a_{41} = 0, & a_{42} = a_{61}, & a_{43} = a_{62}, & a_{11} + a_{44} - a_{63} = 0, \\
& a_{14} = 0, & a_{15} = 0, & a_{13} = a_{65}, & a_{12} + a_{45} - a_{64} = 0.
\end{array}$$

(2.4) At this stage I also want to make a normalisation as described in (1.22), using a transformation of the form $x_1 \mapsto x_1 + \lambda x_0$ to arrange $a_{13} = 0$, and similarly with $x_2, .. x_5$ to get $a_{34} = a_{33} = a_{32} = a_{53} = 0$.

Having done this, the result of the 1st order deformation calculation can be written in human-readable form (well, almost!). I write it down here together with the set-up for the 2nd order calculation:

$$\begin{array}{lcl}
x_1x_3 & = & x_2^2 \\
x_1x_4 & = & x_2x_3 \\
x_1x_5 & = & x_2x_4 \\
x_2x_5 & = & x_3x_4 \\
x_3x_5 & = & x_4^2 \\
x_2x_4 & = & x_3^2
\end{array}
\begin{array}{c} + \\ + \\ + \\ + \\ + \\ + \end{array}
\begin{array}{l}
\left[\begin{array}{lllll}
ax_1 + bx_2 & & & & \\
& ax_2 + bx_3 & & & \\
& fx_2 + (a+g)x_3 + bx_4 & & & \\
& & fx_3 + gx_4 & & \\
& & & fx_4 + gx_5 &
\end{array}\right] x_0 \\
0
\end{array}
\begin{array}{c} + \\ + \\ + \\ + \\ + \\ + \end{array}
\begin{bmatrix} \alpha_1 \\ \alpha_2 \\ \alpha_3 \\ \alpha_4 \\ \alpha_5 \\ \alpha_6 \end{bmatrix} x_0^2$$

Here (a, b, f, g) are free parameters, coordinates on the 4–dimensional vector space

$$\mathbb{T}^1_{(-1)} = \operatorname{coker}\left\{H^0(T_{\mathbb{P}^4}(-1)) \longrightarrow H^0(N_{\mathbb{P}^4|C}(-1))\right\}.$$

(2.5) The 2nd order calculation. Using the same 3 syzygies gives

I.	$\alpha_1 = 0,$	$\alpha_2 = ab,$	$\alpha_6 = -a^2,$	
II.	$\alpha_4 = fg,$	$\alpha_5 = 0,$	$\alpha_6 = -g^2,$	
III.	$\alpha_4 = -af,$	$\alpha_1 = 0,$	$\alpha_3 = (a+g)a + fb,$	$(a+g)b = 0.$

The conclusion: the 1st order deformation with coordinates (a, b, f, g) admits an extension to 2nd order if and only if

$$(a+g)f = (a+g)b = (a+g)(a-g) = 0;$$

and the 2nd order extension is unique if it exists. That is, the locus of $\mathbb{T}^1(\overline{R})$ corresponding to genuine extensions of $\overline{R}$ is the union of the 3–plane $(a+g) = 0$ and the line $f = b = a - g = 0$:

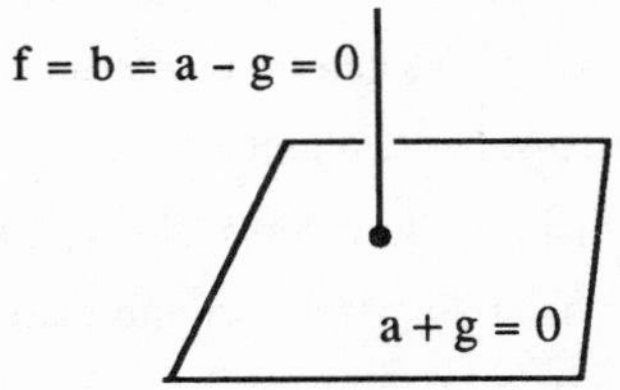

(2.6) Determinantal interpretations. It is not hard to see that if $a = -g$ then the 6 relations can be recast in the determinantal form

$$\operatorname{rk}\begin{bmatrix} x_1 & x_2 & x_3+ax_0 & x_4+fx_0 \\ x_2+bx_0 & x_3-ax_0 & x_4 & x_5 \end{bmatrix} \le 1.$$

That is, the original 2×4 matrix had special or nongeneric entries $a_{12} = a_{21} = x_2$, $a_{13} = a_{22} = x_3$ and $a_{14} = a_{23} = x_4$, and the extension is obtained by allowing these to become general linear forms in 6 variables. These are of course the equations defining a quartic scroll in $\mathbb{P}^5$.

On the other hand, if $b = f = 0$ and $a = g$ then the 6 relations can be written in the form

$$\operatorname{rk}\begin{bmatrix} x_1 & x_2 & x_3+ax_0 \\ x_2 & x_3-ax_0 & x_4 \\ x_3+ax_0 & x_4 & x_5 \end{bmatrix} \le 1.$$

That is, the original 6 relations can be recast in the alternative symmetric determinantal form

$$\operatorname{rk}\begin{bmatrix} x_1 & x_2 & x_3 \\ x_2 & x_3 & x_4 \\ x_3 & x_4 & x_5 \end{bmatrix} \le 1.$$

As a symmetric determinantal, this is nongeneric only in the 'persymmetric' coincidence $a_{13} = a_{22} = x_3$, and the given extension is the least I can do to cure this. When $a \neq 0$, the determinantal equations are of course the equations of the Veronese surface in $\mathbb{P}^5$.

(2.7) The chicken or the egg? Since the answer here (and also in more substantial cases, for example §5) is expressed in terms of determinantal formats, it's tempting to think that the calculations can also. I believe that this view is mistaken, and that the deformation calculation is more fundamental: starting from one of the determinantal formats (as in (2.1)), my experience is that it's not possible to predict the other without carrying out what is in effect a deformation calculation (maybe only in part or in guessed form). In fact, my guess is that in some sense the determinantal format can be seen as arising from an elimination of deformation variables. A ring in determinantal format somehow managing to squeeze out of its straight–jacket under a deformation is an extremely delicate and interesting phenomenon; a similar example provides the main theme of §5, and I believe that there are many other substantial cases in this area of commutative algebra and algebraic geometry.

(2.8) The rest of this section discusses briefly some of the motivation behind my study of the extension problem. As a method, the infinitesimal view has so far only led to convincing results in a small number of cases: the case of numerical quintics treated by Ed Griffin [Griffin2] (worked out in detail in §5), and Dicks' work [Dicks] on canonical rings of surfaces with $p_g = 3, K^2 = 4$ (another class studied by Horikawa). The idea in both cases is that both the geometry and algebra of an individual surface and properties of its moduli space can be treated by viewing the surface as an extension of a general curve $C \in |K|$. Up to now, the computations have been considered to be very hard and long; one reason for the material of §1 is to express these (in so far as possible) as mechanical algorithms.

When we have these computations mechanised, I believe that there will turn out to be dozens of examples where the infinitesimal view can be used to study curves, surfaces, 3-folds and singularities and their moduli spaces. In many cases the defining relations of the graded ring fit into specific formats (for example, involving determinantals), so they give rise to interesting examples in commutative algebra.

(2.9) Mumford's dream. A program advanced by D. Mumford (his problem to the Montreal NATO summer school in 1980) says that questions on the existence of surfaces or the number of components of their moduli space are 'in principle' solvable by computer; roughly, the idea is to define a graded ring by writing down relations (in terms of some chosen generators) as polynomials with indeterminate coefficients, and then to find the subvariety of the set of indeterminates where the graded ring has the required properties. Mumford actually proposed to use this to study surfaces with $p_g = 0, K^2 = 9$. However, I doubt if Mumford's program can be implemented on computers in the foreseeable future (even in simple key cases).

My point is that when detailed information about an ample divisor of the variety under study is available, it is reasonable to study it as an extension problem. This may lead to less intractable problems; the situation is analogous to replacing a general algebraic group by a nilpotent one - it's not far-fetched to see the iterated linear structure of (1.13) as a kind of nilpotent phenomenon.

(2.10) Godeaux surfaces. Surfaces of general type have been the starting point for all of my work in the classification of varieties over the last 13 or 14 years. There is a paradox here, because although the subject as a whole has seen spectacular progress in many directions, some of the basic problems from which I started off seem just as hard now as ever.

A *Godeaux surface* is a minimal surface X of general type with $p_g = 0, K^2 = 1$. These are the smallest possible values of the numerical invariants; the torsion subgroup $(\operatorname{Pic} X)_{\text{Tors}} = \text{Torsion } X = \pi_1^{\text{alg}}(X)$ of a Godeaux surface is one of $\mathbb{Z}/5$, $\mathbb{Z}/4$, $\mathbb{Z}/3$, $\mathbb{Z}/2$ or 0, and the surfaces in the first 3 cases are well understood (at least by me) [Reid1]. Godeaux surfaces with Torsion = $\mathbb{Z}/2$ has been the single most important motivating case for me, and I have put an enormous amount of effort into computing their canonical rings since 1977, before I knew of the link with deformation theory; the ring restricted to the unique curve $C \in |K+\sigma|$ has a nice hyperelliptic description as in Theorem 4.6. The calculation of the ring of the surface itself by the infinitesimal view is a key test of my ideas (see (6.5)), and was originally intended as §6 (*grosse Fuge*) of this paper, but I am reluctantly obliged to leave it to a future occasion.

A Godeaux surface with Torsion = 0 was constructed by R. Barlow in her 1982 Ph.D. thesis (see [Barlow1–2]), and is at present the only known simply connected surface of general type with $p_g = 0$; this has recently been the subject of interest on the part of differential geometers [Kotschik]. The following question is a distant aim.

Question. Is the moduli space of Godeaux surfaces with Torsion = 0 irreducible?

Under certain genericity assumptions, I have a concrete (but fairly complicated) geometric description of the graded ring $R(C) = R(C, K_{X|C})$ associated with a general 2–canonical curve $C \subset X$ in terms of liaisons, involving an elliptic curve $E \subset \mathbb{P}^3$ and a 3–torsion point of E. The ring $R(C)$ can then be written out in terms of generators, relations and syzygies, and the canonical ring $R(X)$ studied by feeding $R(C)$ into the extension methods of §1. Already $R(C)$ is a very big calculation.

(2.11) Remark on 4–manifolds. The current view in 4–dimensional topology is that on the one hand, homotopy theory, surgery and Freedman's work reduces the classification of simply connected 4–manifolds up to homeomorphism to the intersection form on $H^2(M, \mathbb{Z})$; on the other hand, by Donaldson's work, their classification up to diffeomorphism is intimately related to the complex geometry of algebraic surfaces. So for example, although the chain of reasoning is long, and depends on some wild-looking conjectures arising out of Donaldson's work (see [Friedman and Morgan]), one now thinks of the problem of diffeomorphism type of simply connected 4–manifolds with intersection form $(+1, -8)$, as being closely related to my question (2.10). It is quite amazing that there is such a long chain of reasoning, starting at one end with topology and differential geometry, through algebraic geometry and the commutative algebra of complicated graded rings to computer algebra.

(2.12) Speculative applications. There are many of these; to mention only rather substantial ones:

(a) The relative canonical algebra of a fibre space $f: X \longrightarrow B$ of curves of genus g over a base curve. The ultimate aim here is to decide a conjecture of Xiao Gang on 'Morsification': the germ of f around a degenerate fibre can be deformed to a neighbouring fibration having only Morse critical points or nonsingular multiple fibres. The rings (and the calculations) arising here are like those for canonical rings of surfaces. For example, the genus 3 case has been studied in detail by Mendes-Lopes [Mendes-Lopes]: computing the canonical ring $R(F, K_F)$ of a nonreduced fibre of F is a nilpotent extension problem similar to §1, and the rings arising are in some cases similar to the numerical quintics of §5.

(b) Construct new surfaces embedded in $\mathbb{P}^4$, and hence new vector bundles on $\mathbb{P}^4$, starting from a cleverly set up curve $C \subset \mathbb{P}^3$; the point is that the construction and embedding of C can perhaps be done intrinsically and geometrically, even though the commutative algebra of $C \subset \mathbb{P}^3$ (the monad defining the corresponding vector bundle) is certain to be very complicated.

(c) Du Val singularities and 3-fold flip singularities. Another long term aim is an attack on the flip singularities that play a crucial role in Mori's theory of minimal models of 3-folds. A flip singularity $P \in X$ usually contains a Du Val surface singularity $P \in S \in |-K_X|$ as anti-canonical divisor (the 'general elephant'). Technically, the problem is that already the 1st order normal data of X around S is quite awkward to specify: since S is not a Cartier divisor, the normal sheaf may be nontrivial on $S \setminus P$, and moreover, it may not be S_2 at the singularity.

(d) Permanence. A familiar phenomenon in projective geometry is that features of a hyperplane section $C \subset X$ often extend to X. For example, if an ample divisor $C \subset X$ of a surface X is a hyperelliptic curve, one may entertain certain expectations concerning the rational map $\varphi_C: X \dashrightarrow \mathbb{P}^N$; or if C is contained in a scroll $C \subset F \subset \mathbb{P}^{n-1}$, then one may hope to find a bigger scroll in $\mathbb{P}^n$ containing X, etc. (see for example [Serrano]). Similar remarks apply to permanence of features under small deformation. In some cases the infinitesimal view allows these questions to be treated together, and to be explained in terms of 1st order infinitesimal extensions.

Chapter II. Halfcanonical curves and the canonical ring of a regular surface

§3. The canonical ring of a regular surface

(3.1) Theorem. Let X be a canonical surface (that is, the canonical model of a surface of general type). Suppose that

(i) $p_g \geq 2$, $K^2 \geq 3$;

(ii) $q = 0$;

and

(iii) X has an irreducible canonical curve $C \in |K_X|$.

Then the canonical ring $R = R(X, K_X)$ is generated in degrees ≤ 3 and related in degrees ≤ 6.

Standard convention. When R is generated in degrees ≤ 3, I write $R = k[x_i, y_j, z_k]/I$, with $\deg(x_i, y_j, z_k) = (1, 2, 3)$.

(3.2) Counterexample (P. Francia and C. Ciliberto [Ciliberto, §4]). On a minimal surface of general type S, define a *Francia cycle* to be a 2-connected divisor E such that either $KE = 1$ and $E^2 = -1$, or $KE = 2$ and $E^2 = 0$ (think of E as a smooth elliptic or genus 2 curve). A famous theorem of Francia [Francia] says that, with finitely many exceptional families, $2K_X$ is very ample on the canonical model if and only if the minimal model S does not contain a Francia cycle.

Without the assumption (iii), the conclusion of Theorem 3.1 fails infinitely often. The point is just that if the fixed part of $|K_X|$ contains a Francia cycle then the multiplication map

$$S^2R_2 \oplus R_1 \cdot R_3 \to R_4$$

cannot be surjective. In fact φ_{4K} is very ample on E; however, φ_{2K} cannot be very ample on E for reasons of low degree, and the elements of $R_1 \cdot R_3$ vanish along E (by the assumption that E is fixed in $|K_X|$).

Consider the double cover of the quadric cone $Q \subset \mathbb{P}^3$ ramified in the vertex and in a curve $R \in |\mathcal{O}_Q(2m+2n+3)|$ that meets a given generator A in a $(2m+1)$-tuple tacnode and a $(2n+1)$-tuple tacnode (see figure): it is not hard to see that making a minimal resolution X leads

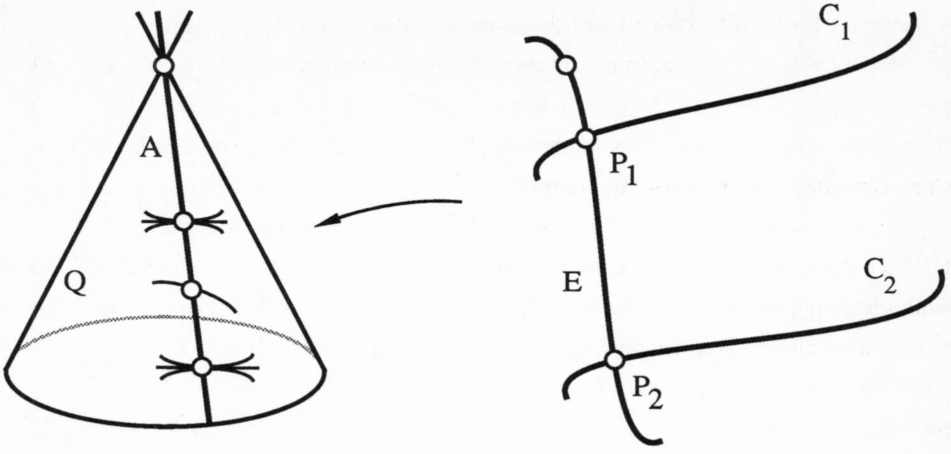

to a elliptic curve E on X with $E^2 = -1$; E is fixed in $|K_X|$ because $K_X E = 1$ and E passes through 2 points P_i on the exceptional curves C_i that are base points of $K_{X|C_i}$.

(3.3) Remarks. (a) The statement of the theorem in terms of the canonical model X allows the possibility that the canonical system $|K_S|$ of the minimal nonsingular model S has (-2)-curves as fixed part; it would be quite unpleasant to have to do the proof in this context.

(b) Two relative versions of the problem are also of interest. If $f: X \longrightarrow Y$ is a resolution of an isolated Gorenstein surface singularity, then the relative canonical algebra $\oplus f_*\omega_X^{\otimes k}$ is generated in degree ≤ 3 (see [Laufer2, (5.2)] and compare [Laufer1, (3.2) and (3.5)]) and I conjecture that it is related in degrees ≤ 6. If $f: X \longrightarrow B$ is a fibre space of curves of genus $g = 2$ or 3 over a base curve then the relative canonical algebra $\oplus f_*\omega_X^{\otimes k}$ is generated in degree ≤ 3 and related in degrees ≤ 6 (a result due to Mendes-Lopes [Mendes-Lopes]); it would be interesting to know if Laufer's argument can be modified to prove this for all g.

(c) Theorem 3.1 may hold even without assumption (ii): that is, although $H^0(X, 2K_X) \longrightarrow H^0(C, 2K_{X|C})$ is not surjective, it might be possible to show that it maps 'onto the bits that matter', as with other questions concerning the 2-canonical map [Francia, Reider]. Also, it seems reasonable to ask about weakening the irreducibility assumption (iii) to $C \in |K_X|$ 3-connected. Thus I conjecture that, with a finite number of exceptional families, the canonical ring of a surface $R(X, K_X)$ is generated in degrees ≤ 3 and related in degrees ≤ 6 if and only if the fixed part of

$|K_S|$ on the minimal model does not contain a Francia cycle.

(d) Exercise. Prove that under the assumptions of (3.1), the 3-canonical model $X^{[3]}$ is projectively normal. [Hint: by (3.1), R_{3d} is spanned by monomials in x_i, y_j, z_k of degree 3d; if such a monomial is a product of two monomials of degree divisible by 3 then OK. Thus $x_0 \cdot R_5$ is in the image of $S^2(R_3) \longrightarrow R_6$. But if x_0 defines an irreducible curve C, the surjectivity of $S^2(R_3) \longrightarrow R_6$ modulo x_0 follows by standard use of the free pencil trick on C.]

(3.4) By the hyperplane section principle (1.2, i–ii), Theorem 3.1 will follow from the following more precise result for curves, applied to the curve $C \in |K_X|$ and the divisor $D = K_{X|C}$.

Theorem. Let C be an irreducible Gorenstein curve of genus $g \geq 2$, and D a Cartier divisor on C such that $2D \sim K_C$; assume that C and D are not in the 4 exceptional cases (i–iv) below. Then the graded ring R(C, D) is generated in degrees ≤ 3 and related in degrees ≤ 6.

Exceptional cases:

(i) C is hyperelliptic of genus $g \neq 2$ and $h^0(\mathcal{O}_C(D)) = 0$; in this case R(C, D) is generated in degrees ≤ 4 and related in degrees ≤ 8.

(ii) $g = 2$, $D = P$ where P is a Weierstrass point; in this case

$$R(C, D) = k[x, y, z]/F, \quad \text{with} \quad \deg(x, y, z, F) = 1, 2, 5, 10.$$

(iii) $g = 3$, $D = g^1{}_2$; in this case

$$R(C, D) = k[x_1, x_2, y]/F, \quad \text{with} \quad \deg(x_1, x_2, y, F) = 1, 1, 4, 8.$$

(iv) $g = 3$, C is nonhyperelliptic and $h^0(\mathcal{O}_C(D)) = 0$; then R(C, D) is generated in degrees ≤ 3, but requires one relation in degree 8 ($2D = K_C$ is very ample, mapping C to a plane quartic $C = C_4 \subset \mathbb{P}^2$, and the relation in degree 8 is the defining equation of C_4).

History. Cases (ii–iv) go back in effect to Enriques; for example, Case (iv) is treated in detail in [Catanese and Debarre].

The proof of Theorem 3.4 is a cheap adaptation of the early stages of the famous *Petri analysis* [4 authors, Chap. III, §2] for the canonical ring $R(C, 2D) = R(C, K_C)$ of a non-hyperelliptic curve C. More closely related to the point of view of [Mumford1] and [Fujita2], similar arguments show that R(C, D) is generated in degrees ≤ 3 and related in degrees ≤ 6 for any divisor D of degree $\geq g+1$ on an irreducible curve C.

(3.5) Set-up for the proof of (3.4). This section *waltzes* through the major case of a nonhyperelliptic curve C; the *trio* section §4 covers the relative minor case when C is hyperelliptic in much more detail; (see (3.11) if you don't know what it means for an irreducible Gorenstein curve C to be hyperelliptic). When $g = 3$ and C is nonhyperelliptic then either (iv) holds, or $h^0(C, \mathcal{O}_C(D)) = 1$; I offer the reader the lovely exercise of seeing that in this case, which corresponds to a plane quartic with a bitangent line, $R(C, \mathcal{O}_C(D))$ is a complete intersection ring

$$R(C, D) = k[x, y_1, y_2, z]/(f, g) \quad \text{with} \quad \deg(f, g) = (4, 6).$$

Thus I suppose throughout this section that C is nonhyperelliptic and $g \geq 4$. Introduce vector space bases as follows:

$$x_1, .. x_a \in H^0(D);$$

$$y_1, .. y_g \in H^0(2D) = H^0(K_C);$$

$$z_1, .. z_{2g-2} \in H^0(3D).$$

Write $I(m, n)$ for the kernel of the natural map

$$\varphi_{m,n}\colon H^0(mD) \otimes H^0(nD) \longrightarrow H^0((m+n)D),$$

and

$$\psi_{\ell;m,n}\colon H^0(\ell D) \otimes I(m, n) \longrightarrow I(\ell+m, n)$$

for the natural map.

(3.6) Main Lemma. (I) $\varphi_{m,2}$ is surjective for every $m \geq 2$;
(II) $I(m+2, 2) = \operatorname{im} \psi_{2;m,2} + \operatorname{im} \psi_{m;2,2}$.

This result is similar to [Fujita2, Lemma 1.8]; the proof occupies (3.8–10) together with a technical appendix.

(3.7) Lemma 3.6 ⟹ Theorem 3.4. (I) implies by induction that if $m = 2\ell \geq 2$ is even, then $H^0(mD)$ is spanned as a vector space by the set $S^\ell(y)$ of monomials of degree ℓ in the y_i; and if $m = 2\ell+1 \geq 3$ is odd then $H^0(mD)$ is spanned as a vector space by the set $z \otimes S^{\ell-1}(y)$ of monomials of the form z_j times a monomial of degree $\ell-1$ in the y_i. This obviously implies that $R(C, D)$ is generated in degrees ≤ 3.

The relations in low degrees can be written

deg 2	$x_ix_j = L_{ij}(y)$		(linear forms)
deg 3	$x_iy_j = M_{ij}(z)$		(linear forms)
deg 4	$x_iz_j = N_{ij}(y)$		(quadratic forms)
deg 6	$z_iz_j = P_{ij}(y)$		(cubic forms).

These relations clearly allow any monomial of degree m in the x_i, y_j, z_k to be expressed as a linear combination of $S^{\ell}(y)$ if $m = 2\ell$ or of $z \otimes S^{\ell-1}(y)$ if $m = 2\ell+1$.

For the relations, suppose that F_m: $f_m(x, y, z) = 0 \in R_m$ is a polynomial relation of degree m between the generators x, y, z of $R(C, D)$. I must show that F_m is a linear combination of products

$$(\text{monomial}) \times (\text{relation in degree } \leq 6).$$

Any term occuring in F_m can be expressed as a linear combination of monomials $S^{\ell}(y)$ or $z \otimes S^{\ell-1}(y)$ by using products of the relations just tabulated. Therefore I need only deal with linear dependence relations between these monomials in R_m (for $m \geq 7$).

By just separating off one y_i in each monomial in an arbitrary way, a linear combination of these monomials in R_m can be written as the image of an element $\xi \in R_{m-2} \otimes R_2$; to say that it vanishes in R_m means that $\xi \in I(m-2, 2)$. But then Lemma 3.6, (II) says that

$$\xi \in \operatorname{im} \psi_{2;m-4,2} + \operatorname{im} \psi_{m-4;2,2}.$$

This means that the relation in degree m corresponding to ξ is a sum of relations in degrees $m-2$ and 4 multiplied up into degree m. By induction, the result follows. Q.E.D.

(3.8) Proof of (3.6, I), and notation. Let $A = P_3 + .. P_g$ be a divisor on C made up of $g-2$ general points. Since C is nonhyperelliptic, K_C is birational, so that $|2D - A| = |K_C - A|$ is a free pencil (by general position [4 authors, p.109]); hence the free pencil trick gives the exact sequence

	$0 \to H^0((m-2)D + A)$	$\longrightarrow$	$H^0(2D - A) \otimes H^0(mD)$	$\longrightarrow$	$H^0((m+2)D - A)$
$m = 2$	1		$2 \times g$		$2g-1$
$m \geq 3$	$(m-2)(g-1)-1$		$2 \times (m-1)(g-1)$		$m(g-1)+1$;

the indicated dimension count shows that the right-hand arrow

$$H^0(2D - A) \otimes H^0(mD) \longrightarrow H^0((m+2)D - A) \to 0$$

is surjective.

Let $t_m \in H^0(mD)$ be an element not vanishing at any of $P_3,.. P_g$, and, as in the Petri analysis, choose the basis $y_1,.. y_g$ of $H^0(K_C)$ such that

$$y_1, y_2 \text{ bases } H^0(2D - A), \text{ and } y_i(P_j) = \delta_{ij} \text{ for } i, j = 3,.. g$$

(Kronecker delta). Then by the free pencil trick,

$$H^0((m+2)D - A) = H^0(mD)y_1 \oplus H^0(mD)y_2,$$

and, obviously, $t_m y_i$ for $i = 3,.. g$ forms a complementary basis of $H^0((m+2)D)$. This proves (3.6, I).

Similarly,

$$H^0((m+4)D-A) = H^0((m+2)D)y_1 + H^0((m+2)D)y_2$$

and $t_{m+2}y_i$ for $i = 3,.. g$ is a complementary basis of $H^0((m+4)D)$.

(3.9) As u runs through $H^0(mD+A)$, the relations

$$\rho(u) = uy_1 \otimes y_2 - uy_2 \otimes y_1 \in I(m+2, 2)$$

express the fact that

$$H^0((m+2)D)y_1 \cap H^0((m+2)D)y_2 = H^0(mD+A)y_1y_2 \subset H^0((m+4)D - A),$$

which is part of the free pencil trick. The key to (3.6, II) is to prove that for $m \geq 3$,

$$\rho(u) \in \operatorname{im} \psi_{2;m,2} \quad \text{for all } u \in H^0(mD+A);$$

since $\xi \otimes \rho(v) = \rho(\xi v)$ for $\xi \in H^0(2D)$ and $v \in H^0((m-2)D+A)$, this follows trivially from

Claim.

$$H^0(2D) \otimes H^0((m-2)D+A) \longrightarrow H^0(mD+A) \to 0$$

is surjective.

(3.10) Proof of (3.6, II). Claim 3.9 is proved in (3.15), and I first polish off (3.6, II) assuming it. Suppose that $m \geq 3$.

Step 1. The subspace $\{\rho(u)\} = \rho(H^0(mD+A))$ is the kernel of

$$H^0((m+2)D) \otimes H^0(2D-A) \longrightarrow H^0((m+4)D-A) \subset H^0((m+4)D),$$

and the $t_m y_i \otimes y_i$ map to a complementary basis. Therefore, a subset

$$S \subset I(m+2, 2) = \ker\{H^0((m+2)D) \otimes H^0(2D) \longrightarrow H^0((m+4)D)\}$$

will span $I(m+2, 2)$ as a k-vector space provided that

(i) S contains the $\rho(u)$;

and (ii) S spans a subspace complementary to

$$H^0((m+2)D) \otimes H^0(2D-A) \oplus \sum k \cdot t_m y_i \otimes y_i,$$

in other words, any $\eta \in H^0((m+2)D) \otimes H^0(2D)$ can be written

$$\eta = \eta_S + \eta_{2D-A} + \eta_t \qquad (*)$$

where $\eta_{2D-A} \in H^0((m+2)D) \otimes H^0(2D-A)$ and η_S, η_t are linear combinations of S and of the $t_m y_i \otimes y_i$ respectively.

Step 2. Now set $S = \operatorname{im} \psi_{2;m,2} + \operatorname{im} \psi_{m;2,2}$. By (3.9), $\operatorname{im} \psi_{2;m,2}$ contains $\rho(u)$ for $u \in H^0(mD+A)$. Therefore, it is enough to verify $(*)$ for any $\eta \in H^0((m+2)D) \otimes H^0(2D)$.

Break up $H^0((m+2)D) \otimes H^0(2D)$ as a direct sum of the following 4 pieces:

$$V_1 = H^0((m+2)D) \otimes H^0(2D-A);$$

$$V_2 = H^0((m+2)D-A) \otimes \sum k \cdot y_i;$$

$$V_3 = \sum k \cdot t_{m+2} y_i \otimes y_j \quad \text{summed over } i, j = 3,.. g \text{ with } i \neq j;$$

$$V_4 = \sum k \cdot t_{m+2} y_i \otimes y_i \quad \text{for } i = 3,.. g.$$

For V_1 and V_4 there's not much to prove. Also since

$$H^0((m+2)D-A) = H^0(mD)y_1 + H^0(mD)y_2$$

and $R(2, 2)$ contains $y_1 \otimes y_i - y_i \otimes y_1$ and $y_2 \otimes y_i - y_i \otimes y_2$ for $i = 3,.. g$, it follows that $V_2 \subset V_1 + \operatorname{im} \psi_{m;2,2}$.

Finally, for the summand V_3, note that for $i, j = 3,.. g$ and $i \neq j$,

$$y_i y_j \in H^0(4D-A) = H^0(2D)y_1 + H^0(2D)y_2,$$

so that $I(2, 2)$ contains the Petri relation

$$y_i \otimes y_j - a_{ij} \otimes y_1 - b_{ij} \otimes y_2 \quad \text{with } a_{ij}, b_{ij} \in H^0(K_C).$$

Therefore also

$$t_{m+2} y_i \otimes y_j \in V_1 + \operatorname{im} \psi_{m;2,2}.$$

This completes the proof of (3.6, II), modulo Claim 3.9.

Coda to §3. 'General' divisors and the proof of (3.9)

(3.11) Lemma (the hyperelliptic dichotomy). Let C be an irreducible Gorenstein curve of genus $g = p_aC \geq 2$.

(i) The canonical linear system $|K_C|$ is free;

(ii) K_C is very ample unless φ_K is a 2–to–1 flat morphism to a normal rational curve.

Proof (See [Catanese, §3] for a discussion of a more general problem; however, my proof of (ii) seems to be new even in the nonsingular case!).

(i) Suppose $P \in C$ is a base point of $|K_C|$; then $h^0(m_P \cdot \mathcal{O}_C(K_C)) = g$ and by RR $h^1(m_P \cdot \mathcal{O}_C(K_C)) = 2$, so by Serre duality the inclusion

$$\mathrm{Hom}(\mathcal{O}_C, \mathcal{O}_C) = k \subset \mathrm{Hom}(m_P, \mathcal{O}_C)$$

is strict. A nonconstant element of $\mathrm{Hom}(m_P, \mathcal{O}_C)$ is a rational function $h \in k(C)$ such that $h \cdot m_P \subset \mathcal{O}_C$. Since $\deg h \cdot m_P = \deg m_P = -1$, it is easy to see that $h \cdot m_P = m_Q$ for some $P \neq Q \in C$, and it follows that P and Q are Cartier divisors on C, hence nonsingular points, and as usual h defines a birational morphism $C \to \mathbb{P}^1$, necessarily an isomorphism.

(ii) If $\varphi_K : C \to \mathbb{P}^{g-1}$ is not birational then it is clearly 2–to–1 to a normal rational curve. Suppose it is birational to a curve of degree $2g-2$. If $A = P_3 + .. P_g$ is a divisor on C made up of $g-2$ general points then $|K_C - A|$ is a free pencil by general position, and arguing as in (3.8), $S^d(H^0(K_C)) \to H^0(dK_C)$ is surjective; thus the ring $R(C, K_C)$ is generated by $H^0(K_C)$. Therefore the ample divisor K_C is very ample. Q.E.D.

(3.12) Claim 3.9 will also follow from the free pencil trick, once I prove that the divisor $A = P_1 + .. P_{g-2}$ made up of $g-2$ general points is 'general enough' for $|D+A|$ to be free and birational.

Proposition. Let C be an irreducible Gorenstein curve of genus g, and D a divisor class such that $2D \sim K_C$. Let $A = P_3 + .. P_g$ be a divisor on C made up of $g-2$ general points. Then

(i) Suppose that $g \geq 3$, and that C is nonhyperelliptic if $g = 3$; then $h^0(\mathcal{O}_C(D)) \leq g - 2$, so that $H^0(C, \mathcal{O}_C(D-A)) = 0$ and $h^0(C, \mathcal{O}_C(D+A)) = g - 2$.

(ii) Suppose that $g \geq 4$, and that C is nonhyperelliptic if $g = 4$; then $|D + A|$ is free; it's a free pencil if $g = 4$.

(iii) Suppose that $g \geq 5$, and that C is nonhyperelliptic if $g = 5$; then φ_{D+A} is birational.

(3.13) Proof of (3.12, ii). It's enough to prove $\mathrm{Hom}(m_P, \mathcal{O}_C(D-A)) = 0$ for every $P \in C$, since then by duality and RR,

$$h^0(m_P \cdot \mathcal{O}_C(D+A)) = g - 3 < h^0(\mathcal{O}_C(D+A)) = g - 2,$$

and $|D + A|$ is free.

Case $H^0(D) = 0$. Then $h^0(m_P \cdot \mathcal{O}_C(D)) = 0$ for every $P \in C$, so by RR, $h^1(m_P \cdot \mathcal{O}_C(D)) = 1$. By Serre duality,

$$\dim \mathrm{Hom}(m_P, \mathcal{O}_C(D)) = 1$$

for every $P \in C$; hence there is just a 1-dimensional family of effective divisors A (of any degree) with $\mathrm{Hom}(m_P, \mathcal{O}_C(D-A)) \neq 0$ for any P. Since A varies in a family of dimension $g-2 \geq 2$, it can be chosen to avoid this set.

Case $H^0(D) \neq 0$. By RR and duality, the inclusion

$$H^0(\mathcal{O}_C(D)) \subset \mathrm{Hom}(m_P, \mathcal{O}_C(D))$$

is strict only for P in the base locus of $|D|$; therefore I can assume that the general divisor A imposes linearly independent conditions on each of the vector spaces $\mathrm{Hom}(m_P, \mathcal{O}_C(D))$ (there are in effect only finitely many of them). So if $\mathrm{Hom}(m_P, \mathcal{O}_C(D-A)) \neq 0$ for $P \in C$ then

$$\dim \mathrm{Hom}(m_P, \mathcal{O}_C(D)) \geq g - 1.$$

Using RR and duality as usual, this is the same as

$$h^0(m_P \cdot \mathcal{O}_C(D)) \geq g - 2.$$

This contradicts (a singular analogue of) Clifford's theorem: by the linear–bilinear trick, the map

$$S^2 H^0(m_P \cdot \mathcal{O}_C(D)) \to H^0(m_P{}^2 \cdot \mathcal{O}_C(K_C))$$

has rank $\geq 2h^0 - 1$ (with equality if and only if the image of C under the rational map defined by $H^0(m_P \cdot \mathcal{O}_C(D))$ is a normal rational curve), so

$$g \geq h^0(m_P{}^2 \cdot \mathcal{O}_C(K_C)) + 1 \geq 2h^0(m_P \cdot \mathcal{O}_C(D)) \geq 2(g-2),$$

that is, $g \leq 4$ and C is hyperelliptic in case of equality. This contradiction proves (ii). The reader can do (i) as an exercise in the same vein.

(3.14) Proof of (3.12, iii). This is very similar: I prove that there exists a nonsingular point Q such that $\mathrm{Hom}(m_P, \mathcal{O}_C(D+Q-A)) = 0$ for every $P \in C$; as before, RR and duality imply that

$$h^0(m_P \cdot \mathcal{O}_C(D+A-Q)) = h^0(\mathcal{O}_C(D+A)) - 2,$$

so that φ_{D+A} is an isomorphism near Q.

Case $h^0(D) \le 1$. Then $h^0(D+Q) = 1$ for a general point Q, and fixing such a point, RR and duality imply that

$$\dim \mathrm{Hom}(m_P, \mathcal{O}_C(D+Q)) = 2$$

for every $P \in C$; hence the family of effective divisors A with $\mathrm{Hom}(m_P, \mathcal{O}_C(D+Q-A)) \neq 0$ for any P has dimension 2, and as A varies in a family of dimension $g-2 \ge 3$, I can choose it to avoid this.

Case $h^0(D) \ge 2$. I pick a general Q, so that $h^0(D+Q) = h^0(D)$; then, as before, the inclusion

$$H^0(\mathcal{O}_C(D+Q)) \subset \mathrm{Hom}(m_P, \mathcal{O}_C(D+Q))$$

is strict only for P a base point of $|D+Q|$; so that there are only finitely many distinct vector spaces $\mathrm{Hom}(m_P, \mathcal{O}_C(D+Q))$, and I can assume that the general divisor A imposes linearly independent conditions on each of them. Thus $\mathrm{Hom}(m_P, \mathcal{O}_C(D+Q-A)) \neq 0$ implies

$$\dim \mathrm{Hom}(m_P, \mathcal{O}_C(D+Q)) \ge g-1, \quad \text{that is,} \quad h^0(m_P \cdot \mathcal{O}_C(D+Q)) \ge g-2.$$

As before, the linear-bilinear trick gives

$$\mathrm{rk}\left\{S^2H^0(m_P \cdot \mathcal{O}_C(D+Q)) \to H^0(m_P{}^2 \cdot \mathcal{O}_C(K_C+2Q))\right\} \ge 2h^0(m_P \cdot \mathcal{O}_C(D+Q)) - 1.$$

Now $|K_C + 2Q|$ is free and $H^0(\mathcal{O}_C(K_C+2Q)) = g+1$, so

$$g+1 \ge h^0(m_P{}^2 \cdot \mathcal{O}_C(K_C+2Q)) + 1 \ge 2h^0(m_P \cdot \mathcal{O}_C(D+Q)) \ge 2(g-2);$$

that is, $g \le 5$ and C is hyperelliptic in case of equality. Q.E.D.

(3.15) Proof of Claim 3.9. $h^0(D+A) = g-2$. If $g \ge 5$ then φ_{D+A} is birational, so that I can choose a divisor $B = Q_1 +.. Q_{g-4}$ made up of general points, and sections $s_i \in H^0(D+A)$ such that $s_i(Q_j) = \delta_{ij}$. Then using the free pencil trick in the usual way shows that

$$H^0(2D) \otimes H^0(D+A-B) \to H^0(3D+A-B)$$

is surjective; if $t \in H^0(2D)$ doesn't vanish at $Q_1,.. Q_4$ then $s_i t$ for $i = 1,.. g-4$ is a complementary basis of $H^0(3D+A)$. The statement for $m \ge 4$ is an easy exercise using the same method. Q.E.D.

§4. Graded rings on hyperelliptic curves

(4.1) Notation, introduction. A nonsingular hyperelliptic curve of genus g is a 2–to–1 cover $\pi\colon C \longrightarrow \mathbb{P}^1$ branched in $2g+2$ distinct points $\{Q_1,.. Q_{2g+2}\} \subset \mathbb{P}^1$, lifting to points $\{P_1,.. P_{2g+2}\} \subset C$ (see the picture below); the $P_i \in C$ are the *Weierstrass points*, the points of C for which $2P_i \in g^1_2$. If $D = \sum d_i P_i$ is a divisor on C made up of Weierstrass points, or equivalently, invariant under the hyperelliptic involution $\iota\colon C \longrightarrow C$, I am going to describe an automatic and painless way of writing down a vector space basis of $H^0(C, \mathcal{O}_C(D))$, and a presentation of the ring $R(C, \mathcal{O}_C(D))$ by generators and relations.

In a nutshell, the method is the following. Fix a basis $(t_1, t_2) \in H^0(\mathbb{P}^1, \mathcal{O}(1)) = H^0(C, g^1_2)$ of homogeneous coordinates on $\mathbb{P}^1$. For each $i = 1,.. 2g+2$, let

$$u_i\colon \mathcal{O}_C \hookrightarrow \mathcal{O}_C(P_i)$$

be the constant section. Since $2P_i \in g^1_2$, it follows that $u_i^2 \in H^0(C, g^1_2)$, so that I can write

$$u_i^2 = \ell_i(t_1, t_2), \qquad (*)$$

where ℓ_i is the linear form in t_1 and t_2 defining the branch point $Q_i \in \mathbb{P}^1$. Now it is more–or–less obvious that any vector space of the form $H^0(C, \mathcal{O}_C(D))$ has a basis consisting of monomials in the u_i, and that the only relations between these are either of a trivial monomial kind or are derived from $(*)$.

(4.2) Easy preliminaries. (i) The decomposition of $\pi_*\mathcal{O}_C$ into the (±1)–eigensheaves of ι is

$$\pi_*\mathcal{O}_C = \mathcal{O}_{\mathbb{P}^1} \oplus \mathcal{O}_{\mathbb{P}^1}(-g-1),$$

and the algebra structure on $\pi_*\mathcal{O}_C$ is given by a multiplication map $f\colon S^2(\mathcal{O}_{\mathbb{P}^1}(-g-1)) = \mathcal{O}_{\mathbb{P}^1}(-2g-2) \longrightarrow \mathcal{O}_{\mathbb{P}^1}$, which is a polynomial $f_{2g+2}(t_1, t_2)$ vanishing at the $2g+2$ branch points Q_i;

(ii) the Weierstrass points add up to a divisor in $|(g+1)g^1_2|$, that is

$$P_1 +.. P_{2g+2} \sim (g+1)g^1_2;$$

(iii) locally near a branch point, $\pi_*\mathcal{O}_C(P_i) = \mathcal{O}_{\mathbb{P}^1} \oplus \mathcal{O}_{\mathbb{P}^1}(Q_i)\cdot u_i$.

Remark. For any partition $\{P_1,..P_a\} \cup \{P_{a+1},..P_{2g+2}\}$ of the Weierstrass points into two sets,

$$P_1 + .. P_a + (g+1-a)g^1{}_2 \sim P_{a+1} + .. P_{2g+2},$$

as follows from (ii) and $2P_i \sim g^1{}_2$. This will be important in what follows (see (4.5)); it corresponds to passing between the (± 1)-eigensheaves of $\pi_* \mathcal{O}_C(P_1 + .. P_a + kg^1{}_2)$.

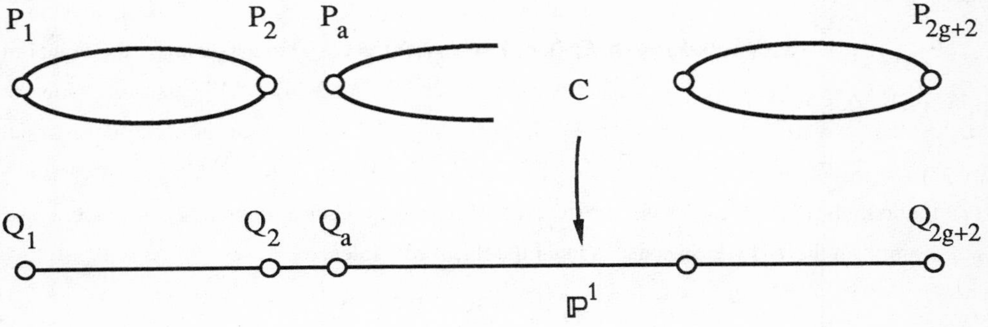

Proof. (i) is standard; one affine piece of C is

$$C: (y^2 = f_{2g+2}(t)).$$

It's easy to see that y/t^{g+1} is a rational function on C with

$$\operatorname{div}(y/t^{g+1}) = P_1 + .. P_{2g+2} - (g+1)\cdot g^1{}_2;$$

this proves (ii). For (iii), if t is a local parameter on $\mathbb{P}^1$ at a branch point $Q \in \mathbb{P}^1$ and $u^2 = t\cdot(\text{unit})$, then u is a local parameter at $P \in C$, so $1/t$ has a simple pole at Q and the (-1)-eigensheaf of $\pi_* \mathcal{O}_C(P)$ is $\mathcal{O}_{\mathbb{P}^1}\cdot u/t = \mathcal{O}_{\mathbb{P}^1}(Q)\cdot u$. Q.E.D.

(4.3) Simplest examples of graded rings. (a) Let $D = g^1{}_2$; then $H^0(\mathcal{O}_C(D)) = (t_1, t_2)$, and

$$H^0(\mathcal{O}_C(kD)) = H^0(\mathbb{P}^1, \mathcal{O}_{\mathbb{P}^1}(k)) \oplus H^0(\mathbb{P}^1, \mathcal{O}_{\mathbb{P}^1}(k-g-1));$$

thus for $k \leq g$ all the sections of $\mathcal{O}_C(kD)$ are in the $(+1)$-eigenspace, so no new generators are needed, and I get the final generator

$$w \in H^0(\mathcal{O}_C((g+1)D))$$

in degree $g+1$ satisfying $w^2 = f_{2g+2}(t_1, t_2)$. So

$$R(C, g^1{}_2) = k[t_1, t_2, w]/F, \quad \text{with} \quad \deg(t_1, t_2, w, F) = 1, 1, g+1, 2g+2,$$

and $C = C_{2g+2} \subset \mathbb{P}(1, 1, g+1)$.

(b) Let $D = P$ with $P \in C$ a Weierstrass point; write $P = P_{2g+2}$ and $P_1,.. P_{2g+1}$ for the remaining Weierstrass points, and

$$u\colon \mathcal{O}_C \hookrightarrow \mathcal{O}_C(P) \quad \text{and} \quad v\colon \mathcal{O}_C \hookrightarrow \mathcal{O}_C(P_1+..P_{2g+1})$$

for the two constant sections. Since $u^2\colon \mathcal{O}_C \hookrightarrow \mathcal{O}_C(2P) = \mathcal{O}_C(g^1{}_2)$ is the constant section, I can choose the coordinates (t_1, t_2) so that $u^2 = t_1$, and $t_2 \in H^0(C, \mathcal{O}_C(2P))$ is a complementary basis element. Now

$$\pi_*\mathcal{O}_C((2k)P) = \pi_*\mathcal{O}_C(kg^1{}_2) = \mathcal{O}_{\mathbb{P}^1}(k) \oplus \mathcal{O}_{\mathbb{P}^1}(k-g-1),$$

and by (4.2, iii),

$$\pi_*\mathcal{O}_C((2k+1)P) = \pi_*\mathcal{O}_C(P) \otimes \mathcal{O}_{\mathbb{P}^1}(k) = \mathcal{O}_{\mathbb{P}^1}(k) \oplus \mathcal{O}_{\mathbb{P}^1}(k-g),$$

so that monomials $u^\ell, u^{\ell-2}t_2,..$ base $H^0(\mathcal{O}_C(\ell P))$ for $\ell \le 2g$; but in degree $2g + 1$ there is a new section z in the (-1)–eigenspace. Under the linear equivalence

$$(2g+1)P \sim P_{2g+2} + g{\cdot}g^1{}_2 \sim P_1 +.. P_{2g+1},$$

z is the constant section $v\colon \mathcal{O}_C \hookrightarrow \mathcal{O}_C(P_1+..P_{2g+1})$; in more detail, if y is chosen as in (4.2, ii) then

$$\operatorname{div}(t_1{}^{g+1}/y) = (g+1)(2P) - (P_1 +.. P_{2g+2}) = (2g+1)P - (P_1 +.. P_{2g+1})$$

so that $z = vt_1{}^{g+1}/y$. If $f = f_{2g+1}(t_1, t_2)$ is the form defining the $2g+1$ branch points in $\mathbb{P}^1$, then $z^2 = f(u^2, t_2)$, so

$$R(C, P) = k[u, t_2, z]/F, \quad \text{with} \quad \deg(u, t_2, z, F) = 1, 2, 2g+1, 4g+2,$$

and $C = C_{4g+2} \subset \mathbb{P}(1, 2, 2g+1)$.

Remark. The ring $R(C, g^1{}_2)$ of (i) can be obtained by eliminating the elements of $R(C, P)$ of odd degree; that is, $R(C, 2P) = R(C, P)^{(2)}$. This means replacing

$$u \text{ by } t_1 = u^2, \quad z \text{ by } w = uz,$$

and

$$F\colon z^2 = f_{4g+2} \text{ by } F'\colon w^2 = u^2 f_{4g+2}(u, t_2) = f'_{2g+2}(t_1, t_2).$$

(4.4) Lemma. Let D be a divisor on C. Equivalent conditions:

(i) the divisor class of D is invariant under ι, that is $D \sim \iota^*D$;

(ii) $D \sim D'$ with $D' = \iota^*D'$;

(iii) D is made up of Weierstrass points, that is (after a possible renumbering),

$$D \sim P_1 +.. P_a + bg^1{}_2.$$

Proof. The implications (ii) $\Rightarrow$ (iii) $\Rightarrow$ (i) are trivial, so assume (i). By adding on a large multiple of $g^1{}_2$ if necessary, I assume that D is effective.

If $\iota^*D \sim D$ but $\iota^*D \neq D$ then $|D|$ is a nontrivial linear system. I pick one Weierstrass point, say P_1; then the divisor class $D - P_1$ is invariant under ι, and $|D - P_1|$ contains an effective divisor, so that induction on $\deg D$ proves (ii). Q.E.D.

Remark. Since $D + \iota^*D \sim (\deg D)\cdot g^1{}_2$ for any divisor D on a hyperelliptic curve, a 4th equivalent condition on D is

(iv) $2D \sim (\deg D)\cdot g^1{}_2$.

This set of divisors includes of course all divisor classes with $2D \sim 0$ or $2D \sim K_C$, etc.

Useful fact: each 2–torsion divisor on a hyperelliptic curve is (up to renumbering) of the form

$$P_1 +.. P_{2a} - a\cdot g^1{}_2 \sim P_{2a+1} +.. P_{2g+2} - (g+1-a)\cdot g^1{}_2.$$

Go on, check for yourself that there are 2^{2g} of these!

(4.5) Theorem. (I) For an invariant divisor $D = P_1 +.. P_a + bg^1{}_2$, set

$$D' = P_{a+1} +.. P_{2g+2} + (a+b-g-1)g^1{}_2,$$

so that $D \sim D'$ by Remark 4.3. Write $u\colon \mathcal{O}_C \hookrightarrow \mathcal{O}_C(P_1 +.. P_a)$ and $v\colon \mathcal{O}_C \hookrightarrow \mathcal{O}_C(P_{a+1} +.. P_{2g+2})$ for the constant sections. Then

$$\pi_*\mathcal{O}_C(D) = \mathcal{O}_{\mathbb{P}^1}(b)\cdot u \oplus \mathcal{O}_{\mathbb{P}^1}(a+b-g-1)\cdot v$$

and

$$H^0(\mathcal{O}_C(D)) = H^0(\mathcal{O}(b))\cdot u \oplus H^0(\mathcal{O}(a+b-g-1))\cdot v.$$

In other words, if I write $S^k(t_1, t_2)$ for the set of $(k+1)$ monomials $t_1{}^k, t_1{}^{k-1}t_2,.. t_2{}^k$ (or $\emptyset$ if $k < 0$) then $H^0(\mathcal{O}_C(D))$ has basis

$$S^b(t_1, t_2)\cdot u, \quad S^{a+b-g-1}(t_1, t_2)\cdot v.$$

(II) Write $f_a(t_1, t_2)$ and $g_{2g+2\ -a}(t_1, t_2)$ for the forms defining $Q_1 +.. Q_a$ and $Q_{a+1} +.. Q_{2g+2}$ in $\mathbb{P}^1$. Then the graded ring $R(C, \mathcal{O}_C(D))$ is generated by monomials in $R(C, \mathcal{O}_C(kD))$ for suitable initial values of k, and related by monomial relations together with relations deduced from

$$u^2 = f_a(t_1, t_2), \quad v^2 = g_{2g+2-a}(t_1, t_2).$$

Proof of (I). $\pi_*\mathcal{O}_C(D)$ has a uniquely determined $\mathbb{Z}/2$ action compatible with the inclusion $\mathcal{O}_C \hookrightarrow \mathcal{O}_C(P_1 + .. P_a)$, and the (+1)-eigensheaf is clearly $\mathcal{O}_{\mathbb{P}^1}(b)\cdot u$. Multiplication by the rational function $y/t_1^{g+1} \in k(C)$ described in the proof of (4.2, ii) induces an isomorphism

$$\mathcal{O}_C(D) \cong \mathcal{O}_C(D'),$$

and since y/t_1^{g+1} is in the (−1)-eigenspace, the isomorphism interchanges the (±1)-eigensheaves. This proves (I).

(4.6) I will regard (II) as a principle, and not go into the long-winded general proof, which involves introducing notation $k_0^{\pm}, k_1^{\pm}$ for the smallest even and odd values of k for which each eigensheaf of $\pi_*\mathcal{O}_C(kD)$ has sections, and a division into cases according to which of these is smaller.

I now give a much more precise statement and proof of (II) in the main case of interest. Suppose that, in the notation of Theorem 4.5,

$$b \geq 0 \quad \text{and} \quad a + 2b < g + 1 \leq 2a + 3b.$$

Note that $2D = (a+2b)\cdot g^1_2$, so that $\pi_*\mathcal{O}_C(2D) = \mathcal{O}(a+2b) \oplus \mathcal{O}(a+2b-g-1)\cdot uv$, where $uv\colon \mathcal{O}_C \hookrightarrow \mathcal{O}_C(P_1 + .. P_{2g+2})$ is the constant section. Write $V^{\pm}$ to denote the (±1)-eigenspaces of a vector space on which ι acts; the point of these inequalities is just to ensure that

$$H^0(D)^+ = H^0(\mathcal{O}(b))\cdot u \neq 0,$$
$$H^0(2D)^- = H^0(\mathcal{O}(a+2b-g-1))\cdot uv = 0, \text{ (so also } H^0(D)^- = 0)$$
$$H^0(3D)^- = H^0(\mathcal{O}(2a+3b-g-1))\cdot v \neq 0.$$

Notice that this case covers all effective halfcanonical divisors on a hyperelliptic curve of genus $g \geq 4$, for which $a + 2b = g - 1$.

Theorem. The graded ring $R(C, D)$ is generated by the following bases:

$$(x_0, x_1,.. x_b) = S^b(t_1, t_2)\cdot u = t_1{}^b u, t_1{}^{b-1}t_2 u,.. t_2{}^b u \in H^0(D)^+;$$
$$(y_0, y_1,.. y_d) = S^d(t_1, t_2) = t_1{}^d, t_1{}^{d-1}t_2,.. t_2{}^d \in H^0(2D)^+;$$
$$(z_0, z_1,.. z_c) = S^c(t_1, t_2)\cdot v = t_1{}^c v, t_1{}^{c-1}t_2 v,.. t_2{}^c v \in H^0(3D)^-.$$

where I set $d = a+2b = \deg D$ and $c = 2a+3b-g-1$ for brevity. The relations are given as follows:

$$\mathrm{rk}\begin{bmatrix} x_0 & x_1 & .. & x_{b-1} & y_0 & y_1 & .. & y_{d-1} & z_0 & z_1 & .. & z_{c-1} \\ x_1 & x_2 & .. & x_b & y_1 & y_2 & .. & y_d & z_1 & z_2 & .. & z_c \end{bmatrix} \le 1$$

(the x or z columns are omitted if $b = 0$ or $c = 0$). And

$$x_i x_j = F_{i+j}(y_0, .. y_d) \quad \text{for all } 0 \le i, j \le b,$$

$$z_i z_j = G_{i+j}(y_0, .. y_d) \quad \text{for all } 0 \le i, j \le c,$$

where

$$F_{i+j} = t_1^{2b-i-j} t_2^{i+j} f_a(t_1, t_2) \quad \text{written out as a linear form in } y_0, .. y_d;$$

$$G_{i+j} = t_1^{2c-i-j} t_2^{i+j} g_{2g+2-a}(t_1, t_2) \quad \text{as a cubic form in } y_0, .. y_d;$$

(4.7) Remarks. (a) Notice that, as promised, the first set consists of monomial relations, and the second of relations deduced from $u^2 = f_a(t_1, t_2)$, $v^2 = g_{2g+2-a}(t_1, t_2)$.

(b) The first set of determinantal relations $\mathrm{rk}\, A \le 1$ says simply that the ratio $(t_1 : t_2)$ defining $\pi\colon C \to \mathbb{P}^1$ is well defined. In fact the projective toric variety defined by $\mathrm{rk}\, A \le 1$ is a *weighted scroll*, that is, a fibre bundle $\varphi\colon F \to \mathbb{P}^1$ with fibre the weighted projective space $\mathbb{P}(1, 2, 3)$: in more detail, $F = \mathrm{Proj}_{\mathbb{P}^1}(\mathcal{A})$, where $\mathcal{A}$ is the graded $\mathcal{O}_{\mathbb{P}^1}$-algebra

$$\mathcal{A} = \bigoplus \varphi_* \mathcal{O}_F(k) = \mathrm{Sym}\left\{\mathcal{O}_{\mathbb{P}^1}(b)_1 \oplus \mathcal{O}_{\mathbb{P}^1}(d)_2 \oplus \mathcal{O}_{\mathbb{P}^1}(c)_3\right\}.$$

This means that the x_i, y_i and z_i can be written simply as

$$(x_0, .. x_b) = S^b(t_1, t_2)\cdot u,$$

$$(y_0, .. y_d) = S^d(t_1, t_2)\cdot w,$$

$$(z_0, .. z_c) = S^c(t_1, t_2)\cdot v,$$

where $u \in H^0(F, \mathcal{O}_F(1) \otimes \varphi^* \mathcal{O}_{\mathbb{P}^1}(-b))$ is a global basis of the summand $\mathcal{A}_1(-b)$ over $\mathbb{P}^1$, and similarly for w and v. In these terms, $C \subset F$ is the codimension 2 complete intersection defined by $u^2 = f_a(t_1, t_2)w$, $v^2 = g_{2g+2-a}(t_1, t_2)w^3$.

(4.8) Proof of the theorem. By (4.5, I), I have an explicit monomial basis of each $H^0(nD)$ in terms of t_1, t_2, u and v; clearly, each monomial in $H^0(2nD)$ equals

$$\text{either} \quad y^{\underline{m}} \text{ with } \deg y^{\underline{m}} = 2n \quad \text{or} \quad y^{\underline{m}'} x_i z_j \text{ with } \deg y^{\underline{m}'} = 2n-4,$$

and similarly each monomial in $H^0((2n+1)D)$ equals

$$\text{either} \quad y^{\underline{m}} x_i \text{ with } \deg y^{\underline{m}} = 2n \quad \text{or} \quad y^{\underline{m}'} z_j \text{ with } \deg y^{\underline{m}'} = 2n-2.$$

Moreover, an expression in x_i, y_j, z_k containing a quadratic term in x_i or z_k can obviously be translated to these using the second set of relations in (4.6). Now make a *first choice* of monomial representative of each such monomial element: for example, ordinary alphanumeric order picks $x_0 y_1 y_d$ ahead of $x_1 y_0 y_d$ or $x_0 y_2 y_{d-1}$ etc., which are equal to it in $R(C, D)$; and it is trivial to go from any $y^{\underline{m}}$ or $y^{\underline{m}'} x_i z_j$ or $y^{\underline{m}} x_i$ or $y^{\underline{m}'} z_j$ to its first choice representative using the monomial relations of the first set of (4.6). Q.E.D.

(4.9) Singular curves and the hyperelliptic case of Theorem 3.1.

For a nonsingular curve, the hyperelliptic case of Theorem 3.1 is included in (4.6). This analysis extends without difficulty to the case of a singular irreducible hyperelliptic curve C; for brevity I restrict myself to the main point, which is to describe the Cartier divisors on C playing the role of the sums of Weierstrass points $P_1 +.. P_a$ in (4.4, iii). The reader may wish to fill in the details as an extended exercise.

The branch points of the cover $C \longrightarrow \mathbb{P}^1$ divide into two types:

Cusp-like points. In local analytic coordinates, $y^2 = x^{2k+1}$. At such a point $P \in C$ there is a unique Cartier divisor $P(1) = \operatorname{div}_P (y/x^k)$ of degree 1 such that $\iota^*P(1) = P(1)$. This has the property that

$$2P(1) = \operatorname{div}_P (x) \sim g^1{}_2,$$

If $k = 0$, then $P(1) = P$ is just an ordinary Weierstrass point of C. In general, the pull-back of $P(1)$ is just the Weierstrass point of the normalised curve, with multiplicity 1 (since y/x^k is a local parameter there), but

$$P(1) + ig^1{}_2 \text{ is effective} \iff i \geq k.$$

The divisor $A_P = P(1) + k{\cdot}g^1{}_2$ is a Cartier divisor of degree of degree $2k+1$ on C, and plays the role of $2k+1$ coincident Weierstrass points.

Node-like points. In local analytic coordinates, $y^2 = x^{2k}$. At such a point $P \in C$ there is a unique nonzero Cartier divisor $P(0) = \operatorname{div}_P (y/x^k)$ of degree 0 such that $\iota^*P(0) = P(0)$. This satisfies $2P(0) = 0$, and the pull-back of $P(0)$ to the normalisation is 0 (since $y/x^k = \pm 1$ is invertible at the two points), but

$$P(0) + ig^1{}_2 \text{ is effective} \iff i \geq k.$$

The divisor $A_P = P(0) + k\cdot g^1{}_2$ is a Cartier divisor of degree of degree $2k$ on C, and plays the role of $2k$ coincident Weierstrass points.

The divisors $\sum A_P$ summed over distinct branch points P and of degree $a \leq g-1$ are characterised as the Cartier divisors on C invariant under ι and with $h^0 = 1$, in complete analogy with sums of distinct Weierstrass points. Now by analogy with Lemma 4.4, it can be seen that any Cartier divisor (or divisor class) on C invariant under ι is a sum of divisors of the form $P(1)$ for cusp-like P, of divisors of the form $P(0)$ for node-like P, and of a multiple of $g^1{}_2$. Any effective Cartier divisor D invariant under ι is of the form

$$D = \sum A_P + bg^1{}_2, \quad \text{with } b \geq 0,$$

summed over a subset of the branch points P, and as in Theorem 4.5, if I set $a = \deg \sum A_P$ and write $\sum' A_P$ for the complementary sum, then $D \sim D'$ where

$$D' = \textstyle\sum' A_P + (a+b-g-1)g^1{}_2,$$

The statement and proof of Theorems 4.5–6 now go through with only minor changes.

Chapter III. Applications

§5. Numerical quintics and other stories

(5.0) Preview. In this *toccata* section I work out in detail the deformation theory in degree ≤ 0 of the ring $\overline{R} = R(C, \mathcal{O}_C(D))$, where C is a nonsingular hyperelliptic curve of genus 6 and $D = 2g^1{}_2 + P \in (1/2)K_C$; in substance, the results are due to Horikawa [Horikawa] and Griffin [Griffin], although my treatment is novel and quite fun.

It turns out that the ring $\overline{R}$ admits two quite different representations in determinantal format. Each of these is *flexible*, in the sense of (1.24) so that changing freely the coefficients of the given matrix format preserves flatness, and thus gives rise to a large family of unobstructed deformations; in fact, I prove that in degrees ≤ 0, deformations from either family have codimension 1 in all 1st order deformations. These two families intersect transversally and their union gives exactly the deformations that extend to 2nd order. (There is an amazingly close analogy with Pinkham's example (2.1–6).) The main results are (5.11) and (5.16).

I conclude the section (5.17) by showing how to apply this to the classification and deformation theory of numerical quintics of dimension ≥ 2; for surfaces, these are of course fundamental classical results of Horikawa.

(5.1) The ring for C. Let C be a nonsingular hyperelliptic curve of genus 6, and $D = 2g^1{}_2 + P$ with $P = P_0$ a Weierstrass point. By the standard hyperelliptic stuff (4.6), $R(C, \mathcal{O}_C(D))$ has generators

$$\begin{aligned} x_1, x_2, x_3 &= t_1{}^2u, t_1t_2u, t_2{}^2u; \\ y &= t_2{}^5; \\ z_1, z_2 &= t_1v, t_2v, \end{aligned}$$

where (t_1, t_2) is a basis of $H^0(g^1{}_2) = H^0(\mathcal{O}_{\mathbb{P}^1}(1))$, and $u: \mathcal{O} \hookrightarrow \mathcal{O}(P_0)$, $v: \mathcal{O} \hookrightarrow \mathcal{O}(P_1 +.. P_{13})$, so

$$u^2 = t_1 \quad \text{and} \quad v^2 = f_{13}(t_1, t_2).$$

By (4.6), the ideal of relations is generated by 9 relations that can be written down as two groups: the 6 relations from

$$\operatorname{rk} A \leq 1 \quad \text{where} \quad A = \begin{bmatrix} x_1 & x_2 & x_3^2 & z_1 \\ x_2 & x_3 & y & z_2 \end{bmatrix},$$

(the vestigial symmetry $a_{12} = a_{21} = x_2$ of A is one crucial ingredient in what follows) and the 3 arising from $v^2 = f_{13}$:

$$\begin{aligned} z_1{}^2 &= t_1{}^2f_{13} &= g_0(x_1,.. y) \\ z_1z_2 &= t_1t_2f_{13} &= g_1(x_1,.. y) \\ z_2{}^2 &= t_2{}^2f_{13} &= g_2(x_1,.. y), \end{aligned}$$

where $\deg g_i = 6$. Write

$$f_{13}(t_1, t_2) = a_0t_1{}^{13} + a_1t_1{}^{12}t_2 +.. a_{10}t_1{}^3t_2{}^{10} + 2b_1t_1{}^2t_2{}^{11} + b_2t_1t_2{}^{12} + b_3t_2{}^{13}$$

(note the names of last 3 coefficients $2b_1, b_2, b_3$); it is easy to see that the nonsingularity of C implies $b_3 \neq 0$. Then

$$\begin{aligned} g_0 &= x_1^2\cdot h \quad +2x_1x_2\cdot b_1y^2 \quad +x_2^2\cdot b_2y^2 \ +(x_3^2)^2\cdot b_3y, \\ g_1 &= x_1x_2\cdot h \ +(x_1x_3+x_2^2)\cdot b_1y^2 \ +x_2x_3\cdot b_2y^2 \ +x_3^2y\cdot b_3y, \\ g_2 &= x_2^2\cdot h \quad +2x_2x_3\cdot b_1y^2 \quad +x_3^2\cdot b_2y^2 \quad +y^2\cdot b_3y, \end{aligned}$$

where

$$\begin{aligned} h = {} & a_0x_1^4+a_1x_1^3x_2+a_2x_1^3x_3+a_3x_1^2x_2x_3+a_4x_1^2x_3^2+a_5x_1^2y \\ & +a_6x_1x_2y+a_7x_1x_3y+a_8x_2x_3+a_9x_3^2y+a_{10}y^2. \end{aligned}$$

More explicitly, I have the following 9 relations.

(5.2) Table of all relations for $\overline{R} = R(C, \mathcal{O}_C(D))$.

$$\begin{aligned} S:&\quad x_1x_3 = x_2^2; \\ T_1:&\quad x_1y = x_2x_3^2; \\ T_2:&\quad x_2y = x_3^3; \\ U_1:&\quad x_1z_2 = x_2z_1; \\ U_2:&\quad x_2z_2 = x_3z_1; \\ V:&\quad x_3^2z_2 = yz_1; \\ -W_0:&\quad z_1^2 = hx_1^2+2b_1y^2\cdot x_1x_2+b_2y^2\cdot x_2^2+b_3y\cdot x_3^4; \\ -W_1:&\quad z_1z_2 = hx_1x_2+b_1y^2\cdot(x_1x_3+x_2^2)+b_2y^2\cdot x_2x_3+b_3y\cdot x_3^2y; \\ -W_2:&\quad z_2^2 = hx_2^2+2b_1y^2\cdot x_2x_3+b_2y^2\cdot x_3^2+b_3y\cdot y^2. \end{aligned}$$

The syzygies holding between these relations are written out explicitly in Table 5.4.

(5.3) First determinantal format. The above notation has been massaged slightly to make the g_i into explicit quadratic expressions in the rows of A; that is,

$$\begin{bmatrix} g_1 & g_2 \\ g_2 & g_3 \end{bmatrix} = A_0M_0{}^tA_0,$$

where

$$A_0 = \begin{bmatrix} x_1 & x_2 & x_3^2 \\ x_2 & x_3 & y \end{bmatrix} \quad \text{and} \quad M_0 = \begin{bmatrix} h & b_1y^2 & 0 \\ b_1y^2 & b_2y^2 & 0 \\ 0 & 0 & b_3y \end{bmatrix}.$$

In other words, the final 3 relations for the z_iz_j can be written in the form

$$AM({}^tA) = 0,$$

where A is as above and M is the symmetric matrix with homogeneous entries

$$M = \begin{bmatrix} M_0 & 0 \\ 0 & -1 \end{bmatrix} \quad \text{of degrees} \quad \begin{matrix} 4 & 4 & 3 & 2 \\ 4 & 4 & 3 & 2 \\ 3 & 3 & 2 & 1 \\ 2 & 2 & 1 & 0 \end{matrix}.$$

(5.4) The following key observation is due (in a slightly harder context) to Duncan Dicks: the syzygies holding between the 9 relations of Table 5.2 are all consequences of the determinantal format.

Proposition. The following are 16 syzygies between the relations $S, T_1,.. V$ of Table 5.2; they generate the module of all syzygies. Moreover, they can be deduced from the determinantal format.

Table of syzygies for $\overline{R} = R(C, \mathcal{O}_C(D))$.

First set:

$x_1T_2 - x_2T_1 + x_3^2S \equiv 0$	$x_2T_2 - x_3T_1 + yS \equiv 0$
$x_1U_2 - x_2U_1 + z_1S \equiv 0$	$x_2U_2 - x_3U_1 + z_2S \equiv 0$
$x_1V - x_3^2U_1 + z_1T_1 \equiv 0$	$x_2V - yU_1 + z_2T_1 \equiv 0$
$x_2V - x_3^2U_2 + z_1T_2 \equiv 0$	$x_3V - yU_2 + z_2T_2 \equiv 0$

Second set:

$$x_2W_0 - x_1W_1 \equiv -(b_1x_1 + b_2x_2)y^2S - b_3x_3{}^2yT_1 + z_1U_1;$$

$$x_3W_0 - x_2W_1 \equiv (x_1h + b_1x_2y^2)S - b_3x_3{}^2yT_2 + z_1U_2;$$

$$yW_0 - x_3{}^2W_1 \equiv (x_1h + b_1x_2y^2)T_1 + (b_1x_1 + b_2x_2)y^2T_2 + z_1V;$$

$$z_2W_0 - z_1W_1 \equiv (x_1h + b_1x_2y^2)U_1 + (b_1x_1 + b_2x_2)y^2U_2 + b_3x_3{}^2yV;$$

$$x_2W_1 - x_1W_2 \equiv -(b_1x_2 + b_2x_3)y^2S - b_3y^2T_1 + z_2U_1;$$

$$x_3W_1 - x_2W_2 \equiv (x_2h + b_1x_3y^2)S - b_3y^2T_2 + z_2U_2;$$

$$yW_1 - x_3{}^2W_2 \equiv (x_2h + b_1x_3y^2)T_1 + (b_1x_2 + b_2x_3)y^2T_2 + z_2V;$$

$$z_2W_1 - z_1W_2 \equiv (x_2h + b_1x_3y^2)U_1 + (b_1x_2 + b_2x_3)y^2U_2 + b_3y^2V.$$

(5.5) Proof. I first show how to derive the 16 syzygies from the determinantal format (which proves they are in fact syzygies). Let $A = (a_{ij})$ and $M = (m_{ij})$ be 2×4 and 4×4 matrixes whose entries are weighted homogeneous polynomials in a polynomial ring of degrees

$$\deg a_{ij} = \begin{matrix} 1 & 1 & 2 & 3 \\ 1 & 1 & 2 & 3 \end{matrix} \quad \text{and} \quad \deg m_{ij} = \begin{matrix} 4 & 4 & 3 & 2 \\ 4 & 4 & 3 & 2 \\ 3 & 3 & 2 & 1 \\ 2 & 2 & 1 & 0 \end{matrix},$$

and I the ideal generated by the 9 polynomial relations

$$\operatorname{rk} A \leq 1, \ AM({}^tA) = 0.$$

There are two ways of deducing syzygies from the format of these relations: first, an obvious determinantal trick is to double a row of A, so that any 3×3 minor vanishes identically. This leads to the first 8 syzygies of Table 5.4. Next, write

$$A^* = \begin{bmatrix} a_{12} & -a_{11} \\ a_{22} & -a_{21} \\ a_{32} & -a_{31} \\ a_{42} & -a_{41} \end{bmatrix},$$

so that $(A^*)A$ is a 4×4 skew matrix with entries the 2×2 minors of A. Then the expression $(A^*)AM({}^tA)$ can be parsed in two different ways: A^* times $AM({}^tA)$ is a linear combination of the second set of relations with coefficients from A^*; whereas $(A^*)A$ times $M({}^tA)$ is a linear combination of the first set of relations (the 2×2 minors of A) with coefficients from $M({}^tA)$.

This leads to the second set of 8 syzygies in Table 5.4.

Finally, why are these all the syzygies? The assertion is that any identity between the 9 relations is a linear combination of the 16 given ones. I sketch a proof by a calculation similar to that of (4.6), but more unpleasant. I can write the identity $\sum \ell_{ij} f_i \equiv 0$, where the ℓ_{ij} are monomials (and the f_i are the 9 relations (5.2), possibly repeated). First of all, since $z_1 S$, $z_1 T_1, .. z_1 V$ all appear on the right-hand side of one of the 16 given syzygies, I can subtract off multiples of them and assume that none of the ℓ_{ij} are divisible by z_1, except possibly if $f_i = W_0, W_1$ or W_2. But an easy argument on the highest power of z_1 then shows that none of the ℓ_{ij} can be divisible by z_1. Similarly for z_2.

Now assuming that none of the ℓ_{ij} are divisible by z_1 or z_2, it's not hard to see that none of W_0, W_1, W_2 can appear in any syzygies at all, and in turn, the same for U_1, U_2, V. Finally, it's not hard to see that the first two syzygies of the table are the only ones between S, T_1, T_2.

I apologise for the above proof *by intimidation*. Here is a proof *by appeal to authority*. The statement is of a kind covered by monomial bases algorithms of Macaulay and Gröbner (so the skeleton of a proof just given is part of an algorithm); in particular, Bayer and Stillman's computer program Macaulay [Bayer and Stillman] can calculate the entire projective resolution of the ring defined by the 9 relations (5.2) in a few seconds (on an obsolete home microcomputer), and confirms the 16 syzygies. (Macaulay assumes working over a prime finite field, preferably $k = \mathbb{Z}/(31{,}991)$, and that constants in k are chosen for the coefficients of h, b_1, b_2 and b_3; however, this proof 'after specialisation' obviously implies the statement I need.) Q.E.D.

(5.6) Remarks. (a) The power of Proposition 5.4, and of the analogous result (5.8) for the second determinantal format, is that varying the entries of the matrixes A and M leads to flat deformations of the ring $\overline{R}$ (since the syzygies only depend on the determinantal format of the equations, this corresponds to varying the relations together with the syzygies). In this case, I say that the format of the equations is *flexible*: since the coefficients of the entries can vary freely in an open set of k^N, leading to large unobstructed families of deformations. It is not clear how to formalise this as a definition, since the expression 'format of the equations' is vague. However, it includes well-known and very useful formats such as generic determinantals.

(b) The relations $\operatorname{rk} A \le 1$, $AM({}^tA) = 0$ for generic matrixes A and M (with M symmetric) are analogous to the defining equations of a Schubert cell. If the weights were all 1, it is easy to see that the corresponding projective variety is just

$$\mathbb{P}^1 \times (\text{universal quadric of } \mathbb{P}^3).$$

Here M gives the quadric $Q \subset \mathbb{P}^3$, the rows of A a point of Q, and the columns a point of

$\mathbb{P}^1$.

(5.7) Second determinantal form. Given a 6×6 skew matrix $N = \{n_{ij}\}$, the condition rank $N \le 2$ is expressed by the vanishing of the 15 (diagonal) Pfaffians of the 4×4 skew matrixes obtained by picking 4 rows and the corresponding columns. More concretely, for $i < j < k < \ell$

$$ij.k\ell = \mathrm{Pfaff}_{ij.k\ell}(N) = n_{ik}n_{j\ell} - n_{i\ell}n_{jk} + n_{ij}n_{k\ell}.$$

Let me while away a happy half-hour by evaluating the 4×4 Pfaffians of the following beauty:

$$N = \begin{bmatrix} 0 & \beta & y_1 & z_1 & x_2 & x_1 \\ & 0 & y_2 & z_2 & x_3 & x_2 \\ & & 0 & q & z_2 & z_1 \\ & & & 0 & py_2 & py_1 \\ & -\mathrm{sym} & & & 0 & p\beta \\ & & & & & 0 \end{bmatrix}, \quad \deg n_{ij} = \begin{matrix} 0&0&2&3&1&1 \\ 0&0&2&3&1&1 \\ 2&2&4&5&3&3 \\ 3&3&5&6&4&4 \\ 1&1&3&4&2&2 \\ 1&1&3&4&2&2 \end{matrix}.$$

(here β, p and q are homogeneous elements of degrees 0, 2 and 5 to be filled in subsequently; the antidiagonal symmetry is part of the format).

The answer, Oh delight! is the following 3 groups:

I.	12.56	$x_1x_3 - x_2^2 + p\beta^2$
	12.36	$x_1y_2 - x_2y_1 + \beta z_1$
	12.35	$x_2y_2 - x_3y_1 + \beta z_2$
	12.46	$x_1z_2 - x_2z_1 + \beta py_1$
	12.45	$x_2z_2 - x_3z_1 + \beta py_2$
	12.34	$y_2z_1 - y_1z_2 + \beta q$
II.	13.46	$x_1q - z_1^2 + py_1^2$
	13.45	$x_2q - z_1z_2 + py_1y_2$
	23.45	$x_3q - z_2^2 + py_2^2$

III. $23.46 = 13.45$

$13.56 = 12.46$

$23.56 = 12.45$

$14.56 = p \times 12.36$

$24.56 = p \times 12.35$

$34.56 = p \times 12.34$

Remarks. (i) If $\beta = 0$, these relations can be put back in the first determinantal form: you just have to express x_1q, x_2q, x_3q as quadratics in the rows of the 2×4 matrix in the top-right.

(ii) On the other hand, if $\beta \neq 0$ then the relations give z_1 and z_2 as polynomials in the other variables. If there is only 1 variable of degree 2, and p, y_1, y_2 are near their values for $\overline{R}$ (given in (5.8) below) then y_1 and y_2 are also functions of x_1, x_2, x_3, and z_1, z_2, y_1 and y_2 can be eliminated to give the single quintic relation 12.34. This discovery is essentially due to Griffin [Griffin1–2].

(iii) Conversely, a set of relations in the first determinantal form given by matrixes A and M as in (5.5) can be put in the Pfaffian form if and only if the first 2×2 minor of A

$$\begin{bmatrix} a_{11} & a_{12} \\ a_{21} & a_{22} \end{bmatrix}$$

can be made symmetric (by row and column operations), or equivalently, the single degree 2 relation $S: (a_{11}a_{22} - a_{12}a_{21})$ is a quadratic form of rank 3.

(5.8) Proposition. (a) If I set

$$y_1 = x_3^2 \text{ and } y_2 = y,\ \beta = 0,\ p = b_3y,\ q = hx_1 + 2b_1x_2y^2 + b_2x_3y^2$$

then the Pfaffian relations of (5.7) generate the same ideal as the 9 relations of (5.2).

(b) All the syzygies holding between the Pfaffian relations of (5.7) can be deduced from the determinantal format. Thus here too, arbitrary (small) variations of y_1, y_2, β, p, q give rise to flat deformations of the ring $\overline{R}$.

Proof. (a) is a trivial substitution: for example, 23.45 becomes

$$-z_2^2 + x_3q + py^2 = -z_2^2 + x_1x_3h + 2b_1x_2x_3y^2 + b_2x_3^2y^2 + b_3y^3 = W_2 + hS.$$

Thus under this specialisation, one can read off

$$\begin{bmatrix} 34.56 & 24.56 & 23.56 & 23.46 & 23.45 \\ & 14.56 & 13.56 & 13.46 & 13.45 \\ & & 12.56 & 12.46 & 12.45 \\ & & & 12.36 & 12.35 \\ & & & & 23.45 \end{bmatrix} = \begin{bmatrix} -b_3yV & b_3yT_2 & U_2 & W_1-b_1y^2S & W_2+hS \\ & b_3yT_1 & U_1 & W_0+b_2y^2S & W_1-b_1y^2S \\ & & S_2 & U_1 & U_2 \\ & & & T_1 & T_2 \\ & & & & -V \end{bmatrix}.$$

(5.9) Syzygies between Pfaffians. Suppose that $B = (b_{ij})$ is a $(2k+1) \times (2k+1)$ skew matrix, and $P = (\mathrm{Pfaff}_{ii})$ the column formed by the $(2k+1)$ diagonal $2k \times 2k$ Pfaffians of B; then $BP \equiv 0$ (or by symmetry $({}^tP)B \equiv 0$). It's useful to know also that the adjugate matrix ($\pm$ maximal minors) of B is $\operatorname{adj} B \equiv P({}^tP)$; and of course, $B(\operatorname{adj} B) \equiv (\det B)\cdot \mathrm{id} \equiv 0$.

This applies to every 5×5 diagonal block of N; thus if I make the skew 6×6 matrix $P = (P_{ij})$ with entries the 4×4 Pfaffians of N,

0	34.56	−24.56	23.56	−23.46	23.45
−34.56	0	14.56	−13.56	13.46	−13.45
24.56	−14.56	0	12.56	−12.46	12.45
−23.56	13.56	−23.56	0	12.36	−12.35
23.46	−13.46	12.46	−12.36	0	12.34
−23.45	13.45	−12.45	12.35	−12.34	0

(that is, $P_{ij} = \pm \mathrm{Pfaff}_{kl.mn}$ with $\pm$ = sign (ijklmn), the *Pfaffian adjugate* of a $(2k+2) \times (2k+2)$ skewsymmetric matrix N), then the off-diagonal elements of NP are identically zero; it's not hard to check that the diagonal elements are all equal to Pf, the 6×6 Pfaffian of N, so that

$$\Sigma = NP \equiv \mathrm{Pf}\cdot \mathrm{id}.$$

Since NP is 6×6, this provides *a priori* 35 identities between the relations of (5.7).

(5.10) Proof of (b). It is clearly enough to prove that after making the specialisation of (5.8, a), the determinantal syzygies just described generate the same module as the 16 syzygies of (5.4). This is a delicious calculation: write $\Sigma = NP = (\sigma_{ij})$, where

$$N = \begin{bmatrix} 0 & 0 & x_3^2 & z_1 & x_2 & x_1 \\ 0 & 0 & y & z_2 & x_3 & x_2 \\ -x_3^2 & -y & 0 & hx_1+2b_1x_2y^2+b_2x_3y^2 & z_2 & z_1 \\ -z_1 & -z_2 & -hx_1-2b_1x_2y^2-b_2x_3y^2 & 0 & b_3y^2 & b_3x_3^2y \\ -x_2 & -x_3 & -z_2 & -b_3y^2 & 0 & 0 \\ -x_1 & -x_2 & -z_1 & -b_3x_3^2y & 0 & 0 \end{bmatrix}$$

and

$$P = \begin{bmatrix} 0 & -b_3yV & -b_3yT_2 & U_2 & -W_1+b_1y^2S & W_2+hS \\ b_3yV & 0 & b_3yT_1 & -U_1 & W_0+b_2y^2S & -W_1+b_1y^2S \\ b_3yT_2 & -b_3yT_1 & 0 & S & -U_1 & U_2 \\ -U_2 & U_1 & -S & 0 & T_1 & -T_2 \\ W_1-b_1y^2S & -W_0-b_2y^2S & U_1 & -T_1 & 0 & -V \\ -W_2-hS & W_1-b_1y^2S & -U_2 & T_2 & V & 0 \end{bmatrix}.$$

Then $\sigma_{14}, \sigma_{13}, \sigma_{15}, \sigma_{16}, \sigma_{24}, \sigma_{23}, \sigma_{25}$ and σ_{26} are identically the 8 syzygies of the first set of (5.4); and $\sigma_{12}, \sigma_{22}-\sigma_{33}, \sigma_{35}, \sigma_{32}, \sigma_{11}-\sigma_{33}, \sigma_{21}, \sigma_{36}, \sigma_{31}$ are the 8 syzygies of the second set plus some multiples of the first. Go on, have a go! Q.E.D.

(5.11) Theorem. (I) In degree 0, every 1st order deformation $R^{(1)}$ of $\overline{R}$ can be put in the second determinantal form (5.7–8).

(II) In degree < 0, every 1st order extension $R^{(1)}$ of $\overline{R}$ can be put in the first determinantal form (5.3–4). (In degree ≤ -2, $R^{(1)}$ can be put in either form.)

Remarks. (a) Fixing a degree $a \le 0$ and making a 1-parameter extension (or deformation), 1st order deformations are thus unobstructed. What happens when extensions (in degree < 0) get mixed up with deformations (in degree 0) is more exciting, and is discussed in (5.15–16) below.

(b) The results here are exactly what one should expect. Plane quintics depend on 12 moduli, whereas hyperelliptic curves of genus 6 depend on 11, and (I) says that the latter can be seen as a smooth codimension 1 degeneration of the former; the only surprise is how complicated the algebra underlying this simple geometry turns out to be. Horikawa's geometric considerations show that a numerical quintic surface having a hyperelliptic canonical curve $C \in |K_X|$ has a genus 2 pencil, and can therefore be written in determinantal form; the result (II) says that this also holds for every 1st order extension 2C of C in degree -1.

(c) To understand the difference between the two cases, note that the deformation as a Pfaffian with $\beta \neq 0$ is impossible in $\deg < 0$ (since β would have to be a polynomial of degree < 0. On the other hand, if you don't add an element x_0 of degree 1 to the polynomial ring, an arbitrary deformation of $S\colon x_1x_3 = x_2^2$ remains a quadric of rank 3; so in the first determinantal format, A must start with the symmetric 2×2 block

$$\begin{bmatrix} x_1 & x_2 \\ x_2 & x_3 \end{bmatrix}$$

(5.12) Setting up the 1st order deformation calculation. By (1.10–11), the Hilbert scheme of 1st order extensions of $\overline{R}$ in degree $-a < 0$ (or deformations in degree $a = 0$) is the vector space

$$\mathbb{H}^{(1)}(\overline{R}, a) = \operatorname{Hom} \overline{R}(, \overline{R})_{-a} = \left\{(f'_i \in \overline{R}_{d_i - a}) \mid \forall j,\ \sum \ell_{ij} f'_i = 0 \in \overline{R}_{s_j - a}\right\}.$$

A useful observation is that although the expression on the right says to use all 16 of the syzygies of (5.4) for $\overline{R}$, they are all implied by the following very convenient subset:

First few syzygies.

$$\sigma_1: \quad x_1T_2 \equiv x_2T_1 - x_3^2S;$$
$$\sigma_2: \quad x_1U_2 \equiv x_2U_1 - z_1S;$$
$$\sigma_3: \quad x_1V \equiv x_3^2U_1 - z_1T_1;$$
$$\sigma_4: \quad x_1W_1 \equiv x_2W_0 + (b_1x_1 + b_2x_2)y^2S + b_3x_3^2yT_1 - z_1U_1;$$
$$\sigma_5: \quad x_1W_2 \equiv x_2W_1 + (b_1x_2 + b_2x_3)y^2S + b_3y^2T_1 - z_2U_1.$$

This is true because every syzygy Σ of (5.4) has a monomial multiple which is a linear combination of these 5, as can be checked by an elementary calculation; for example,

$$x_1(x_2V - x_3^2U_2 + z_1T_2) = x_2\sigma_3 - x_3^2\sigma_2 + z_1\sigma_1.$$

Thus, since each of $x_1, x_2, x_3, y, z_1, z_2$ is a non-zerodivisor of $\overline{R}$, I need only verify the condition $\sum \ell_{ij}f'_i = 0 \in \overline{R}_{s_j-a}$ for these 5 values of j.

For the 1st order calculation, I've got to write down all S', T'_1, U'_1 and $W'_0 \in \overline{R}$ of degrees 2–a, 3–a, 4–a and 6–a such that in turn T'_2, U'_2, V', W'_1 and W'_2 can be found to satisfy

$$\sigma'_1: \quad x_1T'_2 = x_2T'_1 - x_3^2S';$$
$$\sigma'_2: \quad x_1U'_2 = x_2U'_1 - z_1S';$$
$$\sigma'_3: \quad x_1V' = x_3^2U'_1 - z_1T'_1;$$
$$\sigma'_4: \quad x_1W'_1 = x_2W'_0 + (b_1x_1 + b_2x_2)y^2S' + b_3x_3^2yT'_1 - z_1U'_1;$$
$$\sigma'_5: \quad x_1W'_2 = x_2W'_1 + (b_1x_2 + b_2x_3)y^2S' + b_3y^2T'_1 - z_2U'_1.$$

Each of these equalities in $\overline{R}$ is written as a condition of divisibility by x_1; this is a very concrete linear condition on S', T'_1, U'_1 and W'_0, especially since by (4.5, I), it is natural to write down a monomial basis of each $\overline{R}_d$ in alphanumeric order, with x_1 first:

Table of bases of $R_d = H^0(C, \mathcal{O}_C(dD))$:

$H^0(D)$: x_1, x_2, x_3

$H^0(2D)$: $x_1^2, x_1x_2, x_1x_3, x_2x_3, x_3^2, y$

$H^0(3D)$: $x_1^3, x_1^2x_2, x_1^2x_3, x_1x_2x_3, x_1x_3^2, x_1y, x_2y, x_3y; z_1, z_2$

$H^0(4D)$: $x_1^4, x_1^3x_2, x_1^3x_3, x_1^2x_2x_3, x_1^2x_3^2, x_1^2y, x_1x_2y, x_1x_3y, x_2x_3y, x_3^2y, y^2;$ $x_1z_1, x_1z_2, x_2z_2, x_3z_2,$

etc.

(5.13) 1st order deformation calculation in degree 0.

The proof of (5.11) is similar to that of (2.2–6), and I omit some details. *A priori*,

$$S' = \varepsilon_1x_1^2+\varepsilon_2x_1x_2+\varepsilon_3x_1x_3+\varepsilon_4x_2x_3+\varepsilon_5x_3^2+\varepsilon_6y;$$

However, since the relation is

$$x_1x_3 = x_2^2+\lambda S',$$

(with $\deg \lambda = 0$), I can make coordinate changes of the form $x_i \mapsto x_i + \lambda \sum a_{ij}x_j$ to kill all the terms except the last, so assume $S' = \varepsilon_6 y$. Similarly, I can reduce T'_1 to

$$T'_1 = \alpha_7x_2y+\alpha_8x_3y+\beta_1z_1+\beta_2z_2.$$

by changes in y. Now plugging in σ'_1 gives

$$x_1T'_2 = x_2T'_1 - x_3^2S' \in H^0(4D);$$

since x_2x_3y, x_3^2y, x_2z_2 are linearly independent modulo multiples of x_1, I conclude $\varepsilon_6 = \alpha_8 = \beta_2 = 0$, and

$$S' = 0;\ T'_1 = \alpha x_2y+\beta z_1;\ \text{and}\ T'_2 = \alpha x_3y+\beta z_2.$$

Similarly, U'_1 can be reduced to $U'_1 = \gamma_{10}x_3^2y+\gamma_{11}y^2+\delta_4x_3z_2$ by changes in z_1 and z_2, and plugging into σ'_2: $x_1U'_2 = x_2U'_1 - z_1S'$ gives $\gamma_{11} = \delta_4 = 0$, so that

$$U'_1 = \gamma x_3^2y \quad\text{and}\quad U'_2 = \gamma y^2.$$

As an arbitrary element of $H^0(6D)$, W'_0 can be written as

$$\begin{aligned}W'_0 = {} & x_1^2h'+\delta_1x_1x_2y^2+\delta_2x_1x_3y^2+\delta_3x_2x_3y^2+\delta_4x_3^2y^2+\delta_5y^3\\ & +\varepsilon_1x_1x_2x_3z_2+\varepsilon_2x_1x_3^2z_2+\varepsilon_3x_1yz_2+\varepsilon_4x_2yz_2+\varepsilon_5x_3yz_2;\end{aligned}$$

(the ε_i are new, but they too will all die). A coordinate change of the form $z_1 \mapsto z_1 + \delta x_1 Q$, $z_2 \mapsto z_2 + \delta x_2 Q$ can be used to fix up $\varepsilon_1 = \varepsilon_2 = \varepsilon_3 = 0$ (same Q, so as not to alter U'_1 and U'_2). Now the syzygy σ'_4: $x_1 W'_1 = x_2 W'_0 + b_3 x_3{}^2 y(\alpha x_2 y + \beta z_1) - \gamma x_3{}^2 y z_1$ gives

$$\delta_5 = 0 \quad \text{and} \quad \varepsilon_5 = -b_3\beta + \gamma,$$

so that

$$W'_1 = x_1 x_2 h' + \delta_1 x_1 x_3 y^2 + \delta_2 x_2 x_3 y^2 + \delta_3 x_3{}^2 y^2 + (\delta_4 + b_3\alpha) y^3 + \varepsilon_4 x_3 y z_2;$$

in turn, plugging into σ'_5: $x_1 W'_2 = x_2 W'_1 + b_3 y^2(\alpha x_2 y + \beta z_1) - \gamma x_3{}^2 y z_2$ implies $\delta_4 = -2b_3\alpha$, $\varepsilon_4 = 0$ and $\gamma = b_3\beta$. So finally

$$S' = 0,\ T'_1 = \alpha x_2 y + \beta z_1,\ T'_2 = \alpha x_3 y + \beta z_2,\ U'_1 = b_3\beta x_3{}^2 y,\ U'_2 = b_3 \beta y^2,$$

$$V' = -\beta(h x_1 + 2b_1 x_2 y^2 + b_2 x_3 y^2) - \alpha y z_2,$$

$$W'_0 = x_1{}^2 h' + \delta_1 x_1 x_2 y^2 + \delta_2 x_1 x_3 y^2 + \delta_3 x_2 x_3 y^2 - 2b_3 \alpha x_3{}^2 y^2,$$

$$W'_1 = x_1 x_2 h' + \delta_1 x_1 x_3 y^2 + \delta_2 x_2 x_3 y^2 + \delta_3 x_3{}^2 y^2 - b_3 \alpha y^3,$$

$$W'_2 = x_1 x_3 h' + \delta_1 x_2 x_3 y^2 + \delta_2 x_3{}^2 y^2 + \delta_3 y^3.$$

It is now easy to assemble the relations to 1st order into the Pfaffian determinantal format $\operatorname{rk} N^{(1)} \leq 2$, where

$$N^{(1)} = \begin{bmatrix} 0 & \lambda\beta & x_3^2 - \lambda\alpha y & z_1 & x_2 & x_1 \\ & 0 & y & z_2 & x_3 & x_2 \\ & & 0 & q + \lambda q' & z_2 & z_1 \\ & & & 0 & (p+\lambda p')y & (p+\lambda p')(x_3^2 - \lambda\alpha y) \\ & -\mathrm{sym} & & & 0 & \lambda(p+\lambda p')\beta \\ & & & & & 0 \end{bmatrix},$$

with $q' = x_1 h' + \delta_1 x_2 y^2 + \delta_2 x_3 y^2$ and $p' = \delta_3 y$.

(5. 14) 1st order deformation calculation in degree < 0. The computation in degree -2, -3,.. is a straightforward exercise for the reader. (But beware: this kind of conclusion is not at all automatic: Dicks has an example of a ring having obstructed 1st order extensions in degree -4 only.)

I give the computation in degree -1 in skeleton form, since step by step it is almost identical to that in degree 0; I use Latin letters r, s, t, u, v instead of Greeks so that the notations of (5.13) and (5.14) can be added together in the contrapuntal climax (5.15–16).

Set $S' = 0$ (use $x_i \mapsto x_i + {*}x_0$) and $T'_1 = r_3x_2x_3 + r_4x_3^2 + r_5y$ (use $y \mapsto y + x_0({*}x_1 + {*}x_2 + {*}x_3)$). Then

$$\sigma'_1\colon x_1T'_2 = x_2T'_1 - x_3^2S' \implies r_5 = 0 \text{ and } T'_2 = r_3x_3^2 + r_4y.$$

Set $U'_1 = t_7x_3y + s_1z_1 + s_2z_2$ (use $z_1 \mapsto z_1 + {*}x_0q$ and same for z_2). So

$$\sigma'_2\colon x_1U'_2 = x_2U'_1 - z_1S' \implies s_2 = t_7 = 0 \text{ and } U'_2 = s_1z_2.$$

$$\sigma'_3\colon x_1V' = x_3^2U'_1 - z_1T'_1 \implies r_4 = s_1 \text{ and } V' = -r_3x_3z_2.$$

So writing $r_3 = r$, $r_4 = s_1 = s$, this summarises as

$$S' = 0,\ T'_1 = rx_2x_3 + sx_3^2,\ T'_2 = rx_3^2 + sy,\ U'_1 = sz_1,\ U'_2 = sz_2,\ V' = -rx_3z_2,$$

all of which fits together as the x_0 terms in $\operatorname{rk} A^{(1)} \le 1$, where

$$A^{(1)} = \begin{bmatrix} x_1 & x_2 - sx_0 & x_3^2 + rx_0x_3 & z_1 \\ x_2 + sx_0 & x_3 & y & z_2 \end{bmatrix}.$$

Now set

$$-W'_0 = x_1^2h' + t_9x_1x_2x_3y + t_{10}x_1x_3^2y + t_{11}x_1y^2 + t_{12}x_2y^2 + t_{13}x_3y^2 + u_6x_3^2z_2 + u_7yz_2$$

(using $z_1 \mapsto z_1 + x_0x_1m$, $z_2 \mapsto z_2 + x_0x_2m$ as before, and also $z_1 \mapsto z_1 + {*}x_0x_3^2$, $z_2 \mapsto z_2 + {*}x_0y$ to kill the term $u_5x_3^2z_1$). Then

$$\sigma'_4\colon x_1W'_1 = x_2W'_0 + b_3x_3^2y(rx_2x_3 + sx_3^2) - sz_1^2$$

implies $t_{13} = u_7 = 0$ and

$$-W'_1 = -sx_1h + x_1x_2h' + t_9x_1x_3^2y + t_{10}x_2x_3^2y + (t_{11} - 2b_1s)x_2y^2 + (t_{12} - b_2s + b_3r)x_3y^2 + u_6yz_2.$$

Then

$$\sigma'_5\colon x_1W'_2 = x_2W'_1 + b_3y^2(rx_2x_3 + sx_3^2) - sz_1z_2$$

implies $t_{12} = 2b_2s - 2b_3r$, $u_6 = 0$, and

$$-W'_2 = -2sx_2h + x_1x_3h' + t_9x_1y^2 + t_{10}x_2y^2 + (t_{11} - 4b_1s)x_3y^2.$$

Thus, in conclusion, after the same massaging as in (5.2),

$$-W'_0 = x_1{}^2h' + 2v_1x_1x_2x_3y + v_2x_2{}^2x_3y + v_3x_3{}^5 + 2b_1sx_1y^2 + 2b_2sx_2y^2 - 2b_3rx_3{}^3y,$$
$$-W'_1 = -sx_1h + x_1x_2h' + v_1(x_1x_3 + x_2{}^2)x_3y + v_2x_2x_3{}^2y + v_3x_3{}^3y + (b_2s - b_3r)x_3y^2,$$
$$-W'_2 = -2sx_2h + x_2{}^2h' + 2v_1x_2x_3{}^2y + v_2x_3{}^3y + (v_3 - 2b_1s)x_3y^2.$$

The reader familiar with the rules for matrix multiplication will see that to 1st order, these are the x_0 terms of the 3 relations $A^{(1)}M^{(1)}({}^tA^{(1)}) = 0$, where $A^{(1)}$ is given above and

$$M^{(1)} = \begin{bmatrix} h & b_1y^2 & & \\ b_1y^2 & b_2y^2 & & \\ & & b_3y & \\ & & & -1 \end{bmatrix} + x_0\begin{bmatrix} h' & v_1x_3y & \\ v_1x_3y & v_2x_3y & \\ & & \end{bmatrix}.$$

(5.15) Mixing extensions and deformations. I now consider 'extension–deformations' of $\overline{R} = R(C, \mathcal{O}_C(D))$, depending on two variables λ, x_0 of degrees 0 and 1. For example, this situation occurs if I want to study deformations of a given numerical quintic surface S extending C; or equally, if I have a given flat deformation $\overline{R}_\lambda$ of $\overline{R}$ and I want to study simultaneous extensions of the $\overline{R}_\lambda$.

Let A be a graded local Artinian k–algebra (with degree 0 piece maybe bigger than just k). A graded deformation of $\overline{R}$ over A is an A–algebra R_A that is both a flat deformation of $\overline{R}$ over A and graded as k–algebra. If A is generated by elements of different degrees, the ideal of relations defining R_A will be homogeneous in all the variables.

Write $B = k[\lambda, x_0]/(\lambda, x_0)^2$, where $\deg \lambda = 0$, as in (5.13) and $\deg x_0 = 1$. The set of R_B is strictly a 1st order problem, whose solution is just the direct sum of the two vector spaces studied in (5.13) and (5.14): every R_B can be written (after suitable coordinate changes) in the form $B[x_1,.. z_2]/I_B$, where I_B is generated by 9 relations

$${}_BF = F + \lambda F'_{(\lambda)} + x_0F'_{(x_0)},$$

where F is a relation for $\overline{R}$ as in (5.2), $F'_{(\lambda)}$ the bit added on in (5.13), and $F'_{(x_0)}$ the bit added on in (5.14) (examples are given in the proof of (5.16)). I continue to use the notation of

(5.13) and (5.14) without further mention.

(5.16) The 2nd order obstruction. Now let $C = k[\lambda, x_0]/(\lambda^2, x_0^2)$; despite appearances, deformations of $\overline{R}$ over C is no longer a 1st order problem, since the maximal ideal m_C has $m_C^2 \neq 0$.

Theorem. Let R_B be a graded deformation of $\overline{R}$ over B; then

$$R_B \text{ lifts to a deformation } R_C \iff \beta s = 0.$$

Proof. If $\beta = 0$ then R_B can be written in the first determinantal format, and so is unobstructed; if $s = 0$ it can be written in the Pfaffian format, and is likewise unobstructed. So the point is to prove $\Rightarrow$.

By (5.13) and (5.14), the first 4 relations for R_B are

$$\begin{aligned}
{}_BS:&\quad x_1x_3 &=&\ x_2^2,\\
{}_BT_1:&\quad x_1y &=&\ x_2x_3^2 + \lambda(\alpha x_2y + \beta z_1) + x_0(rx_2x_3 + sx_3^2),\\
{}_BT_2:&\quad x_2y &=&\ x_3^3 + \lambda(\alpha x_3y + \beta z_2) + x_0(rx_3^2 + sy),\\
{}_BU_1:&\quad x_1z_2 &=&\ x_2z_1 + \lambda(b_3\beta x_3^2y) + x_0(sz_1).
\end{aligned}$$

The first syzygy $x_1T_2 - x_2T_1 + yS$ for $\overline{R}$ upgrades to one for R_B as follows:

$$x_1({}_BT_2) - x_2({}_BT_1) + y({}_BS) \equiv \lambda(\alpha y({}_BS) + \beta({}_BU_1)) + x_0(rx_3({}_BS) + s({}_BT_1)).$$

A lift of R_B to R_C involves patching the relations ${}_BS, {}_BT_1, {}_BT_2$, etc. to

$${}_CS = {}_BS + \lambda x_0S'', \quad {}_CT_1 = {}_BT_1 + \lambda x_0T''_1, \quad {}_CT_2 = {}_BT_2 + \lambda x_0T''_2 \text{ etc.}$$

in such a way that all the syzygies can be extended (this is exactly the argument of (1.13, (5)). For this it is necessary that

$$x_1({}_CT_2) - x_2({}_CT_1) + y({}_CS) = \lambda(\alpha y({}_CS) + \beta({}_CU_1)) + x_0(rx_3({}_CS) + s({}_CT_1));$$

this is supposed to be an equality in $\overline{R}$ between the λx_0 terms, since the constant and 1st order terms have already been fixed up to vanish.

Now I claim that if $\beta s \neq 0$ this inequality cannot hold for any choice of S'', T''_1, T''_2. In fact the λx_0 term of the right-hand side is already determined by the relations for R_B, and is

$$\lambda\beta x_0(sz_1) + x_0 s\lambda(\alpha x_2 y + \beta z_1)$$

However, the right-hand side consists of assorted multiples of x_1, x_2 and y, so can't possibly hit z_1. Q.E.D.

(5.17) Numerical quintics. A *numerical quintic* is a polarised n-fold X, D such that

(i) D is ample;

(ii) $K_X = (3-n)D$,

(iii) $h^0(\mathcal{O}_X(D)) = n+2$

and

(iv) $D^n = 5$.

One hopes that under suitable nonsingularity conditions, the linear system $|D|$ contains $n-1$ elements whose intersection is a nonsingular curve C. This has been proved by Horikawa if X is a nonsingular surface or 3-fold. (If $|D|$ defines a generically finite map, then easy numerical considerations show that $|D|$ is free and birational, or has a single reduced point as its base locus and is 2-to-1.)

Corollary. Assuming this, the ring $R(X, \mathcal{O}_X(D))$, is either a quintic hypersurface, or of the first determinantal format (5.3); in the latter case, $\varphi_D(X)$ is a quadric of rank 3 or 4; if the rank is 4 then all small deformations of X are given by varying the coefficients in the determinantal format, and 1st order deformations are unobstructed. If the rank is 3 then deformations of X form two components, one of which consists of quintic hypersurfaces.

§6. Six minuets for a mechanical clock

(6.1) Main algorithm. This section outlines routines to mechanise the ideas of §1, intended as a 'pseudocode' computer program to calculate the moduli space of deformations of a ring $\overline{R}$ as the subscheme in $\mathbb{T}^1_{<0}$ defined by the vanishing of an obstruction morphism

$$\texttt{obs}\colon \mathbb{T}^1_{<0} \to \mathbb{T}^2_{<0}.$$

More precisely, I describe an algorithm having the following input and output:

Data: A specification of a ring $\overline{R}$ in terms of generators, syzygies and second syzygies, that is, the ring $A = k[x_1,.. x_n]$, the generators (f_i) of $\overline{I}$ and the resolution (1.14)

$$.. \xrightarrow{(m_{jn})} \oplus A(-s_j) \xrightarrow{(\ell_{ij})} \oplus A(-d_i) \xrightarrow{(f_i)} \overline{I} \to 0.$$

Result: The graded vector spaces $\mathbb{T}^1_{<0}$, $\mathbb{T}^2_{<0}$ and the polynomial map `obs`: $\mathbb{T}^1_{<0} \to \mathbb{T}^2_{<0}$ between them.

The input and output are objects of the same type: the f_i are homogeneous polynomials, and so are the components $\mathtt{obs}_\iota$ of `obs`. If the coefficients of the f_i are symbolic (expressions involving indeterminates), then so are the $\mathtt{obs}_\iota$, whereas if the f_i are numerical (rational numbers or elements of a finite field), then so are the $\mathtt{obs}_\iota$.

Some practical considerations on implementing this algorithm are discussed in (6.5) below.

(6.2) Monomial basis routine. First of all, it's clear that the whole computation will only involve $\overline{R}_{\leq d}$ for some d that is readily determined *a priori* from the data (in fact the minimum degree of the second syzygies $\min t_k$ will do). A monomial basis for $\overline{R}$ is a set of monomials $x^{\underline{m}_\lambda}$ in the variables $\{x_1,.. x_n\}$ whose images in $\overline{R}$ form a basis; because of what I just said, I will only need to deal with a finite set $\{x^{\underline{m}_\lambda}\}_{\lambda \in \Lambda}$ forming a basis of $\overline{R}_{\leq d}$. I need the following:

(i) at the outset, a choice of a monomial basis $\{x^{\underline{m}_\lambda}\}$ of $\overline{R}$;

(ii) a general-purpose procedure `general_polynomial(name, deg)` that writes out a general polynomial of given degree

$$f = \sum \mathtt{c}_{(\mathtt{name})\lambda} \cdot x^{\underline{m}_\lambda},$$

with indeterminate coefficients, where $\mathtt{c}_{(\mathtt{name})}$ is a suitable name for the coefficients, and the sum takes place over all basic monomials of given degree.

(iii) a reduction procedure that takes any polynomial h into its normal form $hh \mod \overline{I}$, with hh a linear combination of the $x^{\underline{m}_\lambda}$;

(iv) the *toll*, a vector $\mathtt{toll} = \{\mathtt{toll}_i\}$ such that

$$h \equiv hh + \sum \mathtt{toll}_i \cdot f_i.$$

Here the normal form hh is the final outcome of a reduction process, and is sufficient for 1st order purposes; `toll` is a record of how each relation f_i is used in the reduction process. It's used to replace equalities in $\overline{R}$ by identities in $k[x_1,.. x_n]$; in particular, to determine $\ell^{(k)}{}_{ij}$ when going back from (1.12, (6)) to (1.12, (5)) for higher order work.

In concrete cases, the required monomial basis may be fixed up in an *ad hoc* way; for example, in §5 it's not hard to see how to use the relations (5.2) to pass from any element of the ring $\overline{R}$ to a combination of basic monomials as in (5.12).

(6.3) Computation of $\mathbb{T}^1$ and $\mathbb{T}^2$. Since $H^{(1)}$, $\mathbb{T}^1$ and $\mathbb{T}^2$ are graded vector spaces, the 1st order computation breaks up into independent routines for the graded pieces of each degree $-k$.

The two vector spaces $H^{(1)}$ and $\mathbb{T}^2$ are the homology of the conormal complex

$$L_0 = \oplus \overline{R}(d_i) \xrightarrow{\delta_0} L_1 = \oplus \overline{R}(s_j) \xrightarrow{\delta_1} ..$$

where δ_0 and δ_1 are matrixes with polynomial entries, the transpose of (ℓ_{ij}), (m_{jk}) given in the data. An element of L_0 of degree $-k$ is a vector $\varphi = \{f'_i\}$ with $f'_i \in \overline{R}_{d_i-k}$. First write out $f'_i = \sum cf_{*i\lambda}\cdot x^{\underline{m}_\lambda}$ using the procedure (6.2, ii) (putting enough information into the coefficient names $cf_{*i\lambda}$ to distinguish them from all previous names).

To calculate $\mathbb{T}^1$ in any degree $-k<0$:

Step 0. Generate $\{f'_i\}$ for each i, and make a list of their coefficients.

Step 1. Divide by the group of coordinate transformations $x_\ell \mapsto x_\ell + c_\ell x_0$, where $c_\ell \in \overline{R}_{a_i-k}$. The effect is to replace f'_i by $f'_i + \sum c_\ell \partial f_i/\partial x_\ell$, so that the quotient by a group action is in this case just a quotient vector space. As in (2.4) and (5.13), this simply means using the coordinate changes to assign values (usually 0) to as many of the coefficients $cf_{*i\lambda}$ of the f'_i as possible. (To calculate $H^{(1)}$ this step would be omitted.)

Start a loop on j, ranging from 1 up to the number of syzygies: for each of the syzygies σ_j, carry out the following 3 steps.

Step 2. Evaluate the jth entry of $\delta_0(\{f'_i\})$; that is, calculate $\sum \ell_{ij}f'_i$ in the polynomial ring and carry out the reduction

$$\texttt{to\_kill}_j := \text{normal form of } \sum \ell_{ij}f'_i$$

using the procedure (6.2, iii). The kernel of δ_0 is of course obtained by setting $\texttt{to\_kill}_j = 0$

for each j.

Step 3. To equate coefficients in `to_kill`$_j = 0$, start an internal loop through each monomial $x^{\underline{m}\mu}$ of degree $s_j - k$. Set to zero the coefficient of $x^{\underline{m}\mu}$ in `to_kill`$_j$ to get a homogeneous linear equation

$$\texttt{eqh}(j, x^{\underline{m}\mu}) := ((\text{coefficient of } x^{\underline{m}\mu} \text{ in } \texttt{to\_kill}_j) = 0)$$

in the list of indeterminate coefficients $\{cf_{*i\lambda}\}$. Solve this, to obtain a new relation of the form

$$\text{dependent coefficient} = \text{combination of others},$$

for one coefficient $cf_{*i\lambda}$, and assign this value.

Keep side-effect. The subscript $*i\lambda$ is *a priori* not know; for the higher order computation, it is crucial to remember the point at which the coefficients $cf_{*i\lambda}$ are solved for; so at this point, make a table `Table`$_1$ relating $*i\lambda$ to $(j, x^{\underline{m}\mu})$.

(end of $x^{\underline{m}\mu}$ loop, end of j loop).

After all this, some of the coefficients $cf_{*i\lambda}$ have values assigned. Define `defvar[k]` to be the set of remaining independent (unassigned) coefficients; these are coordinates on $\mathbb{T}^1{}_{-k}$.

The calculation of $\mathbb{T}^2{}_{-k}$ is similar, and I only sketch it: start from the vector space

$$\{g'_j = \sum cg_{*j\lambda} \cdot x^{\underline{m}\lambda} \in \overline{R}_{s_j - k}\},$$

with a monomial basis, divide by $\operatorname{im} \delta_0$ (assigning values to some of the coefficients $cg_{*j\lambda}$), then use the linear equations defining $\ker \delta_1$ to assign values to more of the $cg_{*j\lambda}$. Here I again make a table `Table`$_2$ to remember when a coefficient $cg_{*j\lambda}$ is solved for (that is, assigned a value) by the linear equations defining $\ker \delta_1$. The meaning of this table is that δ_1 (the second syzygies) expresses the $x^{\underline{m}\mu}$ term of the syzygy s_j (corresponding to $cg_{*j\lambda}$) as a linear combination of terms of the other syzygies, and does not give rise to new independent obstructions. At the end, the unassigned coefficients $cg_{*j\lambda}$ are coordinates on $\mathbb{T}^2{}_{-k}$.

Short-cuts. (1) It may not be necessary to write out all the $\{f'_i\}$ as general polynomials: if one entry ℓ_{ij} of the matrix δ_0 is a non-zerodivisor of $\overline{R}$ then the jth syzygy is equivalent to

$$\ell_{ij} \text{ divides } \sum_{\iota \neq i} \ell_{\iota j} f'_\iota \quad \text{and} \quad f'_i := \{\sum_{\iota \neq i} \ell_{\iota j} f'_\iota\} / \ell_{ij}.$$

For example, in (5.12) all 5 of the syzygies used were treated in this way.

(2) The 1st order calculation of (6.3) can be done as a self-contained, purely linear routine. However, to avoid repeating all the 1st order work when proceeding to 2nd order, it is desirable to get hold of the `toll` in passing to the normal form in Step 2, which gives ℓ'_{ij}.

(6.4) Higher order theory. The higher order theory is done by induction, and I start off the induction assuming that the calculation of $\mathbb{T}^1{}_{-\kappa}$ and $\mathbb{T}^2{}_{-\kappa}$ have been carried out for all $\kappa < k$, resulting in an array `defvar[1 .. k-1]`, where each `defvar[κ]` consists of coordinates on $\mathbb{T}^1{}_{-\kappa}$. Write `defvars` for the union of these. The obstructions will be kept in a set `obs` of relations between the `defvars` (more usefully, as an ideal in k[`defvars`]), and I initialise `obs := {}` to be the empty set (or the zero ideal).

I fix $\deg x_0 = 1$. Now I work by induction on k, starting with $k = 2$ because the 1st order part has already been done in (6.3). Consider kth order deformations of the form

$$f_i + x_0 f'_i + .. x_0{}^k f^{(k)}{}_i \quad \text{and} \quad \ell_{ij} + x_0 \ell'_{ij} + .. x_0{}^k \ell^{(k)}{}_{ij}$$

as in (1.13). Assuming that everything is known up to order (k−1), the new unknowns $f^{(k)}{}_i$ must satisfy

$$\sum \ell_{ij} f^{(k)}{}_i = \psi_j \in \overline{R}_{s_j - k a_0}, \qquad (*)$$

where $\psi_j = -\sum_{a=1}^{k-1} \sum \ell^{(a)}{}_{ij} f^{(k-a)}{}_i$ is as in (1.12, (6)).

I calculate all possibilities for the kth order terms at the same time as writing out the new obstructions.

Step 1. Write out the general form of $\varphi = \{f^{(k)}{}_i\}$ with $f^{(k)}{}_i \in \overline{R}_{d_i - k}$, and divide by the coordinate change $x_\ell \mapsto x_\ell + c_\ell x_0$, where $c_\ell \in \overline{R}_{a_i - k}$. This is exactly the same calculation as (6.3, Steps 0–1).

Now for each syzygy j, carry out the following steps, parallel to those in (6.3):

Step 2. Evaluate $\sum \ell_{ij} f^{(k)}{}_i - \psi_j$ where ψ_j is as in $(*)$ above, and reduce

$$\texttt{to\_kill}_j := \text{normal form of } \sum \ell_{ij} f^{(k)}{}_i - \psi_j$$

using the reduction procedure (6.2, iii); remember to set $\ell^{(k)}{}_{ij} :=$ `-toll`$_j$. This is similar to (6.3, Step 2), except that carrying forward the ψ_j adds nonlinear terms from the order $<k$ calculations to the purely linear kth order part $\sum \ell_{ij} f^{(k)}{}_i$.

Step 3. Equate coefficients of each monomial $x^{\underline{m}_\mu}$ in `to_kill`$_j$, to get an equation `eqin`$(j, x^{\underline{m}_\mu})$ in the kth order indeterminate coefficients $\{cf_{*i\lambda}\}$ which is inhomogeneous linear in these, but with coefficients involving the defvar[κ] with $\kappa < k$.

Step 4. In deciding how to handle each equation `eqin`$(j, x^{\underline{m}_\mu})$ there is a division into 3 cases, depending on the information remembered in the 'side-effect' tables $\mathtt{Table}_1, \mathtt{Table}_2$ constructed in (6.3, Step 3).

Case A. If $\mathtt{Table}_1$ remembers that $\mathtt{eqh}(j, x^{\underline{m}_\mu})$ was used in (6.3, Step 3) of the 1st order problem to assign a value to $cf_{*i\lambda}$ in the 1st order problem, then $\mathtt{eqin}(j, x^{\underline{m}_\mu})$ can also be used to assign a value to $cf_{*i\lambda}$ (not the same value). This is true because the kth order indeterminates $cf_{*i\lambda}$ appear with the same coefficients in the two equations.

Case B. If $\mathtt{Table}_2$ remembers that in the 1st order problem the $x^{\underline{m}_\mu}$ term of the syzygy s_j was a linear combination of terms of the other syzygies, then just ignore $\mathtt{eqin}(j, x^{\underline{m}_\mu})$.

Case C. If neither of Cases A–B hold then $\mathtt{eqin}(j, x^{\underline{m}_\mu})$ is a new obstructions, a component of $\mathtt{obs}\colon \mathbb{T}^1{}_{<0} \to \mathbb{T}^2{}_{-k}$.

(end of $x^{\underline{m}_\mu}$ loop, end of j loop).

(6.5) Considerations of space and time. Concerning the feasibility of implementing this algorithm, I have a version of it running (written in Maple, [Maple]) to compute the deformation theory of a very specific example related to Godeaux surfaces with torsion $\mathbb{Z}/2$; in this example the ring needs 8 generators, 20 relations, 64 syzygies and 90 second syzygies, and the relations f_i have 10 indeterminates among their coefficients. My program works with symbolic coefficients, and polishes off easily the computation of $\mathbb{T}^1{}_{<0}$, which is 17-dimensional; at present I don't have the computation of $\mathbb{T}^2{}_{<0}$ implemented, although this does not present special difficulties (it should be approximately 10-dimensional). My computation of obstructions eventually grinds to a halt, growing too large for computer memory and the tolerance of my fellow-users (typically taking 80% of 16Mb memory, about 30 hours of CPU time, for an unfinished calculation). The obstructions are polynomials in 17 variables, with hundreds of nonzero symbolic coefficients (the first two take half a page each to print out, and after that 3 or 4 pages each), so that for example using the standard Gröbner basis package provided in Maple is not feasible to control them. I nevertheless expect that a modified version of this program will eventually run to completion, and decide the irreducibility of the moduli space of Godeaux surfaces with torsion $\mathbb{Z}/2$.

As described in (2.10), my preliminary notes on this calculation formed §6 (17 pages) of the preprint of this paper, and pending a more definitive composition these are still available on request, together with some version of my Maple routines.

On the other hand, there is the alternative approach in the spirit of Macaulay [Bayer and Stillman] in which all the rings are defined by polynomials with one word coefficients; I guess the advantage is that polynomials can be viewed as a much simpler data type (the array of its coefficients) instead of the recursive structure of symbolic computation packages such as Maple, thus saving on all the overheads of simplification. It seems clear to me that in these terms my algorithm will involve the same order of magnitude of complexity as the existing Macaulay package, so that rings of the size of my example (but defined by polynomials with coefficients in a prime field $\mathbb{Z}/(p)$) can be dealt with easily.

References

[4 authors] E. Arbarello, M. Cornalba, P.A. Griffiths and J. Harris, Geometry of algebraic curves, vol. I, Springer, 1985.

[Artin] M. Artin, Lectures on deformations of singularities, Tata Inst. lecture notes **54**, Bombay 1976.

[Bădescu] L. Bădescu, On ample divisors, I, Nagoya Math J. **86** (1982), 155–171; and II, in Proceedings of the week of algebraic geometry (at Bucharest 1980), Teubner Texte zur Mathematik **40**, Leipzig 1981, 12–32.

[Barlow1] R. Barlow, Some new surfaces with $p_g = 0$, Warwick Ph.D. thesis, 1982.

[Barlow2] R. Barlow, A simply connected surface of general type with $p_g = 0$, Invent. Math **79** (1985), 293–301.

[Bayer and Stillman] D. Bayer and P. Stillman, Macaulay, computer program, version 2.0 for Apple Macintosh, distributed Nov. 1986 from Columbia Univ. and Brandeis Univ.

[Catanese] F. Catanese, Pluricanonical Gorenstein curves, in Enumerative geometry and classical algebraic geometry (at Nice, 1981), Birkhaüser, 1982, 51–95.

[Catanese and Debarre] F. Catanese and O. Debarre, Surfaces with $K^2 = 2$, $p_g = 1$, $q = 0$, Orsay preprint, 1988, to appear in Crelle's J.

[Ciliberto] C. Ciliberto, Sul grado dei generatori dell'anello canonico di una superficie di tipo generale, Rend. Sem. Mat. Univ. Politec. Torino **41**:3 (1983/84), 83–111.

[Dicks] D. Dicks, Surfaces with $p_g = 3$, $K^2 = 4$ and extension–deformation theory, Warwick Ph.D. thesis, 1988.

[Francia] P. Francia, The bicanonical map for surfaces of general type, Univ. of Genova preprint (1983), to appear.

[Friedman and Morgan] R. Friedman and J. Morgan, Algebraic surface and 4–manifolds: some conjectures and speculation, Bull. Amer. Math. Soc. **18** (1988), 1–19.

[Fujita1] T. Fujita, Impossibility criterion of being an ample divisor, J. Math Soc. Japan **34** (1982), 355–363.

[Fujita2] T. Fujita, Defining equations for certain types of polarized varieties, in Complex analysis and algebraic geometry, (Kodaira volume), W.L. Baily, Jr. and T. Shioda eds., Iwanami and C.U.P. 1977, pp.165–173.

[Griffin1] E.E. Griffin, Families of quintic surfaces and curves, Harvard Ph.D. thesis, 1982.

[Griffin2] E.E. Griffin, Families of quintic surfaces and curves, Compositio Math **55** (1985), 33–62.

[Grothendieck] A. Grothendieck, Catégories cofibrées additives et complexe cotangent relatif, LNM **79** (1968).

[Horikawa] E. Horikawa, On deformations of quintic surfaces, Invent. Math **31** (1975), 43–85.

[Illusie1] L. Illusie, Cotangent complex and deformations of torsors and group schemes, in Toposes, algebraic geometry and logic, LNM **274** (1971), 159–189.

[Illusie2] L. Illusie, Complexe cotangent et déformations, I, II, LNM **239** (1971) and **283** (1972).

[Kotschik] D. Kotschik, On manifolds homeomorphic to $\mathbb{CP}2 \# 8\overline{\mathbb{CP}^2}$, Invent. Math 95 (1989), 591-600, 1988.

[Laufer1] H.B. Laufer, Weak simultaneous resolution for deformations of Gorenstein surface singularities, in Singularities (at Arcata, 1981), Proc. of Symposia in Pure Math **40**, A.M.S., 1983, Part 2, 1–29.

[Laufer2] H.B. Laufer, Generation of 4-pluricanonical forms for surface singularities, Amer. J. Math **109** (1987), 571–589.

[Lichtenbaum and Schlessinger] S. Lichtenbaum and M. Schlessinger, The cotangent complex of a morphism, Trans. Amer. Math Soc. **128** (1967), 41–70.

[L'vovskii1] S.M. L'vovskii, Criterion for nonrepresentability of a variety as a hyperplane section, Vestnik Mosk. Univ. (1) Mat. Mekh, 1985, No.3, 25–28.

[L'vovskii2] S.M. L'vovskii, On extensions of varieties defined by quadratic equations, Mat. Sbornik **135** (1988), 312–324.

[Maple] Maple, Computer algebra package, v.4.2 for SUN/Unix system, Symbolic computation group, Univ. of Waterloo, Canada c. 1988.

[Mendes-Lopes] M. Mendes-Lopes, The relative canonical algebra for genus 3 fibrations, Warwick Ph.D. thesis, 1988.

[Mumford1] D. Mumford, Varieties defined by quadratic equations, in Questions on algebraic varieties (CIME conference proceedings, Varenna, 1969), Cremonese, Roma, 1970, pp.29–100.

[Mumford2] D. Mumford, Lectures on curves on an algebraic surface, Ann. of Math Studies **59**, Princeton, 1966.

[Pinkham] H. Pinkham, Deformations of algebraic varieties with G_m-action, Astérisque **20**, Société Math de France, Paris, 1974.

[Reid1] M. Reid, Surfaces with $p_g = 0, K^2 = 1$, J. Fac. Science, Univ. of Tokyo, **25** (1978), 75-92.

[Reider] I. Reider, Vector bundles of rank 2 and linear systems on algebraic surfaces, Ann. of Math **127** (1988), 309-316.

[Saint-Donat] B. Saint-Donat, Projective models of K3 surfaces, Amer. J. Math **96** (1974), 602-639.

[Schlessinger] M. Schlessinger, Functors on Artin rings, Trans. Amer. Math Soc. **130** (1968), 41-70.

[Scorza1] Sopra una certa classe di varietà razionali, Opere scelte, Roma 1960, vol.1, 372-375.

[Scorza2] Sulle varietà di Segre, same Opere scelte, vol.1, 376-386.

[Serrano] F. Serrano, Extension of morphisms defined on a divisor, Math Ann. **277** (1987), 395-413.

[Sommese1] A.J. Sommese, On manifolds that cannot be ample divisors, Math Ann. **221** (1976), 55-72.

[Sommese2] A.J. Sommese, On the birational theory of hyperplane sections of projective threefolds, Notre Dame Univ. preprint, c. 1981.

[Sommese3] A.J. Sommese, Configurations of −2 rational curves on hyperplane sections of projective threefolds, in Classification of algebraic and analytic manifolds, K. Ueno (ed.), Birkhäuser, 1983, 465-497.

[Swinnerton-Dyer] H.P.F. Swinnerton-Dyer, An enumeration of all varieties of degree 4, Amer. J. Math **95** (1973), 403-418.

[XXX] anonymous correspondence (rumoured to be by A. Weil), Amer. J. Math **79** (1957), 951-952.

Miles Reid, Math Inst., Univ. of Warwick, Coventry CV4 7AL, England
Electronic mail: miles @ UK.AC.Warwick.Maths

Toward Abel–Jacobi theory for higher dimensional varieties

IGOR REIDER[1]

School of Mathematics, Institute for Advanced Study

Princeton, New Jersey, 08540, U.S.A.

§1. Introduction and a summary of the results

The purpose of this paper is to outline a construction of a non-abelian candidate for Abel–Jacobi theory in higher dimensions. Our approach is based on the following point of view:

(1.1) A "Jacobian" for an n-dimensional variety X should be an object associated to a family of rank n-bundles on X naturally attached to X.

A good candidate of the above nature should possess:

(1.2)

1) A theory analogous to the theory of special divisors for curves (this could mean that "Jacobian" should detect "special" vector bundles;

2) A suitable Torelli Theorem (this should be a meaning of naturality of a family in (1.1)).

Our construction parallels the classical situation. Recall: if X is a curve, then one has an object $\mathcal{J}(X)$, the jacobian of X, and an Abel–Jacobi map

$$\alpha : X^{(d)} \longrightarrow \mathcal{J}(X), \tag{1.3}$$

where $X^{(d)}$ is the d-th symmetric product of X. We describe the construction of an analogue for (1.3) in the case X is an algebraic surface.

Fix a nef and big line bundle $\mathcal{O}_X(L)$ and a positive integer d. As in the case of curves we consider $X^{(d)}$, the Hilbert scheme of 0-dimensional subschemes of X of length d. The next step is to relate $X^{(d)}$ to a family of $rk\,2$ sheaves on X. (In case $\dim X = 1$ the elements of $X^{(d)}$ automatically give rank 1 sheaves on X, so this step is a higher dimensional phenomenon). This is done by defining the strata $\Gamma_d^r(L)$ of $X^{(d)}$ together with the sheaf $\mathbf{Ext}^1$. Then we put

$$Y_d^r(L) = \mathbf{P}(\mathbf{Ext}^1)$$

[1] Research partially supported by NSF Postdoctoral Fellowship and NSF Grant DMS–8702588 and University of Oklahoma Research Council.

which, by construction, is a family of torsion free sheaves on X whose Chern invariants are $\mathcal{O}_X(L)$ and d. This is the source of our Abel–Jacobi map.

Turning to the target for an analogue of (1.3) we observe that there is the distinguished filtration of $H^0(K_X + L)$ for points in $Y_d^r(L)$. This gives rise to the map (defined on some Zariski open subset of $Y_d^r(L)$):

$$\alpha : Y_d^r(L) \dashrightarrow F\Big(H^0(K_X + L)\Big) \tag{1.4}$$

where F is an appropriate variety of flags of $H^0(K_X + L)$. This is our non-abelian version of Abel–Jacobi map.

Now we seek a data associated to (α, Y_d^r, F) as in (1.4) which would play a role of polarization in the classical theory. Such data is essentially contained in the definition of the map α in (1.4). Namely, there is a torsion free sheaf denoted $\mathcal{T}_Y$ on Y (here and below $Y = Y_d^r$; we omit indicies when there is no ambiguity) such that $\tilde{Y} = \mathbf{P}(\tilde{\mathcal{T}}_Y)$ has the destinguished divisor denoted $D_{Y/\Gamma}$, where $\tilde{\mathcal{T}}_Y = \mathcal{T}_Y \oplus \mathcal{O}_Y$.

Definition 1.5. The pair $(\tilde{Y}, D_{Y/\Gamma})$ is called the polarization of (α, Y, F).

The main point is that the above polarization is analogous to the classical one with respect to the Torelli type theorems. To make this precise let $\mathcal{Z} \xrightarrow{p_1} \Gamma$ be the incidence cycle over Γ (i.e. $\mathcal{Z} = \{([Z], x) \in \Gamma \times X \mid x \in Z\}$ and p_1 is the projection on the first factor). Consider the fibre product

$$\begin{array}{ccc} \tilde{\mathcal{Z}} = \mathcal{Z} \times_\Gamma Y & \xrightarrow{t_2} & Y \\ {\scriptstyle t_1}\downarrow & & \downarrow{\scriptstyle \pi} \\ \mathcal{Z} & \longrightarrow & \Gamma \end{array} \tag{1.6}$$

We observe that there is the canonical map

$$\begin{array}{ccc} \kappa : \tilde{\mathcal{Z}} & \longrightarrow & \mathbf{P}(\tilde{\mathcal{T}}_Y^*) \stackrel{\text{def}}{=} \tilde{Y}^* \\ & {\scriptstyle t_2}\searrow \quad \swarrow{\scriptstyle q^*} & \\ & Y & \end{array} \tag{1.7}$$

commuting with the projections t_2 and q^*. Let $\tilde{\mathcal{Z}}'$ be the image of κ. It defines the divisor $D_{Y/\Gamma}$ in $\tilde{Y} = \mathbf{P}(\tilde{\mathcal{T}}_Y)$ via the "point-hyperplane" correspondence in $\tilde{Y}^* \times_\Gamma \tilde{Y} = \mathbf{P}(\tilde{\mathcal{T}}_Y^*) \times_\Gamma \mathbf{P}(\tilde{\mathcal{T}}_Y)$. Our first result is the following

THEOREM 1. *The polarization $(\tilde{Y}, D_{Y/\Gamma})$ determines the image $\tilde{\mathcal{Z}}'$ of κ in (1.7).*

COROLLARY 2. *If κ is generically $1-1$, then $(\tilde{Y}, D_{\Gamma/Y})$ determines $\tilde{\mathcal{Z}}$ and hence the incidence cycle $\mathcal{Z}$ (see (1.6) for notations).*

The analogy of our constructions with the classical ones persists on the level of IVHS (Infinitesimal Variation of Hodge structure). By an IVHS type data we mean any natural algebraic invariants associated to the differential of the period map (see [**2**], which is our basic reference on the subject). Since our non-abelian version of Abel–Jacobi map in (1.4) is a "period map" it is natural to look for IVHS type data associated to (α, Y, F) as well as to state Torelli problems of (α, Y, F) in terms of IVHS associated. Indeed, the algebraic invariants associated to α appear quite naturally: Set $\mathcal{M}_Y = \{\mathcal{M}_Y^0 \supset \mathcal{M}_Y^1 \supset \dots \mathcal{M}_Y^k\}$ to be the filtration of sheaves associated to α in (1.4) (the sheaves in $\mathcal{M}_Y^i$'s are the pull-back of the universal bundles on Grassmann varieties involved in the definition of F; see (3.14)).

It turns out that there is a pair of sheaf morphisms:

$$\begin{aligned} &1) \quad p^0 : \Theta_{Y/\Gamma} \longrightarrow (\mathcal{M}_Y^0)^* \otimes \mathcal{T}_Y^* \otimes \mathcal{O}_Y(1) \\ &2) \quad p = \underset{\ell \geq 1}{\oplus} p^\ell : \Theta_{Y/\Gamma} \otimes \mathcal{O}_U \longrightarrow \underset{\ell \geq 1}{\oplus} \mathcal{H}om_{\mathcal{O}_U}\left(\frac{\mathcal{M}_Y^\ell}{\mathcal{M}_Y^{\ell+1}}, \frac{\mathcal{M}_Y^{\ell-1}}{\mathcal{M}_Y^\ell}\right) \end{aligned} \tag{1.8}$$

where $\mathcal{O}_Y(1)$ is a tautological sheaf on Y, U is Zariski open subset of Y where $\mathcal{M}_Y^\ell$'s are locally free, $\Theta_{Y/\Gamma}$ the relative tangent sheaf of $\pi : Y \to \Gamma$.

Definition 1.9. The data $(\mathcal{T}_Y, \mathcal{M}_Y, p^0, p)$ will be called the Infinitesimal polarization of (α, Y, F).

Remark 1.10. The morphism p as in (1.8) 2), is the differential of α in (1.4) restricted to the relative tangent sheaf of $\pi : Y \to \Gamma$.

It is possible to show that there is a morphism

$$h_\Gamma : \mathcal{T}_Y \longrightarrow \Theta_{Y/\Gamma}$$

Moreover, h_Γ is an isomorphism on $\overset{\circ}{Y}$, the Zariski open subset of Y, corresponding to the locally free sheaves.

An introduction of the Infinitesimal polarization is in accordance with the Griffiths' philosophy: an IVHS type data should contain geometry of the underlying geometric object. Before stating a result illustrating this principle we need the following:

Definition 1.12. A 0-dimensional subscheme Z of X is said to have special decomposition with respect to $\mathcal{O}_X(L)$ if $Z = \sum_{i=1}^{k} Z_i$ for $k \geq 2$, where $[Z_i] \in \Gamma_{d_i}^{r_i}(L)$, $r_i \geq 0$, $d_i > 0$.

THEOREM 2. *Let* $\Gamma(i) = \{[Z] \in \Gamma_d^r(L) \mid \ker(p_y^i) \neq 0$ *for some* $y \in \pi^{-1}([Z]) \cap \mathring{Y}\}$, *where* p_y^i *is* p^i *in (1.8) at the point* $y \in Y$, $\mathring{Y}$ *is the Zariski open subset of* Y, *corresponding to the locally free sheaves. Then every* $[Z] \in \Gamma(i)$ *has a special decomposition with respect to* $\mathcal{O}_X(L)$.

This result can be viewed as the Infinitesimal Torrelli for 0-cycles as well as an illustration of 1) in (1.2).

The remainder of this paper is as follows: in §2 we define the strata $\Gamma_d^r(L)$ of $X^{(d)}$ and build $Y_d^r(L)$, the destinguished family of torsion free sheaves on X; in §3 we define the distinguished filtration of $H^0(K_X + L)$ for points in Y_d^r, the polarization $(\tilde{Y}, D_{Y/\Gamma})$ and the infinitesimal polarization; in §4 the geometry of the Infinitesimal polariazation is discussed.

More complete versions of the above constructions, their applications to the study of algebraic surfaces and higher dimensional varieties will appear elsewhere.

It is a pleasure to thank all organizers of the algebraic geometry conference in L'Aquila, where the first version of this paper was presented.

§2. A stratification of the Hilbert scheme of 0-cycles and the destinguished family of sheaves

Let X be an algebraic surface and $X^{(d)}$ the Hilbert scheme of 0-dimensional subschemes of length d on X. There is the incidence cycle

$$\begin{array}{ccc} & \mathcal{Z}_d & \\ {\scriptstyle p_1}\swarrow & & \searrow{\scriptstyle p_2} \\ X^{(d)} & & X \end{array} \tag{2.1}$$

where $p_{2*}\left(p_1^*([Z])\right) = Z$ for every $[Z] \in X^{(d)}$.

Fix a line bundle $\mathcal{O}_X(L)$ and consider the morphism of sheaves on $X^{(d)}$:

$$H^0(K_X + L) \otimes \mathcal{O}_{X^{(d)}} \xrightarrow{p} p_{1*}\left[\mathcal{O}_{\mathcal{Z}_d} \otimes p_2^* \mathcal{O}_X(K_X + L)\right], \tag{2.2}$$

where $\mathcal{O}_X(K_X)$ is the canonical line bundle of X.

Define

$$\Gamma_d^r(L) := \{\, [Z] \in X^{(d)} \mid \dim \left(\mathrm{coker}\ p([Z])\right) \geq r+1\}. \tag{2.3}$$

This stratifies $X^{(d)}$:

$$X^{(d)} \supset \Gamma_d^0 \supset \Gamma_d^1 \supset \cdots \supset \Gamma_d^r \supset \Gamma_d^{r+1} \supset \cdots \tag{2.4}$$

(we omit the index L above if a line bundle is specified).

Also denote $\overset{\circ}{\Gamma}{}_d^r = \Gamma_d^r - \Gamma_d^{r+1}$.

Fix $\mathcal{O}_X(L)$ nef and big line bundle on X and consider the sheaf $\mathbf{Ext}^1 = \ker(p^*)$, where p is as in (2.2). By **[1]**, **[3]** there is a natural identification

$$\mathbf{Ext}^1 \otimes \mathcal{O}_{X^{(d)},[Z]} = \mathrm{Ext}^1\left(\mathcal{J}_Z(L), \mathcal{O}_X\right) \otimes \mathcal{O}_{X^{(d)},[Z]} \tag{2.5}$$

for every $[Z] \in X^{(d)}$, where $\mathcal{J}_Z$ is the ideal sheaf of Z on X.

Remark 2.6. By definition $\mathbf{Ext}^1$ is supported on Γ_d^0 and $rk\left(\mathbf{Ext}^1 \otimes \mathcal{O}_{\overset{\circ}{\Gamma}{}_d^r}\right) = r+1$.

Define:

$$\begin{aligned} Y_d^r(L) &= \mathbf{P}\left(\mathbf{Ext}^1 \otimes \mathcal{O}_{\Gamma_d^r}\right), \\ Y_d^r(L) &\xrightarrow{\ \pi\ } \Gamma_d^r(L) \end{aligned} \tag{2.7}$$

where π is the natural projection.

By (2.5) the scheme $Y_d^r(L)$ parametrizes a family of torsion free sheaves whose Chern invariants are $\mathcal{O}_X(L)$ and d. This completes the first part of the outline in §1.

§3. Main constructions

1. A non-abelian version of Abel-Jacobi map

For simplicity we give our construction for a fixed cycle (the construction for a family of cycles is straightforward).

Let $[Z] \in \overset{\circ}{\Gamma}{}_d^r(L)$. Denote $\mathbf{P}_Z = \mathbf{P}\left(\mathbf{Ext}^1 \otimes \mathcal{O}_{X^{(d)},[Z]}\right)$. By (2.5) we have

$$\mathbf{P}_Z = \mathbf{P}\left(\mathrm{Ext}^1(\mathcal{J}_Z(L), \mathcal{O}_X)\right) \tag{3.1}$$

On $\mathbf{P}_Z \times X$ there is $\mathcal{E}$, a sheaf of rank 2, such that $\mathcal{E}|_{\{\alpha\}\times X}$ corresponds to the extension class defined by $\alpha \in \mathbf{P}_Z$, for every $\alpha \in \mathbf{P}_Z$. More explicitly, $\mathcal{E}$ is the extension

$$0 \longrightarrow p_1^*\mathcal{O}_{\mathbf{P}_Z}(1) \longrightarrow \mathcal{E} \longrightarrow p_2^*\left(\mathcal{J}_Z(L)\right) \longrightarrow 0 \tag{3.2}$$

defined by id $\in$ Hom $\left(\mathrm{Ext}^1(\mathcal{J}_Z(L), \mathcal{O}_X), \mathrm{Ext}^1(\mathcal{J}_Z(L), \mathcal{O}_X)\right) \subset \mathrm{Ext}^1\left(p_2^*\mathcal{J}_Z(L), p_1^*\mathcal{O}_{\mathbf{P}}(1)\right)$, where $p_i, i = 1, 2$, is the projection of $\mathbf{P}_Z \times X$ on the i-th factor.

Tensoring (3.2) with $p_2^*\mathcal{O}_X(-L)$ we obtain

$$0 \longrightarrow p_1^*\mathcal{O}_{\mathbf{P}_Z}(1) \otimes p_2^*\mathcal{O}_X(-L) \longrightarrow \mathcal{E}' \longrightarrow p_2^*\mathcal{J}_Z \longrightarrow 0 \tag{3.3}$$

where $\mathcal{E}' \stackrel{\mathrm{def}}{=} \mathcal{E} \otimes p_2^*\mathcal{O}_X(-L)$. Combining this with the standard sequence

$$0 \longrightarrow p_2^*\mathcal{J}_Z \longrightarrow \mathcal{O}_{\mathbf{P}_Z \times X} \longrightarrow \mathcal{O}_{p_2^*Z} \longrightarrow 0 \tag{3.4}$$

yields the following diagram

$$\begin{array}{ccccc}
 & & 0 & & \\
 & & \downarrow & & \\
 & & H^0(\mathcal{O}_X) \otimes \mathcal{O}_{\mathbf{P}_Z} = \mathcal{O}_{\mathbf{P}_Z} & & \\
 & & \downarrow & & \\
 & & H^0(\mathcal{O}_Z) \otimes \mathcal{O}_{\mathbf{P}_Z} & & \\
 & & \downarrow & & \\
0 \longrightarrow R^1 p_{1*}(\mathcal{E}') & \longrightarrow & H^1(\mathcal{J}_Z) \otimes \mathcal{O}_{\mathbf{P}_Z} & \longrightarrow & H^2(-L) \otimes \mathcal{O}_{\mathbf{P}_Z}(1) \\
 & & \downarrow & & \\
 & & H^1(\mathcal{O}_X) \otimes \mathcal{O}_{\mathbf{P}_Z} & & \\
 & & \downarrow & & \\
 & & 0 & &
\end{array} \tag{3.5}$$

This diagram induces the homomorphism

$$\mathrm{Res}_Z : H^0(\mathcal{O}_Z) \otimes \mathcal{O}_{\mathbf{P}_Z} \longrightarrow H^2(-L) \otimes \mathcal{O}_{\mathbf{P}_Z}(1) \tag{3.6}$$

($\mathrm{Res}_Z(\alpha) : H^0(\mathcal{O}_Z) \longrightarrow H^2(-L)$ for every $\alpha \in \mathbf{P}_Z$, is the residue map as defined in[**3**]).

Set $\tilde{\mathcal{T}}_Z = \ker(\mathrm{Res}_Z)$. Observe the inclusion

$$i : \mathcal{O}_{\mathbf{P}_Z} \longrightarrow \tilde{\mathcal{T}}_Z \tag{3.7}$$

which follows from (3.5).

Now we are ready to define the filtration of $H^0(K_X + L) \otimes \mathcal{O}_{\mathbf{P}_Z}$ mentioned in §1. By definition of $\tilde{\mathcal{T}}_Z$ we have the morphism

$$\mathcal{S}^{\cdot}\tilde{\mathcal{T}}_Z = \bigoplus_{m \geq 1} \mathcal{S}^m \tilde{\mathcal{T}}_Z \xrightarrow{e} H^0(\mathcal{O}_Z) \otimes \mathcal{O}_{\mathbf{P}_Z} \tag{3.8}$$

Define:

1) $\tilde{\lambda}_m : \mathcal{S}^m \tilde{\mathcal{T}}_Z \longrightarrow H^2(-L) \otimes \mathcal{O}_{\mathbf{P}_Z}(1)$ to be the composition of $e_m = e|_{\mathcal{S}^m \tilde{\mathcal{T}}_Z}$ and Res_Z in (3.6).

(3.9)

2) $\mathcal{M}_Z^{m-1} = \ker\left(\tilde{\lambda}_m^* \otimes \mathcal{O}_{\mathbf{P}_Z}(1)\right)$

Remark 3.10. The inclusion i in (3.7) implies the inclusion

$$i_m : \mathcal{S}^m \tilde{\mathcal{T}}_Z \longrightarrow \mathcal{S}^{m+1} \tilde{\mathcal{T}}_Z$$

which yields the inclusion

$$\mathcal{M}_Z^{m-1} \hookleftarrow \mathcal{M}_Z^m$$

for every $m \geq 1$. Observe, by definition of $\tilde{\mathcal{T}}_Z$, we have

$$\mathcal{M}_Z^0 = H^0(K_X + L) \otimes \mathcal{O}_{\mathbf{P}_Z} \tag{3.11}$$

Definition 3.12. Let $[Z] \in \overset{\circ}{\Gamma}{}_d^r$ and $\mathbf{P}_Z$ as in (3.1). The filtration $\mathcal{M}_Z := \{\mathcal{M}_Z^0 \supset \mathcal{M}_Z^1 \supset \cdots \supset \mathcal{M}_Z^k\}$ will be called the period filtration for Z with respect to $\mathcal{O}_X(L)$. The index of the last piece of $\mathcal{M}_Z$ will be called the weight of the filtration.

Remark 3.13. By definition, $\mathcal{M}_Z^i$'s are torsion free sheaves on $\mathbf{P}_Z$ (in fact, they are locally free in codim ≥ 3, since they are 2-nd syzygy sheaves (see [4])). Put $m_i = rk(\mathcal{M}_Z^i)$ and consider the flag variety:

$$\begin{aligned} F(k, \vec{m};\, H^0(K_X + L)) = \Big\{ [M] \in \prod_{i=0}^{k} Gr\big(m_i, H^0(K_X + L)\big) \Big| \\ M = \{ H^0(K_X + L) = M^0 \supset M^1 \supset \cdots \supset M^k\}, \quad \text{where} \\ [M^i] \in Gr\big(m_i, H^0(K_X + L)\big) \Big\} \end{aligned} \tag{3.14}$$

where k is the weight of $\mathcal{M}_Z$ and $\vec{m} = (m_0, m_1, \ldots, m_k)$. If there is no ambiguity we omit the indices in the notation $F\big(k, \vec{m};\, H^0(K_X + L)\big)$. So we obtain the desired map

$$\alpha_Z : \mathbf{P}_Z \dashrightarrow F\big(k, \vec{m};\, H^0(K_X + L)\big) \tag{3.15}$$

by sending $\xi \in \mathbf{P}_Z$ to the point of the flag variety defined by $\mathcal{M}_{Z,\xi}$.

Remark 3.16. 1) It is straightforward to define the period filtration and the corresponding map into the flag variety for families of 0-cycles.

2) If $[Z] \in \mathring{\Gamma}^0_d$, then the filtration $\mathcal{M}_Z = \{\mathcal{M}^0_Z\}$ is trivial and hence is α_Z.

2. A polarization of (α, Y, F)

We turn now to the canonical map κ whose existence was asserted in (1.7). Recall the morphism $e : \mathcal{S}^{\cdot}\tilde{\mathcal{T}}_Z \longrightarrow H^0(\mathcal{O}_Z) \otimes \mathcal{O}_{\mathbf{P}_Z}$ as in (3.8). This induces the morphism

$$\begin{array}{ccc} \kappa_Z : Z \times \mathbf{P}_Z & \xrightarrow{\qquad\qquad} & \mathbf{P}(\tilde{\mathcal{T}}^*_Z) \\ & {\scriptstyle t_2(Z)} \searrow \quad \swarrow {\scriptstyle q^*(Z)} & \\ & \mathbf{P}_Z & \end{array} \tag{3.17}$$

where $t_2(Z)$ and $q(Z)$ are the natural projections. The map κ can be defined by setting

$$\kappa\Big|_{t_1^{-1}([Z])} = \kappa_Z$$

for every $[Z] \in \Gamma$ (refer to (1.6) and (1.7) for notations).

Finally, to define the divisor D_Z in $\mathbf{P}(\tilde{\mathcal{T}}_Z)$ we dualize the image of κ_Z. More precisely, consider

$$\begin{array}{ccc} & T = Z \times \mathbf{P}_Z \times_{\mathbf{P}_Z} \mathbf{P}(\tilde{\mathcal{T}}_Z) & \\ {\scriptstyle r_1} \swarrow & & \searrow {\scriptstyle r_2} \\ Z \times \mathbf{P}_Z & & \mathbf{P}(\tilde{\mathcal{T}}_Z) \end{array} \tag{3.18}$$

Consider the morphism of sheves on T:

$$r_1^*\left(\kappa_Z^*\mathcal{O}_{\mathbf{P}(\tilde{\mathcal{T}}^*_Z)}(-1)\right) \otimes r_2^*\mathcal{O}_{\mathbf{P}(\tilde{\mathcal{T}}^*_Z)}(-1) \xrightarrow{\ s\ } \mathcal{O}_T$$

where $\mathcal{O}_{\mathbf{P}(\tilde{\mathcal{T}}^*_Z)}(1)$ (resp. $\mathcal{O}_{\mathbf{P}(\tilde{\mathcal{T}}_Z)}(1)$) is the tautological sheaf on $\mathbf{P}(\tilde{\mathcal{T}}^*_Z)$ (resp. on $\mathbf{P}(\tilde{\mathcal{T}}_Z)$) such that $(q^*_Z)_*(\mathcal{O}_{\mathbf{P}(\tilde{\mathcal{T}}^*_Z)}(1)) = \tilde{\mathcal{T}}_Z$ (resp. $(q_Z)_*(\mathcal{O}_{\mathbf{P}(\tilde{\mathcal{T}}_Z)}(1)) = \tilde{\mathcal{T}}_Z$. Put $D_T = (s = 0)$ and define $D_Z = (r_2)_*(D_T)$. Again we can define $D_{Y/\Gamma}$ as in Definition 1.5 by setting $D_{Y/\Gamma}|_{q^{-1}(\mathbf{P}_Z)} = D_Z$, where $q : \tilde{Y} = \mathbf{P}(\tilde{\mathcal{T}}_Z) \longrightarrow Y$ is the natural projection.

3. The infinitesimal polarization of (α, Y, F)

We begin by explaining the morphism p in (1.8), 2). The differential of α_Z in (3.15) induces the morphism

$$\mathcal{M}^{\ell+1}_Y \otimes \mathcal{O}_U \longrightarrow \Omega^1_U \otimes \frac{\mathcal{M}^0_Y}{\mathcal{M}^{\ell+1}_Y}$$

where U is the Zariski open subset of Y where $\mathcal{M}_Y^\ell$'s are locally free. This morphism factors

$$\mathcal{M}_Y^{\ell+1} \otimes \mathcal{O}_U \longrightarrow \Omega_U^1 \otimes \frac{\mathcal{M}_Y^\ell}{\mathcal{M}_Y^{\ell+1}} \tag{3.19}$$

(see §4, for details).

Combining (3.19) with inclusions $\mathcal{M}_Y^{\ell+1} \hookrightarrow \mathcal{M}_Y^\ell$ we deduce

$$\frac{\mathcal{M}_Y^\ell}{\mathcal{M}_Y^{\ell+1}} \otimes \mathcal{O}_U \longrightarrow \Omega_U^1 \otimes \frac{\mathcal{M}_Y^{\ell-1}}{\mathcal{M}_Y^\ell}$$

which is equivalent to

$$\Theta_U \longrightarrow \mathrm{Hom}_{\mathcal{O}_U}\left(\frac{\mathcal{M}_Y^\ell}{\mathcal{M}_Y^{\ell+1}}, \frac{\mathcal{M}_Y^{\ell-1}}{\mathcal{M}_Y^\ell}\right) \tag{3.20}$$

The restriction of the morphism in (3.20) to $\Theta_{U/\Gamma}$ is the morphism p^ℓ as in (1.8), 2). To describe the morphism p^0 in (1.8) 1), we recall the morphism Res_Z in (3.6) which gives rise to the following diagram

$$\begin{array}{ccc}
 & & 0 \\
 & & \downarrow \\
 & & H^0(K_X+L)^* \otimes \Omega_{\mathbf{P}_Z}(1) \\
 & & \downarrow \\
H^0(\mathcal{O}_Z)\otimes\mathcal{O}_{\mathbf{P}_Z} & \xrightarrow{r} & H^0(K_X+L)^* \otimes \dfrac{H^0(\mathcal{O}_Z(K_X+L))}{H^0(K_X+L)} \otimes \mathcal{O}_{\mathbf{P}_Z} \\
 & {\scriptstyle \mathrm{Res}_Z}\searrow & \downarrow \\
 & & H^0(K_X+L)^* \otimes \mathcal{O}_{\mathbf{P}_Z}(1) \\
 & & \downarrow \\
 & & 0
\end{array} \tag{3.21}$$

This induces the morphism

$$\tilde{\mathcal{T}}_Z \longrightarrow H^0(K_X+L)^* \otimes \Omega_{\mathbf{P}_Z}(1) \tag{3.22}$$

Observe: the subsheaf $\mathcal{O}_{\mathbf{P}_Z} \hookrightarrow H^0(\mathcal{O}_Z)\otimes\mathcal{O}_{\mathbf{P}_Z}$ (see (3.5)) is in $\ker(r)$ of (3.21). This implies that the morphism in (3.22) factors through $\mathcal{T}_Z = \frac{\tilde{\mathcal{T}}_Z}{\mathcal{O}_{\mathbf{P}_Z}}$ to give the morphism

$$\mathcal{T}_Z \longrightarrow H^0(K_X+L)^* \otimes \Omega_{\mathbf{P}_Z}(1)$$

which is equivalent to

$$p_Z^0 : \Theta_{\mathbf{P}_Z} \longrightarrow H^0(K_X+L)^* \otimes \mathcal{T}_Z^* \otimes \mathcal{O}_{\mathbf{P}_Z}(1)$$

We can define $\overset{\circ}{p}$ as in (1.8) 1), by setting

$$p^0\big|_{\pi^{-1}([Z])} = p^0\big|_{\mathbf{P}_Z} = p_Z^0$$

for every $[Z] \in \Gamma$.

§4. Geometry of 0-cycles and the Infinitesimal polarization

A reason that the Infinitesimal polarization is related to the geometry of the underlying 0-cycles comes from being able to identify $\mathcal{T}_Y$ with $\Theta_{Y/\Gamma}$ on the Zariski open subset $\overset{\circ}{Y}$ of Y corresponding to the locally free sheaves.

LEMMA 4.1. *There is a natural morphism*

$$h_\Gamma : \mathcal{T}_Y \longrightarrow \Theta_{Y/\Gamma}$$

which is an isomorphism on $\overset{\circ}{Y}$, *the subset of* Y *corresponding to the locally free sheaves.*

Proof: We show how to define $h_Z = h_\Gamma|_{\mathbf{P}_Z}$. From (3.5) we have

$$\mathcal{T}_Z \simeq \ker\left(R^1p_{1*}(\mathcal{E}') \xrightarrow{\gamma} H^1(\mathcal{O}_X)\otimes\mathcal{O}_{\mathbf{P}_Z}\right) \tag{4.2}$$

where $\mathcal{T}_Z = \frac{\tilde{\mathcal{T}}_Z}{\mathcal{O}_{\mathbf{P}_Z}}$ (see (3.6) and (3.7)). Tensoring (3.2) with $p_2^*\mathcal{O}_X(K_X)$ and taking its direct image under p_1 we deduce

$$H^1(K_X)\otimes\mathcal{O}_{\mathbf{P}_Z}(1) \longrightarrow R^1p_{1*}(\mathcal{E}\otimes p_2^*K_X) \longrightarrow H^1\big(\mathcal{J}_Z(K_X+L)\big)\otimes\mathcal{O}_{\mathbf{P}_Z} \longrightarrow$$
$$\longrightarrow H^2(K_X)\otimes\mathcal{O}_{\mathbf{P}_Z}(1) \longrightarrow 0 \tag{4.3}$$

Dualizing (4.3) yields

$$0 \longrightarrow \mathcal{O}_{\mathbf{P}_Z}(-1) \longrightarrow H^1\big(\mathcal{J}_Z(K_X+L)\big)^*\otimes\mathcal{O}_{\mathbf{P}_Z} \longrightarrow \big(R^1p_{1*}(\mathcal{E}\otimes p_2^*K_X)\big)^* \longrightarrow$$
$$\longrightarrow H^1(\mathcal{O}_X)\otimes\mathcal{O}_{\mathbf{P}_Z}(-1) \tag{4.4}$$

The last two terms on the left in (4.4) are part of the Euler sequence twisted by $\mathcal{O}_{\mathbf{P}_Z}(-1)$ (recall: $H^1\big(\mathcal{J}_Z(K_X+L)\big)^* = \mathrm{Ext}^1\big(\mathcal{J}_Z(L),\mathcal{O}_X\big)$ and (3.1) for definition of $\mathbf{P}_Z$). So (4.4) implies

$$\Theta_{\mathbf{P}_Z} = \ker\big(R^1p_{1*}(\mathcal{E}\otimes p_2^*K_X)^*\otimes\mathcal{O}_{\mathbf{P}_Z}(1) \xrightarrow{\delta} H^1(\mathcal{O}_X)\otimes\mathcal{O}_{\mathbf{P}_Z}\big) \tag{4.5}$$

Consider the pairing

$$\begin{array}{ccc} R^1p_{1*}(\mathcal{E}')\otimes R^1p_{1*}(\mathcal{E}\otimes p_2^*K_X) & \longrightarrow & R^2p_{1*}\left(\det\mathcal{E}\otimes p_2^*\mathcal{O}_X(-L+K_X)\right) \\ \Vert \text{def} & & \Vert \\ R^1p_{1*}(\mathcal{E}\otimes p_2^*\mathcal{O}_X(-L)) & & H^2(K_X)\otimes\mathcal{O}_{\mathbf{P}_Z}(1) \end{array} \tag{4.6}$$

This pairing is non-degenerate at $y\in\mathbf{P}_Z$ such that $\mathcal{E}_y=\mathcal{E}|_{\{y\}\times X}$ is locally free: $\mathcal{E}'_y=\mathcal{E}_y\otimes\mathcal{O}_X(-L)=\mathcal{E}^*_y$ and the pairing (4.6) at y becomes $H^1(\mathcal{E}^*_y)\otimes H^1(\mathcal{E}_y\otimes\mathcal{O}_X(K_X))\longrightarrow H^2(K_X)$ which is the duality pairing.

Rewrite (4.6) as follows:

$$R^1p_{1*}(\mathcal{E}')\xrightarrow{\phi}\left(R^1p_{1*}(\mathcal{E}\otimes p_2^*K_X)\right)^*\otimes\mathcal{O}_{\mathbf{P}_Z}(1) \tag{4.7}$$

and consider the diagram

$$\begin{array}{ccc} R^1p_{1*}(\mathcal{E}') & \xrightarrow{\phi} & R^1p_{1*}\left(\mathcal{E}\otimes p_2^*(K_X)\right)^*\otimes\mathcal{O}_{\mathbf{P}_Z}(1) \\ \gamma\downarrow & & \downarrow\delta \\ H^1(\mathcal{O}_X)\otimes\mathcal{O}_{\mathbf{P}_Z} & = & H^1(\mathcal{O}_X)\otimes\mathcal{O}_{\mathbf{P}_Z} \end{array}$$

where γ and δ as in (4.2) and (4.5) respectively.

This yields the morphism

$$h_Z:\mathcal{T}_Z\overset{(4.2)}{=}\ker\gamma\longrightarrow\ker\delta\overset{(4.5)}{=}\Theta_{\mathbf{P}_Z}$$

which is an isomorphism on $\overset{\circ}{Y}_Z$, the Zariski open subset of $\mathbf{P}_Z$ corresponding to the locally free sheaves, because ϕ is an isomorphism on $\overset{\circ}{Y}_Z$.

Q.E.D.

Remark 4.8. Let $\overset{\circ}{Y}_Z$ be the Zariski open subset of $\mathbf{P}_Z$ where h_Z as in the proof of Lemma 4.1 is an isomorphism. This induces an isomorphism

$$\tilde{h}_Z:\tilde{\mathcal{T}}_Z\otimes\mathcal{O}_{\overset{\circ}{Y}_Z}\longrightarrow\left[\frac{H^0\left(\mathcal{O}_Z(K_X+L)\right)}{H^0(K_X+L)}\right]^*\otimes\mathcal{O}_{\overset{\circ}{Y}_Z}(1) \tag{4.9}$$

which follows from the diagram

$$\begin{array}{ccccccccc} 0 & \longrightarrow & \mathcal{O}_{\mathbf{P}_Z} & \longrightarrow & \tilde{\mathcal{T}}_Z & \longrightarrow & \mathcal{T}_Z & \longrightarrow & 0 \\ & & & & & & \downarrow & & \\ 0 & \longrightarrow & \mathcal{O}_{\mathbf{P}_Z} & \longrightarrow & H\otimes\mathcal{O}_{\mathbf{P}_Z}(1) & \longrightarrow & \Theta_{\mathbf{P}_Z} & \longrightarrow & 0 \end{array}$$

where $H = \mathrm{Ext}^1(\mathcal{J}_Z(K_X+L)) = H^1(\mathcal{J}_Z(K_X+L))^* = \left[\frac{H^0(\mathcal{O}_Z(K_X+L))}{H^0(K_X+L)}\right]^*$. For $y \in \mathring{Y}_Z$ the isomorphism

$$\tilde{h}_Z(y) : \tilde{\mathcal{T}}_{Z,y} \xrightarrow{\sim} \left[\frac{H^0(\mathcal{O}_Z(K_X+L))}{H^0(K_X+L)},\right]^*$$

where $\tilde{\mathcal{T}}_{Z,y} \stackrel{\mathrm{def}}{=} \tilde{\mathcal{T}}_Z \otimes k(y)$, is given by the residue map as in **[2]** (also **[1]**).

Using (4.9) we can reinterpret the morphisms p^i's in (1.8). Consider the pairing

$$\tilde{\mathcal{T}}_Z \otimes H^0(K_X+L) \longrightarrow H^0(\mathcal{O}_Z(K+L)) \otimes \mathcal{O}_{\mathbf{P}_Z}$$

defined by the multiplication. By (4.9) and definitions of $\mathcal{M}_Z^i$'s in (3.9) 2), we have

$$\tilde{\mathcal{T}}_Z \otimes \mathcal{M}_Z^i \longrightarrow \frac{\mathcal{M}_Z^{i-1}}{H^0(\mathcal{J}_Z(K_X+L)) \otimes \mathcal{O}_{\mathring{Y}_Z}}$$

This combined with inclusions $\mathcal{M}_Z^{i+1} \hookrightarrow \mathcal{M}_Z^i$ yields

$$\tilde{\mathcal{T}}_Z \otimes \frac{\mathcal{M}_Z^i}{\mathcal{M}_Z^{i+1}} \longrightarrow \frac{\mathcal{M}_Z^{i-1}}{\mathcal{M}_Z^i}$$

which factors through

$$\mathcal{T}_Z \otimes \frac{\mathcal{M}_Z^i}{\mathcal{M}_Z^{i+1}} \longrightarrow \frac{\mathcal{M}_Z^{i-1}}{\mathcal{M}_Z^i}$$

and is equivalent to

$$\mathcal{T}_Z \longrightarrow \mathcal{H}om\left(\frac{\mathcal{M}_Z^i}{\mathcal{M}_Z^{i+1}}, \frac{\mathcal{M}_Z^{i-1}}{\mathcal{M}_Z^i}\right)$$

We claim (but omit a proof here) that this coincides with the differential of α_Z in (3.15) via the isomorphism (4.9). The above gives a simple way to compute $p^i|_{\mathring{Y}_Z}$ as in (1.8) 2).

Turning to the morphism $p^0|_{\mathring{Y}_Z}$ we recall

$$\tilde{\lambda}_2 : S^2\tilde{\mathcal{T}}_Z \longrightarrow H^2(-L)^* \otimes \mathcal{O}_{\mathbf{P}_Z}(1) = (\mathcal{M}_Z^0)^* \otimes \mathcal{O}_{\mathbf{P}_Z}(1)$$

which factors through

$$S^2\mathcal{T}_Z \longrightarrow (\mathcal{M}_Z^0)^* \otimes \mathcal{O}_{\mathbf{P}_Z}(1)$$

This yields

$$\mathcal{T}_Z \longrightarrow (\mathcal{M}_Z^0)^* \otimes \mathcal{T}_Z^* \otimes \mathcal{O}_{\mathbf{P}_Z}(1)$$

Using (4.9) we arrive to (1.8), 1).

The meaning of the fact that the Infinitesimal polarization is geometric is illustrated by the fact that it detects the decompositions of cycles in $\Gamma_d^r(L)$ into special subcycles. More precisely, we have the following

Proposition 4.10 (= **Theorem** 2). *Let $[Z] \in \Gamma(i)$, then Z has special decomposition with respect to $\mathcal{O}_X(L)$ (see Definition 1.12; for $\Gamma(i)$, see the statement of Theorem 2).*

Proof: Let $y \in \mathbf{P}_Z$ be such that $h_Z(y)$ is an isomorphism, where h_Z as in the proof of Lemma 4.1 and $h_Z(y)$ its value at y. Set $T_{Z,y} = \mathcal{T}_Z \otimes k(y)$, the fibre of $\mathcal{T}_Z$ at y, and $T^{(i)}_{Z,y}$, the subspace of $T_{Z,y}$ isomorphic to $\ker(p^i_y)$ (p^i_y as in the statement of Theorem 2) via $h_Z(y)$. Put $\tilde{T}_{Z,y} = \tilde{\mathcal{T}}_Z \otimes k(y)$, the fibre of $\tilde{\mathcal{T}}_Z$ at y, and $\tilde{T}^{(i)}_{Z,y}$, the inverse image of $T^{(i)}_{Z,y}$ under the projection $\tilde{T}_{Z,y} \longrightarrow T_{Z,y}$.

Observe: $\tilde{T}^{(i)}_{Z,y} \supsetneqq H^0(\mathcal{O}_X)$ and it is a subring of $H^0(\mathcal{O}_Z)$. To see this is enough to show that $\tilde{T}^{(i)}_{Z,y}$ is closed under the multiplication: let $f, g \in \tilde{T}^{(i)}_{Z,y}$; to show that $f \cdot g \in \tilde{T}^{(i)}_{Z,y}$ is equivalent to $(f \cdot g) \cdot m \in \frac{M^i_{Z,y}}{H^0(\mathcal{J}_Z(K_X+L))}$ for every $m \in M^i_{Z,y}$, where $M^i_{Z,y}$ is the fibre of $\mathcal{M}^i_Z$ at y and we use the interpretation of p^i_y discussed in Remark 4.8.

Since $g \in \tilde{T}^{(i)}_{Z,y}$ we have: $g \cdot m \in \frac{M^i_{Z,y}}{H^0(\mathcal{J}_Z(K+L))}$ for every $m \in M^i_{Z,y}$. Putting $m_g = g \cdot m$ and using the fact that $f \in \tilde{T}^{(i)}_{Z,y}$ we obtain:

$$(f \cdot g) \cdot m = f(g \cdot m) = f \cdot m_g \in \frac{M^i_{Z,y}}{H^0\left(\mathcal{J}_Z(K+L)\right)}.$$

The inclusion of rings $\tilde{T}^{(i)}_{Z,y} \subset H^0(\mathcal{O}_Z)$ induces surjective morphism of schemes

$$f : Z \longrightarrow Z' = \operatorname{Spec}\left(\tilde{T}^{(i)}_{Z,y}\right).$$

Take Z'_1, a proper subscheme of Z', and $Z'_2 = Z' \setminus Z'_1$. Seting $Z_i = f^*(Z'_i)$ we obtain the decomposition

$$Z = Z_1 + Z_2.$$

This decomposition is special (in a sense of Definition 1.12) since

$$p_{Z_i} : H^0(K_X + L) \longrightarrow H^0\left(\mathcal{O}_{Z_i}(K_X + L)\right),$$

the restriction map, is not surjective for $i = 1, 2$ (this can be seen as follows: there exists $g_i \in \tilde{T}^{(i)}_{Z,y}$ such that $g_i \notin \mathcal{J}_{Z_i}$ and $g_i \in H^0(\mathcal{O}_{Z \setminus Z_i} \otimes \mathcal{J}_{Z \setminus Z_i})$; this implies

$$0 = \langle \operatorname{Res}_Z(g_i), \phi \rangle = \langle \operatorname{Res}_{Z_i}(g_i), \phi \rangle \tag{4.11}$$

for every $\phi \in H^0(K_X + L)$, where Res_Z as in (3.6). Since $(\operatorname{Res}_{Z_i})^* = p_{Z_i}$ it follows from (4.11)

$$g_i \in \ker(\operatorname{Res}_{Z_i}) = [\operatorname{coker}(p_{Z_i})]^*$$

Q.E.D.

Remark 4.12. If Z in Proposition 4.10 is reduced, then for every $p' \in Z'$ the subcycle $Z_{p'} = f^{-1}(p')$ is special with respect to $\mathcal{O}_X(L+K_X)$ (i.e. $h^1\left(\mathcal{J}_{Z_{p'}}(K_X+L)\right) \neq 0$) for every $p' \in Z'$ and the special decomposition for Z is as follows:

$$Z = \sum_{p' \in Z'} Z_{p'}.$$

In particular, if $\mathcal{O}_X(L+K_X)$ is base point free, then $\deg(Z_{p'}) > 1$, for every $p' \in Z'$ (since $\Gamma_1^0(L) = \emptyset$; see (2.3) for notations); if $\mathcal{O}_X(L+K_X)$ is very ample, then $\deg(Z_{p'}) > 2$, for every $p' \in Z'$ (since $\Gamma_2^0(L) = \emptyset$).

References

[1] F. Catanese, *Footnotes on a theorem of I. Reider*, preprint.

[2] P. Griffiths, et al, *Topics in transcendental algebraic geometry*, Annals of Math. Studies, Princeton University Press (1984).

[3] P. Griffiths, J. Harris, *Principles of algebraic geometry*, John Wiley, New York (1978).

[4] C. Okoneck, M. Schneider and H. Spindler, *Vector bundles on complex projective spaces*, Progress in Math. **3** (1980), Brikhauser.

REIDER-SERRANO'S METHOD ON NORMAL SURFACES

Fumio Sakai

Department of Mathematics, Saitama University, Urawa 338, Japan

Dedicated to Prof. Dr. F.Hirzebruch on his 60th birthday

Let Y be a normal projective surface over $\mathbb{C}$, and let D be a Weil divisor on Y. Generalizing the methods of Reider and Serrano, we prove a criterion for the very ampleness of the adjoint linear system $|K_Y + D|$.

Introduction.

Recently, Reider([Rd]) and Serrano [Se] have discovered new methods to study linear systems on a smooth projective surface. Reider's argument goes back to Mumford's proof [Mu] of the Ramanujam vanishing theorem. On the other hand, Serrano has applied Miyaoka's version [Mi] of the Ramanujam vanishing theorem. We will observe that by either method, one can prove the following:

Proposition 1. *Let* D *be a big divisor with* $D^2 > 0$ *on a smooth projective surface* X. *If* $H^1(X,\mathcal{O}(-D)) \neq 0$, *then there is a nonzero effective divisor* E *such that* (i) $(D - E)E \leq 0$, (ii) $D - 2E$ *is a big divisor.*

With the aid of this result, we study adjoint linear systems on a normal surface Y. Let $\mathcal{L}$ be a line bundle on Y. The vector space $H^0(Y,\mathcal{L})$ defines a complete linear system $|\mathcal{L}|$ and a rational map $\Phi_{\mathcal{L}}$ of Y. For a point $y \in Y$, let $\mathfrak{m}_y$ denote the maximal ideal sheaf of y. We say that $|\mathcal{L}|$ *has no base points* if the maps $H^0(Y,\mathcal{L}) \to \mathcal{L}\otimes(\mathcal{O}/\mathfrak{m}_y)$ $\forall y \in Y$ are surjective. We say that $|\mathcal{L}|$ *separates two distinct points* if the maps $H^0(Y,\mathcal{L}) \to \mathcal{L}\otimes(\mathcal{O}/\mathfrak{m}_y \oplus \mathcal{O}/\mathfrak{m}_{y'})$ $\forall y \neq \forall y' \in Y$ are surjective. We also say that

$|\mathcal{L}|$ *separates tangent vectors everywhere* if the maps $H^0(Y,\mathcal{L}) \to \mathcal{L}\otimes(\mathcal{O}/m_y^2)$ $\forall y \in Y$ are surjective. We know that (i) $|\mathcal{L}|$ has no base points $\Leftrightarrow \Phi_{\mathcal{L}}$ is a morphism, (ii) $|\mathcal{L}|$ separates two distinct points $\Leftrightarrow \Phi_{\mathcal{L}}$ is a one-to-one morphism, (iii) $|\mathcal{L}|$ separates tangent vectors everywhere $\Leftrightarrow \Phi_{\mathcal{L}}$ is a local embedding. Finally, $\mathcal{L}$ (or $|\mathcal{L}|$) is *very ample* if $\Phi_{\mathcal{L}}$ gives an embedding of Y, i.e., $\Phi_{\mathcal{L}}$ is a one-to-one morphism and is a local embedding. Our main result is the following:

Theorem 1. *Let* Y *be a normal projective surface, and let* D *be a nef divisor on* Y. *Assume that the adjoint divisor* $K_Y + D$ *is a Cartier divisor. If* $D^2 > 8 + \eta(Y) + \gamma(Y)$, *then* $|K_Y + D|$ *is very ample unless there exists a nonzero effective divisor* E *on* Y *such that*

$$0 \leq DE < 4 + (\eta(Y) + \gamma(Y))/2$$

$$DE - 2 - (\eta(Y) + \gamma(Y))/4 \leq E^2 \leq (DE)^2/D^2, \text{ and}$$

$$E^2 < 0 \text{ if } DE = 0.$$

Compared with the smooth case (cf. Theorem 2 in Sect.3), this result involves the nonnegative compensation terms $\eta(Y)$, $\gamma(Y)$. The term $\eta(Y)$ measures the sum of the contributions from the non-Gorenstein singularities, while $\gamma(Y)$ is the contribution from the worst non-rational singularity. We sketch here how we prove the above theorem. Set $\mathcal{L} = \mathcal{O}(K_Y + D)$. Take an ideal sheaf $\mathcal{A}$ with 0-dimensional support on Y. Suppose that the map $H^0(Y,\mathcal{L}) \to \mathcal{L}\otimes(\mathcal{O}/\mathcal{A})$ is not surjective. In order to construct a nonzero effective divisor E on Y, we argue as follows. Choose a smooth projective surface X, a birational morphism $\pi: X \to Y$ and an effective divisor J on X such that $\mathcal{J} = \pi_*\mathcal{O}(-J) \subset \mathcal{A}$. Consider the cohomology sequence:

$$\begin{array}{ccccccc} H^0(X,\pi^*\mathcal{L}\otimes\mathcal{O}(-J)) & \to & H^0(X,\pi^*\mathcal{L}) & \to & H^0(X,\pi^*\mathcal{L}\otimes(\mathcal{O}/\mathcal{O}(-J))) & \to & H^1(X,\pi^*\mathcal{L}\otimes\mathcal{O}(-J)) \\ \wr\!\! | & & \wr\!\! | & & \cup & & \\ H^0(Y,\mathcal{L}\otimes\mathcal{J}) & \to & H^0(Y,\mathcal{L}) & \to & \mathcal{L}\otimes(\mathcal{O}/\mathcal{J}) & & \end{array}$$

Since $\mathcal{O}/\mathcal{J} \to \mathcal{O}/\mathcal{A}$ is surjective, the map $H^0(Y,\mathcal{L}) \to \mathcal{L}\otimes(\mathcal{O}/\mathcal{J})$ cannot be surjective. It follows from the above sequence that $H^1(X,\pi^*\mathcal{L}\otimes\mathcal{O}(-J)) \neq 0$. If we write $\pi^*\mathcal{L}\otimes\mathcal{O}(-J) = \mathcal{O}(K_X + \tilde{D})$, by duality, we see that $H^1(X,\mathcal{O}(-\tilde{D})) \neq 0$. If $\tilde{D}$ is big, e.g., if $\tilde{D}^2 > 0$ (cf.Lemma 3), then one can apply Proposition 1 to $\tilde{D}$ and one finds a nonzero effective divisor $\tilde{E}$ on X such that $(\tilde{D} - \tilde{E})\tilde{E} \leq 0$ and $\tilde{D} - 2\tilde{E}$ is big. If we choose J carefully, we can make $E = \pi_*\tilde{E}$ nonzero. As an application of Theorem 1, we show the following result. Let Y be a normal $\mathbf{Q}$-Gorenstein projective surface of index r. If K_Y is ample, then $|nrK_Y|$ is very ample for $n \geq 4 + r^{-1} + (\eta(Y) + \gamma(Y))/2$.

The content of this paper is the following. Sect.1 is devoted to prepare notation and technical results about divisors on normal surfaces. In Sect.2 we define several local invariants connected with the maximal ideal sheaf of a normal surface singularity. In Sect.3 we provide two proofs of Proposition 1. Sect.4 contains the proof of Theorem 1. In Sect.5 we apply Theorem 1 to pluri-canonical and pluri-anticanonical systems on normal surfaces. In Sect.6 we discuss the relative version of Proposition 1.

I would like to thank A.Sommese for encouragement, helpful discussions and correspondence on this topics, and for sending me the preprints of his papers [AS], [SV]. I also thank T.Fujita for inspiring me by his preprint [F].

1. Preliminaries.

We refer to [Sa3] for the basic results on normal surfaces. Let Y be a normal Moishezon surface. By a *divisor* we mean a Weil divisor. There is a $\mathbf{Q}$-valued intersection theory on divisors. We denote by $\equiv$ the numerical equivalence. A divisor D is *nef* if $DC \geq 0$ for all irreducible curves C on Y, and is *pseudoeffective* if $DP \geq 0$ for all nef divisors P on Y. Define $\kappa(D,Y) = \text{tr.deg.} \bigoplus_{m\geq 0} H^0(Y,\mathcal{O}(mD)) - 1$, and $-\infty$ if $H^0(Y,\mathcal{O}(mD)) = 0$ for all $m > 0$. We say that D is *big* if $\kappa(D,Y) = 2$.

Lemma 1. *The following are equivalent:*

(i) D *is big and* $D^2 > 0$.

(ii) D *belongs to the positive cone (i.e.,* $D^2 > 0$, $PD > 0$ *for a nef divisor* P).

Lemma 2 (cf.[Ra], p.44). *Let* D *be a nef divisor with* $D^2 > 0$ on Y. *If* $D \equiv D_1 + D_2$ *where both* D_i *are pseudoeffective and* $D_i \not\equiv 0$, *then we have* $D_1D_2 > 0$.

Lemma 3. *Let* $f:\tilde{Y} \to Y$ *be a birational morphism of normal Moishezon surfaces. Let* $\tilde{D}$ *be a divisor on* $\tilde{Y}$, *and set* $D = f_*\tilde{D}$. *Then*

(i) *If* $\tilde{D}$ *is big, then so is* D.

(ii) *If* $\tilde{D}^2 > 0$ *and if* D *is big, then* $\tilde{D}$ *is big.*

Proof. (i) Write $f^*D = \tilde{D} + G$ and $G = G_+ - G_-$ with $G_+ \geq 0$, $G_- \geq 0$. Since $f^*(D) + G_- = \tilde{D} + G_+$, we infer from Theorem (6.2) in [Sa3] that $\kappa(D,Y) = \kappa(\tilde{D} + G_+,\tilde{Y}) \geq \kappa(\tilde{D},\tilde{Y})$. (ii) Since $\tilde{D}^2 > 0$, either $\tilde{D}$ or $-\tilde{D}$ is big. If $-\tilde{D}$ were big, then G would be big, which is absurd. Q.E.D.

Proposition 2. *Let* $f:\tilde{Y} \to Y$ *be a birational morphism of two normal Moishezon surfaces. Let* $\tilde{D}$ *be a divisor on* $\tilde{Y}$ *with* $\tilde{D}^2 > 0$. *Suppose there is a nonzero effective divisor* $\tilde{E}$ *such that* (a) $(\tilde{D} - \tilde{E})\tilde{E} \leq 0$, (b) $\tilde{D} - 2\tilde{E}$ *is big. Set* $D = f_*\tilde{D}$, $\alpha = (D^2 - \tilde{D}^2)$, $E = f_*\tilde{E}$. *If* D *is nef and* $E > 0$, *then*

$$0 \leq DE < \alpha/2$$

$$DE - \alpha/4 \leq E^2 \leq (DE)^2/D^2, \text{ and}$$

$$E^2 < 0 \text{ if } DE = 0.$$

Proof. It follows from Lemma 3, (i) that D - 2E is big. Note that $D^2 = \tilde{D}^2 + \alpha > 0$. By Lemma 2, $DE \geq 0$, $(D - 2E)E > 0$. Write $f^*D = \tilde{D} + G$, $f^*E = \tilde{E} + \Gamma$. Then $(\tilde{D} - \tilde{E})\tilde{E} = (f^*(D - E) - G + \Gamma)(f^*E - \Gamma) = (D - E)E - \Gamma^2 + G\Gamma$. Thus

$$(D - E)E \leq (\Gamma - \frac{1}{2} G)^2 - \frac{1}{4} G^2 \leq - \frac{1}{4} G^2 = \alpha/4.$$

We have the inequality: $DE - \alpha/4 \leq E^2 < DE/2$, so that $0 \leq DE < \alpha/2$. The Hodge index theorem on Y gives the inequality: $E^2 \leq (DE)^2/D^2$. Since $D^2 > \alpha$, in the above range of DE, $(DE)^2/D^2 \leq DE/2$. If $DE = 0$, then $E^2 < 0$. Q.E.D.

2. Local invariants.

Let (V,y) be a germ of a normal surface singularity. Let $\mathfrak{m}$ be the maximal ideal sheaf of y. We now prepare some local invariants, which will be used in Sect.4. Let $\pi: U \to V$ be the minimal resolution of y. There exists an effective $\mathbb{Q}$-divisor Δ supported on $\pi^{-1}(y)$ such that $\pi^*K_V = K_U + \Delta$. Note that $\mathrm{Supp}(\Delta) = \pi^{-1}(y)$ unless y is a rational double point. Cf.[Sa3]. Define

$$\delta = -\Delta^2,$$

The *numerical index* of y is defined to be the least integer s such that $s\Delta$ is integral. Recall that y is *Gorenstein* if $\mathcal{O}(K_V)$ is trivial near y. In this case, s = 1. Let Z be the fundamental cycle of y. One knows then that $\mathfrak{m} \simeq \pi_*\mathcal{O}(-Z)$ and $\mathcal{O}/\mathfrak{m} \simeq \pi_*(\mathcal{O}/\mathcal{O}(-Z))$. If y is a rational singularity, Artin [A] showed that $\mathfrak{m}^n \simeq \pi_*\mathcal{O}(-nZ)$ $\forall$ $n \geq 1$.

To cover the non-rational case, consider the following conditions of a nonzero effective divisor Γ on $\pi^{-1}(y)$.

(*) $\Delta - \Gamma$ is π-nef, i.e.,$(\Delta - \Gamma)C \geq 0$ $\forall$ C on $\pi^{-1}(y)$,

(**) $\pi_*\mathcal{O}(-\Gamma) \subset \mathfrak{m}^2$.

If Γ satisfies (*), then $R^1\pi_*\mathcal{O}(-\Gamma) = 0$. So $\mathcal{O}/\pi_*\mathcal{O}(-\Gamma) \simeq \pi_*(\mathcal{O}/\mathcal{O}(-\Gamma))$. There is a unique minimal divisor Z^a satisfying (*). We call this Z^a the *adjoint fundamental cycle* of y. It then turns out that $h^1(\mathcal{O}_{Z^a}) = \dim R^1\pi_*\mathcal{O}_U$, $Z^a \geq Z$, $s\Delta \geq Z^a \geq \Delta$. In particular, $Z^a = \Delta$ if y is Gorenstein. Set $m^a = \pi_*\mathcal{O}(-Z^a)$. One has a canonical surjection $\mathcal{O}/m^a \to \mathcal{O}/m$. Define

$$\mu = -(Z^a - \Delta)^2.$$

Among effective divisors Γ satisfying (*) and (**), let Z^t be the one such that $-(Z^t - \Delta)^2$ takes the minimum. In view of (**), there is a surjection $\mathcal{O}/\pi_*\mathcal{O}(-Z^t) \to \mathcal{O}/m^2$. Define

$$\nu = -(Z^t - \Delta)^2 - \mu.$$

In general, it is not easy to find Z^t. To estimate μ, ν, set $\mathcal{J}_n = \pi_*\mathcal{O}(-ns\Delta)$. If y is not a rational double point, then $\mathcal{J}_n \subset m$. Let e be the least positive integer e such that $\mathcal{J}_e \subset m^2$. Then $es\Delta$ satisfies (*) and (**), hence

$$\mu \leq (s - 1)^2\delta, \qquad \mu + \nu \leq (es - 1)^2\delta.$$

To indicate y, we write as m_y, m_y^a, Δ_y, Z_y, Z_y^a, Z_y^t, δ_y, μ_y, ν_y.

Lemma 4. *If* y *is not a rational double point, then*

(i) $e \leq 2$ *if* $s \geq 3$, $e \leq 3$ *if* $s = 2$ *and* $e \leq 5$ *if* $s = 1$,

(ii) *if* y *is Gorenstein, then* $e \leq 4$.

Proof. By Theorem 3.2 in [L2], if both $-ps\Delta - 2K_U$ and $-qs\Delta - 3K_U$ are π-nef, then the map $\mathcal{J}_p \otimes \mathcal{J}_q \to \mathcal{J}_{p+q}$ is surjective, and $\mathcal{J}_{p+q} = \mathcal{J}_p\mathcal{J}_q \subset m^2$. Since $\Delta \equiv -K_U$ on $\pi^{-1}(y)$ and $-\Delta$ is π-nef, the assertion (i) follows from this. Suppose y is Gorenstein. We have $\pi_*\mathcal{O}(nK_U) = \pi_*\mathcal{O}(\pi^*(nK_V) - n\Delta) = \mathcal{O}(K_V)\otimes\mathcal{J}_n \simeq \mathcal{J}_n$. The result of Laufer [L3] then asserts that $\mathcal{J}_4 = \mathcal{J}_1\mathcal{J}_3 + \mathcal{J}_2\mathcal{J}_2$. Hence $\mathcal{J}_4 \subset m^2$. Q.E.D.

Example 1. Suppose that y is resolved by a single smooth curve C of genus $g \geq 2$. Set $\mathcal{L} = \mathcal{O}_C(-C)$, $d = \deg \mathcal{L} = -C^2$. If $d \geq 2g - 2$ and if $\mathcal{L}$ is normally generated, e.g., if $d \geq 2g + 1$, then $2C$ satisfies (*), (**). In this case, $Z = C$, $Z^a = Z^t = 2C$, $\Delta = ((d + 2g - 2)/d)C$, and $s\Delta = (d + 2g - 2)C$ (if $d > 2g - 2$), $= 2C$ (if $d = 2g - 2$). Cf. [MR].

Remark 1. If y is Gorenstein, then $\mu = 0$, $\nu \leq (e - 1)^2\delta$, so $\nu \leq 9\delta$. In case y is minimally elliptic, if $\delta \geq 3$, then $e = 2$ and so $\nu = \delta$ (Laufer [L1], Reid [Re]).

3. Reider's method and Serrano's method.

We give two proofs of Proposition 1 in the Introduction.

Proof by Mumford-Reider's method. The non-vanishing of $H^1(X,\mathcal{O}(-D))$ gives rise to a vector bundle $\mathcal{F}$ with a nontrivial extension:

$$0 \to \mathcal{O} \to \mathcal{F} \to \mathcal{O}(D) \to 0.$$

Clearly, $c_1(\mathcal{F})^2 = D^2$, $c_2(\mathcal{F}) = 0$, so that $c_1(\mathcal{F})^2 > 4c_2(\mathcal{F})$. One deduces from the Bogomolov theory that there is an extension:

$$0 \to \mathcal{O}(Q)\otimes\mathcal{L} \to \mathcal{F} \to I_Z\mathcal{L} \to 0,$$

where $\mathcal{L}$ is an invertible sheaf, I_Z is an ideal sheaf of a 0-dimensional scheme Z and Q is a big divisor with $Q^2 > 0$. Since the composition map $\mathcal{O} \to \mathcal{F} \to I_Z\mathcal{L}$ is not zero, one finds an effective divisor E such that $\mathcal{L} \simeq \mathcal{O}(E)$. It follows that $D = c_1(\mathcal{F}) = Q + 2E$, $0 = c_2(\mathcal{F}) = (E + Q)E + \deg Z$ and hence $(D - E)E = (E + Q)E \leq 0$. Q.E.D.

Proof by Miyaoka-Serrano's method. Consider the Zariski decomposition: $D = P + N$. Since D is big, it is known that $P^2 > 0$. Write $N = \sum\alpha_i E_i$, $\alpha_i \geq 0$, $\alpha_i \in \mathbf{Q}$, and let $[N]$ denote the integral part of N. It follows from the Miyaoka-Ramanujam vanishing theorem [Mi] that $H^1(X,\mathcal{O}(-(D-[N]))) = 0$. So if $H^1(X,\mathcal{O}(-D)) \neq 0$, we must have $[N] > 0$. Consider a sequence: $D_0 = D - [N], \ldots, D_k = D_{k-1} + E_{j_k}, \ldots, D_n = D$. There is an exact sequence:

$$0 \to \mathcal{O}(-D_k) \to \mathcal{O}(-D_{k-1}) \to \mathcal{O}_{E_{j_k}}(-D_{k-1}) \to 0.$$

If $D_{k-1}E_{j_k} > 0$ for all k, one inductively obtains the vanishing: $H^1(X, \mathcal{O}(-D)) = 0$. One can therefore construct a sequence $D_0, \ldots, D_k$ with $k < n$ such that $D_kE_j \le 0$ for all irreducible components E_j of $D - D_k \le [N]$. Set $E = D - D_k$. By construction, $(D - E)E \le 0$. One finds that $(D - 2E)^2 \ge D^2 > 0$, because $D^2 = (D - 2E)^2 + 4(D - E)E$. One has also $P(D - 2E) = P^2 > 0$. By Lemma 1, D - 2E is big. Q.E.D.

Remark 2. We further obtain the following properties of E:

(i)* $(D - E)E_j \le 0$ for all components E_j of E,

(iii) the intersection matrix E_iE_j of all irreducible components of E is negative definite.

For smooth projective surfaces, we have the following criterion.

Theorem 2 (Reider [Rd], cf. [BL], [SV], [F], [Se]). *Let* X *be a smooth projective surface, and let* D *be a nef divisor on* X.

(i) *If* $D^2 > 4$, *then* $|K_X + D|$ *has no base points unless there exists a nonzero effective divisor* E *such that*

$$DE = 0,\ E^2 = -1,\ \textit{or}\ DE = 1,\ E^2 = 0.$$

(ii) *If* $D^2 > 8$, *then* $|K_X + D|$ *is very ample unless there exists a nonzero effective divisor* E *satisfying one of the following:*

$DE = 0,\ E^2 = -1,\ \textit{or}\ -2,$

$DE = 1,\ E^2 = 0\ \textit{or}\ -1,$

$DE = 2,\ E^2 = 0,$

$D \equiv 3E,\ E^2 = 1.$

Proof. This is now a corollary of Theorem 3 in Sect.4. By proving (i) we illustrate how Proposition 1 yields this kind of results. Assume that $D^2 > 4$ and that $|K_X + D|$ has a base point x. Let $\varphi:\tilde{X} \to X$ be the blowing up of X at x, and let L denote the exceptional curve of φ. We must have $H^1(\tilde{X},\mathcal{O}(K_{\tilde{X}} + \varphi^*D - 2L)) \neq 0$. Set $\tilde{D} = \varphi^*D - 2L$. By duality, we get $H^1(\tilde{X},\mathcal{O}(-\tilde{D})) \neq 0$. The hypothesis: $D^2 > 4$ implies that $\tilde{D}^2 > 0$, so that $\tilde{D}$ is big (Lemma 3). Applying Proposition 1 to $\tilde{D}$, we find a non-zero effective divisor $\tilde{E}$ such that $\tilde{D} - 2\tilde{E}$ is big and $(\tilde{D} - \tilde{E})\tilde{E} \leq 0$. Set $E = \varphi_*\tilde{E}$. We see easily that $E > 0$. Then Proposition 2 applies, and gives : $DE-1 \leq E^2 < DE/2$, $DE - 1 \leq E^2 \leq (DE)^2/D^2$. So $DE = 0$ or 1. If $DE = 0$, then $E^2 = -1$, and if $DE = 1$, then $E^2 = 0$. Q.E.D.

4. Adjoint linear systems on a normal surface.

Let Y be a normal Moishezon surface and let D be a divisor on Y. We consider the situation in which the adjoint divisor $K_Y + D$ is a Cartier divisor. Let Rat(Y) (resp. Irr(Y)) denote the locus of rational (resp. non-rational) singularities on Y. Define

$$\eta(Y) = \sum_{y \in \mathrm{Rat}(Y)} \delta_y + \sum_{y \in \mathrm{Irr}(Y)} \mu_y$$

$$\gamma(Y) = \begin{cases} 0 & \text{if } \mathrm{Irr}(Y) = \emptyset \\ \max\limits_{y \in \mathrm{Irr}(Y)} \{ (\nu_y - 8)^+ \} & \text{if } \mathrm{Irr}(Y) \neq \emptyset \end{cases}$$

Here $\alpha^+ = \max\{\alpha, 0\}$ for $\alpha \in \mathbb{Q}$. For the definition of δ_y, μ_y and ν_y, see Sect.2. Note that $8 + \gamma(Y) = \max\{8, \nu_y\}$ where the maximum is taken over all $y \in \mathrm{Irr}(Y)$. If Y is Gorenstein, then $\eta(Y) = 0$.

Theorem 3. *Let Y be a normal Moishezon surface. Let D be a nef divisor on Y such that $K_Y + D$ is a Cartier divisor.*

(i) *If* $D^2 > 4 + \eta(Y)$, *then* $|K_Y + D|$ *has no base points unless there exists a nonzero effective divisor* E *such that*

$$0 \leq DE < 2 + \frac{1}{2}\eta(Y)$$
$$DE - 1 - \frac{1}{4}\eta(Y) \leq E^2 \leq (DE)^2/D^2, \text{ and}$$
$$E^2 < 0 \text{ if } DE = 0.$$

(ii) *If* $D^2 > 8 + \eta(Y)$, *then* $|K_Y + D|$ *separates two distinct points on* Y *and separates tangent vectors everywhere off* Irr(Y) *unless there exists a nonzero effective divisor* E *such that*

$$0 \leq DE < 4 + \frac{1}{2}\eta(Y)$$
$$DE - 2 - \frac{1}{4}\eta(Y) \leq E^2 \leq (DE)^2/D^2, \text{ and}$$
$$E^2 < 0 \text{ if } DE = 0.$$

(iii) *If* $D^2 > 8 + \eta(Y) + \gamma(Y)$, *then* $|K_Y + D|$ *separates tangent vectors everywhere unless there exists a nonzero effective divisor* E *such that*

$$0 \leq DE < 4 + \frac{1}{2}(\eta(Y) + \gamma(Y))$$
$$DE - 2 - \frac{1}{4}(\eta(Y) + \gamma(Y)) \leq E^2 \leq (DE)^2/D^2, \text{ and}$$
$$E^2 < 0 \text{ if } DE = 0.$$

Proof of Theorem 1 in the Introduction. The assertion follows from (ii) and (iii) of Theorem 3.

Proof of Theorem 3. We omit the proof of (i). Write $\mathcal{L} = \mathcal{O}(K_Y + D)$. Let $\pi: X \to Y$ be the minimal resolution of the singularities of Y. Set $\Delta = \sum \Delta_y$, and

$$\Lambda = \sum_{y \in Irr(Y)} Z_y^a .$$

Then we have $\eta(Y) = -(\Delta - \Lambda)^2$, because

$$\Delta - \Lambda = \sum_{y \in Rat(Y)} \Delta_y - \sum_{y \in Irr(Y)} (Z_y^a - \Delta_y).$$

We first prove the following:

Lemma 5. *Let* $\tilde{E}$ *be a nonzero effective divisor on* X *such that* $(\Delta - \Lambda - \tilde{E})\tilde{E} \leq 0$. *Then* $E = \pi_*\tilde{E} > 0$.

Proof. Assume to the contrary that $\tilde{E}$ is exceptional for π. Decompose as $\tilde{E} = \tilde{E}_R + \tilde{E}_I$ where $\tilde{E}_R$ (resp. $\tilde{E}_I$) are supported on $\pi^{-1}(\mathrm{Rat}(Y))$ (resp. $\pi^{-1}(\mathrm{Irr}(Y))$). We see that $(\Delta - \Lambda - \tilde{E})\tilde{E}_R = -(K_X + \tilde{E}_R)\tilde{E}_R \geq 2$ (if $\tilde{E}_R > 0$). We also have $(\Delta - \Lambda - \tilde{E})\tilde{E}_I > 0$ (if $\tilde{E}_I > 0$), because $\Delta - \Lambda$ is π-nef on $\pi^{-1}(\mathrm{Irr}(Y))$. We infer from these that $(\Delta - \Lambda - \tilde{E})\tilde{E} > 0$, a contradiction. Q.E.D.

Now we prove the first part of (ii). Assume that $D^2 > 8 + \eta(Y)$. Let y, y′ be two distinct points on Y. Suppose the the canonical map $H^0(Y,\mathcal{L}) \to \mathcal{L}\otimes(\mathcal{O}/m_y \oplus \mathcal{O}/m_{y'})$ is not surjective. We divide into six cases: (a) both y and y′ are smooth, (b) y is smooth but $y' \in \mathrm{Rat}(Y)$, (c) y is smooth but $y' \in \mathrm{Irr}(Y)$, (d) $y, y' \in \mathrm{Rat}(Y)$, (e) $y \in \mathrm{Rat}(Y)$, $y' \in \mathrm{Irr}(Y)$ (f) $y, y' \in \mathrm{Irr}(Y)$. We here deal with the cases (a), (d), and (f). The other cases can be done similarly.

In case (a), let $\varphi:\tilde{X} \to X$ be the blowing ups at y and y′, and let L, L′ be the exceptional curves, respectively. Set $\Phi = \pi \circ \varphi$. Letting $\mathcal{I} = \Phi_*\mathcal{O}(-\Lambda-L-L')$, we have

$$\begin{array}{ccc} H^0(\tilde{X},\Phi^*\mathcal{L}) & \to & H^0(\tilde{X},\Phi^*\mathcal{L}\otimes(\mathcal{O}/\mathcal{O}(-\Lambda-L-L'))) \\ \wr\!\parallel & & \cup \\ H^0(Y,\mathcal{L}) & \to & \mathcal{L}\otimes(\mathcal{O}/\mathcal{I}) \end{array}$$

Since $\mathcal{O}/\mathcal{I} \to \mathcal{O}/m_y \oplus \mathcal{O}/m_{y'}$ is surjective, it follows therefore that $H^1(\tilde{X},\Phi^*\mathcal{L}\otimes\mathcal{O}(-\Lambda-L-L')) \neq 0$. Set $\tilde{D} = \pi^*D + \Delta - \Lambda - 2L - 2L'$. Then we have $H^1(\tilde{X},\mathcal{O}(-\tilde{D})) \neq 0$. Since $\tilde{D} = D^2 - \eta(Y) - 8$, under our hypothesis $\tilde{D}^2 > 0$. We see from Lemma 3 that $\tilde{D}$ is big. We apply Proposition 1 to $\tilde{D}$, and we obtain a nonzero effective divisor $\tilde{E}$ on $\tilde{X}$ such that $(\tilde{D} - \tilde{E})\tilde{E} \leq 0$ and $\tilde{D} - 2\tilde{E}$ is big. It is easy to see that $\varphi_*\tilde{E} > 0$, and hence by Lemma 5, we find that $E > 0$. The properties of E then follows from Proposition 2.

In case (d), let Z, Z' be the fundamental cycles of y, y', respectively. As in case (a), we see that $H^1(X, \pi^*\mathcal{L}\otimes\mathcal{O}(-\Lambda-Z-Z')) \neq 0$. Set $\tilde{D} = \pi^*D + \Delta - \Lambda - Z - Z'$. Then we have $H^1(X, \mathcal{O}(-\tilde{D})) \neq 0$. Since $\tilde{D}^2 = D^2 - \eta(Y) - 8$, we can proceed as in case (a).

In case (f), since $\mathfrak{m}_y^a \subset \mathfrak{m}_y$, the map $H^0(Y, \mathcal{L}) \to \mathcal{O}/\mathfrak{m}_y^a \oplus \mathcal{O}/\mathfrak{m}_{y'}^a$ cannot be surjective. Since $\pi_*(\mathcal{O}/\mathcal{O}(-\Lambda)) \simeq \bigoplus_{y \in \mathrm{Irr}(Y)} \mathcal{O}/\mathfrak{m}_y^a$, we deduce that $H^1(X, \pi^*\mathcal{L}\otimes\mathcal{O}(-\Lambda)) \neq 0$. If we set $\tilde{D} = \pi^*D + \Delta - \Lambda$, $\tilde{D}^2 = D^2 - \eta(Y)$. The rest is the same as in case (a).

In order to prove the rest of (ii) and (iii), we proceed to consider the separation of tangent vectors. Take a point $y \in Y$. Assume that $D^2 > \eta(Y) + \nu_y$ if $y \in \mathrm{Irr}(Y)$, $> \eta(Y) + 8$ otherwise. Suppose that the map $H^0(Y, \mathcal{L}) \to \mathcal{L}\otimes(\mathcal{O}/\mathfrak{m}_y^2)$ is not surjective.

We first deal with the case in which y is a smooth point. We regard y as a point on X. We can choose local coordinates (u,v) at y, so that $H^0(Y, \mathcal{L}) \to \mathcal{L}\otimes\mathcal{O}/\mathcal{A}$ is not surjective, where $\mathcal{A} = (u, v^2)$. Let $\varphi_1 : X_1 \to X$ be the blowing up of Y at y, and let L_1 be the exceptional curve over y. Let $\varphi_2 : \tilde{X} \to X_1$ be the blowing up of X_1 at a point y_1 on L_1, which corresponds to the direction $\partial/\partial v$ and let L_2 be the exceptional curve of φ_2. Let $\bar{L}_1$ denote the strict transform of L_1. Set $\Phi = \pi \circ \varphi_1 \circ \varphi_2$. In this case, we have $\Phi_*\mathcal{O}(-\bar{L}_1 - 2L_2) \simeq \mathcal{A}$. Since $\Phi_*\mathcal{O}(-\Lambda-\bar{L}_1-2L_2) \subset \mathcal{A}$, we must have $H^1(\tilde{X}, \Phi^*\mathcal{L}\otimes\mathcal{O}(-\Lambda-\bar{L}_1-2L_2)) \neq 0$. Set $\tilde{D} = \Phi^*D + \Delta - \Lambda - 2\bar{L}_1 - 4L_2$. By duality, $H^1(\tilde{X}, \mathcal{O}(-\tilde{D})) \neq 0$. Since $\tilde{D}^2 = D^2 - \eta(Y) - 8$, the assumption implies that $\tilde{D}^2 > 0$, so that $\tilde{D}$ is big. By Proposition 1, we find an effective divisor $\tilde{E} > 0$ such that (i) $(\tilde{D} - \tilde{E})\tilde{E} \leq 0$, (ii) $\tilde{D} - 2\tilde{E}$ is big. Set $E = \varphi_*\tilde{E}$. By using Lemma 5, we can prove that $E > 0$. We infer from Proposition 2 that $0 \leq DE < 4 + \eta(Y)/2$, and $DE - 2 - \eta(Y)/4 \leq E^2 \leq (DE)^2/D^2$.

We pass to the case in which y is singular. Let $Z = Z_y$, $Z^a = Z_y^a$, $Z^t = Z_y^t$. Set $\Gamma = 2Z$ if $y \in \mathrm{Rat}(Y)$, $= Z^t - Z^a$ if $y \in \mathrm{Irr}(Y)$. Set $\mathcal{J} = \pi_*\mathcal{O}(-\Lambda-\Gamma)$. Since $\mathcal{J} \subset \mathfrak{m}_y^2$, the map $H^0(Y, \mathcal{L}) \to \mathcal{L}\otimes\mathcal{O}/\mathcal{J}$ cannot be surjective. From the cohomology sequence:

$$\begin{array}{ccc} H^0(X,\pi^*\mathcal{L}) & \to & H^0(X,\pi^*\mathcal{L}\otimes(\mathcal{O}/\mathcal{O}(-\Lambda-\Gamma))) \\ \wr\!\! & & \cup \\ H^0(Y,\mathcal{L}) & \to & \mathcal{O}/\mathcal{I} \end{array}$$

we deduce that $H^1(X,\pi^*\mathcal{L}\otimes\mathcal{O}(-\Lambda-\Gamma))) \neq 0$. Letting $\tilde{D} = \pi^*D + \Delta - \Lambda - \Gamma$, by duality, we have $H^1(\tilde{Y},\mathcal{O}(-\tilde{D})) \neq 0$. If $y \in \mathrm{Rat}(Y)$, then we have

$$D^2 - \tilde{D}^2 = -(\Delta - \Lambda - \Gamma)^2 = -(\Delta - \Lambda)^2 - 4(K_X + Z)Z = \eta(Y) + 8.$$

If $y \in \mathrm{Irr}(Y)$, then we have

$$D^2 - \tilde{D}^2 = -(\Delta - \Lambda)^2 - \{(Z^t - \Delta_y)^2 - (Z^a - \Delta_y)^2\} = \eta(Y) + \nu_y.$$

In either case, we get $\tilde{D}^2 > 0$, so that $\tilde{D}$ is big. We then apply Proposition 1 to this $\tilde{D}$, and we find a nonzero effective divisor $\tilde{E}$ with $(\tilde{D} - \tilde{E})\tilde{E} \leq 0$. As in the former case, we get $E = \pi_*\tilde{E} > 0$. By Proposition 2, we assert that E satisfies the conditions in (ii) if y is rational, and the conditions in (iii) if y is non-rational. Q.E.D.

Remark 3. In the above proof, I learned the trick of using the ideal (u,v^2) from A.Sommese. See [BFS] for its systematic study and for the higher order analysis.

If D is a Cartier divisor, one can study $|K_Y + D|$ on Gor(Y), which is the Gorenstein locus Y.

Theorem 4. *Let* Y *be a normal Moishezon surface, and let* D *be a nef Cartier divisor on* Y.

(i) *If* $D^2 > 4$, *then* $|K_Y + D|$ *has no base points on* Gor(Y) *unless there exists a nonzero effecitve divisor* E *such that*

$$DE = 0,\ -1 \leq E^2 < 0, \text{ or } DE = 1,\ 0 \leq E^2 \leq 1/D^2.$$

(ii) *If* $D^2 > 8$, *then* $|K_Y + D|$ *separates two distinct points on* Gor(Y) *unless there exists a nonzero effective divisor* E *satisfying one of the following conditions:*

$DE = 0, \quad -2 \le E^2 \le 0,$

$DE = 1, \quad -1 \le E^2 \le 1/D^2,$

$DE = 2, \quad 0 \le E^2 \le 4/D^2,$

$D \equiv 3E, \quad E^2 = 1.$

Corollary. *Let* Y *be a normal projective surface. Let* H *be an ample Cartier divisor on* Y. *Then* $|K_Y + nH|$ *separates two distinct points on* Gor(Y) *for* $n \geq 4$.

6. Pluricanonical and pluri-anticanonical systems.

We study pluricanonical and pluri-anticanonical systems. When Y is $\mathbb{Q}$-Gorenstein, there is a positive integer r such that rK_Y is a Cartier divisor. The least such r is called the *index* of Y.

Theorem 5. *Let* Y *be a normal* $\mathbb{Q}$*-Gorenstein projective surface with index* r, *such that* K_Y *is ample. Then*

(i) $|nrK_Y|$ *has no base points for* $n \geq 2 + r^{-1} + \eta(Y)/2$, *and for* $n > 1 + 3/(2r) + \eta(Y)/4$ *if* $K_Y^2 \geq 2/r$.

(ii) $|nrK_Y|$ *separates two distinct points on* Y *and is very ample off* Irr(Y) *for* $n \geq 4 + r^{-1} + \eta(Y)/2$, *and for* $n > 2 + 3/(2r) + \eta(Y)/4$ *if* $K_Y^2 \geq 2/r$.

(iii) $|nrK_Y|$ *is very ample for* $n \geq 4 + r^{-1} + (\eta(Y) + \gamma(Y))/2$, *and for* $n > 2 + 3/(2r) + (\eta(Y) + \gamma(Y))/4$ *if* $K_Y^2 \geq 2/r$.

Proof. (i) Set $D = (nr - 1)K_Y$. Since rK_Y is Cartier and ample, we have $K_YE \geq 1/r$ for all divisors $E > 0$ on Y, and we get $DE \geq n - r^{-1}$. If $n \geq r^{-1} + 2 + \eta(Y)/2$, then $D^2 > 4 + \eta(Y)$ and there would be no E satisfying (i) of Theorem 3. If $n > 1 + 3/(2r) + \eta(Y)/4$ and $K_Y^2 \geq 2/r$, then $D^2 > 4 + \eta(Y)$. We divide into two cases: (a) $K_Y E \geq 2/r$, (b) $K_Y E = 1/r$. In case (a), we would have $DE \geq 2 + \eta(Y)/2$. In case (b), we would have $DE - 1 - \eta(Y)/4 > (DE)^2/D^2$. In either case, E with (i) of Theorem 3 cannot exist. We omit the proofs of (ii), (iii). Q.E.D.

Corollary (cf.[B], [Sa1]). *Let* Y *be a normal Gorenstein projective surface, such that* K_Y *is ample. Then*

(i) $|nK_Y|$ *separates two distinct points for* $n \geq 5$.

(ii) $|nK_Y|$ *is very ample for* $n \geq 5 + \frac{1}{2}\gamma(Y)$.

We can show an analogous result for the anticanonical divisor. We here state the very ampleness part.

Theorem 6. *Let* Y *be a normal* **Q**-*Gorenstein projective surface with index* r, *such that* $-K_Y$ *is ample. Then* $|-nrK_Y|$ *is very ample for* $n \geq 4 - r^{-1} + (\eta(Y) + \gamma(Y))/2$, *and for* $n > 2 - 1/(2r) + (\eta(Y) + \gamma(Y))/4$ *if* $K_Y^2 \geq 2/r$.

Corollary. *Let* Y *be a normal Gorenstein projective surface, such that* $-K_Y$ *is ample. Then* $|-nK_Y|$ *is very ample for* $n \geq 3$, *and for* $n \geq 1$ *if* $K_Y^2 \geq 3$.

Proof. In this case, it is known that either (a) Y has only rational double points, or (b) Y is an elliptic cone. In case (a), $|-nK_Y|$ is very ample if $(n + 1)^2K_Y^2 \geq 10$. Cf. [SV]. For the case (b), see Example 2 below. Q.E.D.

Example 2. We consider the Gorenstein cones. Let X be a $\mathbf{P}^1$-bundle $\mathbf{P}(\mathcal{O}\oplus\mathcal{O}(-\mathfrak{d}))$ over a smooth curve B of genus g, where $\mathfrak{d}$ is a divisor of degree $d > 0$. There is a unique section b with $b^2 < 0$. Let $\pi: X \to Y$ be the contraction of b to a vertex z of the cone Y. We find that $\delta_z =$

$(2g - 2 + d)^2/d$, $K_Y^2 = (2g - 2 - d)^2/d$. We see that Y is Gorenstein in the following cases: (a) $g \geq 2$, $d|2g - 2$, $K_B \sim q\mathfrak{b}$, where $q = (2g - 2)/d$, (b) $g = 1$, $d \geq 1$, (c) $g = 0$, $d = 2$.

In case (a), K_Y is ample if $q > 1$, $K_Y \sim 0$ if $q = 1$. Suppose $q > 1$. Set $D = (n - 1)K_Y$. If $n \geq 5$, $|nK_Y|$ is very ample off z (Theorem 5). Since there are no curves with negative self-intersection other than b on X, we infer from Theorem 3 and Remark 2 that $|nK_Y|$ separates tangent vectors at z if $D^2 > 8 + \gamma(Y) = \max\{8, \nu_z\}$. Since $\nu_z \leq 9d(1 + q)^2$ (Remark 1), $D^2 = d(n - 1)^2(q - 1)^2$, we conclude that $|nK_Y|$ is very ample for $n \geq 11$, for $n \geq 5$ and $g \geq 4d + 1$.

In case (b), set $D = -(n + 1)K_Y$. Suppose $D^2 \geq 10$. Then $|-nK_Y|$ is very ample off z unless there is an effective divisor E satisfying the conditions (ii) in Theorem 3. Suppose such E exists. Since $DE \geq n + 1$ we have $n = 1$, $K_YE = -1$. It is easy to see that $C^2 \geq 1/d$ for every irreducible curve C on Y, and if $C^2 = 1/d$, C is a ruling line, and so $K_YC = -2$. Thus $E^2 \geq 1/d$, and we get $E^2 = 1/d$, since $E^2 \leq 4/D^2 = 1/d$. This is a contradiction. Thus $|-nK_Y|$ is very ample off z. By the same reason as in the case (a), $|-nK_Y|$ separates tangent vectors at z if $D^2 > 8 + \gamma(Y) = \max\{8, \nu_z\}$. Since $D^2 = (n + 1)^2d$, $\nu_z \leq 9d$ (= d if $d \geq 3$), we see that $|-nK_Y|$ is very ample for $n \geq 3$, for $n \geq 1$ and $d \geq 3$.

In case (c), $|-nK_Y|$ is very ample for $n \geq 2$.

7. Relative case.

Let (V,y) be a germ of a normal surface singularity, $\pi: U \to V$ a resolution of y. An invertible sheaf $\mathcal{L}$ on U is *π-generated* if the canonical map $\pi^*\pi_*\mathcal{L} \to \mathcal{L}$ is surjective.

Proposition 3. *Let* D *be a divisor on* U, *if* $R^1\pi_*\mathcal{O}(K_U + D) \neq 0$, *then there exists a nonzero effective divisor* E *supported on* $\pi^{-1}(y)$ *such that* $(D - E)E \leq 0$.

Proof. According to the relative version of the Zariski decomposition ([Sa2]), one can write $D = P + N$ where (i) N is an effective $\mathbf{Q}$-divisor on $\pi^{-1}(y)$, (ii) P is π-nef and $PE_j = 0$ for all irreducible components E_j of E. One has the vanishing theorem $R^1\pi_*\mathcal{O}(K_U + D - [N]) = 0$. Cf. [Sa2], Appendix. One can prove the assertion in a similar manner to that in the second proof of Proposition 1. Q.E.D.

Theorem 7. *Let* V, y, U, π, D *be the same as above. Then* $\mathcal{O}(K_U + D)$ *is* π*-generated unless there exists a nonzero effecitve divisor* E *on* $\pi^{-1}(y)$ *such that* $(D - E)E \leq 1$.

Proof. Assume V is Stein, and consider a point x on U. Let $\varphi:\tilde{U} \to U$ be the blowing up of U at x, and let L be the exceptional curve. Set $\Phi = \pi\circ\varphi$. If x is a base point of $|K_U + D|$, then we infer easily that $H^1(\tilde{U},\mathcal{O}(K_{\tilde{U}} + \varphi^*D - 2L)) \neq 0$. Applying Proposition 3 to the divisor $\tilde{D} = \varphi^*D - 2L$, we find an effective divisor $\tilde{E}$ supported on $\Phi^{-1}(y)$ such that $(\tilde{D} - \tilde{E})\tilde{E} \leq 0$. Then $E = \varphi_*\tilde{E} > 0$, $(D - E)E \leq 1$ (Proposition 2). Q.E.D.

Corollary ([Sh]). *In particular,* $\mathcal{O}(K_U)$ *is* π*-generated unless there exists an effective divisor* E *with* $E^2 = -1$.

References:

[AS] Andreatta,M., Sommese,A.:On the adjunction mapping for singular projective varieties. Preprint

[A] Artin,M.:On isolated rational singularities of surfaces. Amer. J. Math. **88**, 129-136 (1966)

[BFS] Beltrametti,M., Francia,P., Sommese,A.:On Reider's method and higher order embeddings. Preprint

[BL] Beltrametti,M., Lanteri,A.:On the 2 and 3-connectedness of ample divisors on a surface. Manuscripta Math. **58**, 109-128 (1987)

[B] Bombieri,E.:Canonical models of surfaces of general type. Publ. Math. IHES **42**, 172-219 (1973)

[F] Fujita,T.: Remarks on adjoint bundles of polarized surfaces. Preprint

[I] Ionescu,P.:Ample and very ample divisors. Preprint

[L1] Laufer,H.:On minimally elliptic singularities. Amer.J.Math. **99**, 1257-1295 (1977)

[L2] Laufer,H.:Weak simultaneous resolution for deformations of Gorenstein surface singularities. Proc. Symp. Pure Math. **40**, 1-29 (1983)

[L3] Laufer,H.:Generation of 4-pluricanonical forms for surface singularities. Amer.J.Math. **109**, 571-590 (1987)

[Mi] Miyaoka,Y.:On the Mumford-Ramanujam vanishing theorem on a surface. In :Journées de Géom. Alg. d'Angers, pp. 239-247. Sijthoff and Noordhoff, 1980

[MR] Morrow,J., Rossi,H.:Canonical embeddings. Trans. A.M.S. **261**, 547-565 (1980)

[Mu] Mumford,D.:Some footnotes to the work of C.P. Ramanujam. In: C.P.Ramanujam - a tribute, pp. 247-262, Springer-Verlag, 1978

[Ra] Ramanujam,C.P.:Remarks on the Kodaira vanishing theorem. J. Indian Math. Soc. **36**, 41-51 (1972)

[Re] Reid,M.:Elliptic Gorenstein singularities of surfaces. Preprint (1975)

[Rd] Reider,I.:Vector bundles of rank 2 and linear systems on algebraic surfaces. Ann. Math. **127**, 309-316 (1988)

[Sa1] Sakai,F.:Enriques classification of normal Gorenstein surfaces. Amer.J.Math. **104**, 1233-1341 (1982)

[Sa2] Sakai,F.:Anticanonical models of rational surfaces. Math. Ann. **269**, 389-410 (1984)

[Sa3] Sakai,F.:Weil divisors on normal surfaces. Duke Math. J. **51**, 877-887 (1984)

[Se] Serrano,F.:Extension of morphisms defined on a divisor. Math. Ann. 277, 395-413 (1987)

[Sh] Shepherd-Barron,N.I.:Some questions on singularities in 2 and 3 dimensions. Thesis, Univ.of Warwick, 1980

[SV] Sommese,A., Van de Ven, A.:On the adjunction mapping. Math.Ann. 278, 593-603 (1987)

List of seminars held during the conference

F.Catanese: Components of the moduli spaces of surfaces.

M.Chang: Buchsbaum subvarieties of codimension 2 in P^n.

C.Ciliberto: Hyperplane sections of K3 surfaces.

H.Clemens: The use of *D* - modules in the study of deformations of submanifolds: I - II.

L.Ein: Some special Cremona transformations.

T.Fujita: Classification of polarized varieties by sectional genus and Δ-genus: I - II - III.

K.Hulek: Abelian surfaces in $\mathbf{P}^4$ and their moduli.

J.Murre: Height pairing of algebraic cycles.

C.Peskine: Remarks on the normal bundle of smooth threefolds in $\mathbf{P}^5$ - Remarks on Noether theorem for smooth surfaces in $\mathbf{P}^3$.

Z.Ran: Monodromy of plane curves.

M.Reid: Infinitesimal view of extending a hyperplane section - The quadrics through a canonical surface.

I.Reider: Toward Abel-Jacobi theory for higher dimensional varieties and Torelli Theorem.

F.Sakai: Reider-Serrano's method on normal surfaces.

M.Schneider: Compactifications of $\mathbf{C}^3$.

A.J.Sommese: The Classical Adjunction Mapping - The Adjunction Theoretic Structure of Projective Varieties - Some Recent Results On Hyperplane Sections - Projective Classification of Varieties.

Vol. 1259: F. Cano Torres, Desingularization Strategies for Three-Dimensional Vector Fields. IX, 189 pages. 1987.

Vol. 1260: N.H. Pavel, Nonlinear Evolution Operators and Semi-groups. VI, 285 pages. 1987.

Vol. 1261: H. Abels, Finite Presentability of S-Arithmetic Groups. Compact Presentability of Solvable Groups. VI, 178 pages. 1987.

Vol. 1262: E. Hlawka (Hrsg.), Zahlentheoretische Analysis II. Seminar, 1984–86. V, 158 Seiten. 1987.

Vol. 1263: V.L. Hansen (Ed.), Differential Geometry. Proceedings, 1985. XI, 288 pages. 1987.

Vol. 1264: Wu Wen-tsün, Rational Homotopy Type. VIII, 219 pages. 1987.

Vol. 1265: W. Van Assche, Asymptotics for Orthogonal Polynomials. VI, 201 pages. 1987.

Vol. 1266: F. Ghione, C. Peskine, E. Sernesi (Eds.), Space Curves. Proceedings, 1985. VI, 272 pages. 1987.

Vol. 1267: J. Lindenstrauss, V.D. Milman (Eds.), Geometrical Aspects of Functional Analysis. Seminar. VII, 212 pages. 1987.

Vol. 1268: S.G. Krantz (Ed.), Complex Analysis. Seminar, 1986. VII, 195 pages. 1987.

Vol. 1269: M. Shiota, Nash Manifolds. VI, 223 pages. 1987.

Vol. 1270: C. Carasso, P.-A. Raviart, D. Serre (Eds.), Nonlinear Hyperbolic Problems. Proceedings, 1986. XV, 341 pages. 1987.

Vol. 1271: A.M. Cohen, W.H. Hesselink, W.L.J. van der Kallen, J.R. Strooker (Eds.), Algebraic Groups Utrecht 1986. Proceedings. XII, 284 pages. 1987.

Vol. 1272: M.S. Livšic, L.L. Waksman, Commuting Nonselfadjoint Operators in Hilbert Space. III, 115 pages. 1987.

Vol. 1273: G.-M. Greuel, G. Trautmann (Eds.), Singularities, Representation of Algebras, and Vector Bundles. Proceedings, 1985. XIV, 383 pages. 1987.

Vol. 1274: N. C. Phillips, Equivariant K-Theory and Freeness of Group Actions on C*-Algebras. VIII, 371 pages. 1987.

Vol. 1275: C.A. Berenstein (Ed.), Complex Analysis I. Proceedings, 1985–86. XV, 331 pages. 1987.

Vol. 1276: C.A. Berenstein (Ed.), Complex Analysis II. Proceedings, 1985–86. IX, 320 pages. 1987.

Vol. 1277: C.A. Berenstein (Ed.), Complex Analysis III. Proceedings, 1985–86. X, 350 pages. 1987.

Vol. 1278: S.S. Koh (Ed.), Invariant Theory. Proceedings, 1985. V, 102 pages. 1987.

Vol. 1279: D. Ieşan, Saint-Venant's Problem. VIII, 162 Seiten. 1987.

Vol. 1280: E. Neher, Jordan Triple Systems by the Grid Approach. XII, 193 pages. 1987.

Vol. 1281: O.H. Kegel, F. Menegazzo, G. Zacher (Eds.), Group Theory. Proceedings, 1986. VII, 179 pages. 1987.

Vol. 1282: D.E. Handelman, Positive Polynomials, Convex Integral Polytopes, and a Random Walk Problem. XI, 136 pages. 1987.

Vol. 1283: S. Mardešić, J. Segal (Eds.), Geometric Topology and Shape Theory. Proceedings, 1986. V, 261 pages. 1987.

Vol. 1284: B.H. Matzat, Konstruktive Galoistheorie. X, 286 pages. 1987.

Vol. 1285: I.W. Knowles, Y. Saitō (Eds.), Differential Equations and Mathematical Physics. Proceedings, 1986. XVI, 499 pages. 1987.

Vol. 1286: H.R. Miller, D.C. Ravenel (Eds.), Algebraic Topology. Proceedings, 1986. VII, 341 pages. 1987.

Vol. 1287: E.B. Saff (Ed.), Approximation Theory, Tampa. Proceedings, 1985–1986. V, 228 pages. 1987.

Vol. 1288: Yu. L. Rodin, Generalized Analytic Functions on Riemann Surfaces. V, 128 pages, 1987.

Vol. 1289: Yu. I. Manin (Ed.), K-Theory, Arithmetic and Geometry. Seminar, 1984–1986. V, 399 pages. 1987.

Vol. 1290: G. Wüstholz (Ed.), Diophantine Approximation and Transcendence Theory. Seminar, 1985. V, 243 pages. 1987.

Vol. 1291: C. Mœglin, M.-F. Vignéras, J.-L. Waldspurger, Correspondances de Howe sur un Corps p-adique. VII, 163 pages. 1987

Vol. 1292: J.T. Baldwin (Ed.), Classification Theory. Proceedings, 1985. VI, 500 pages. 1987.

Vol. 1293: W. Ebeling, The Monodromy Groups of Isolated Singularities of Complete Intersections. XIV, 153 pages. 1987.

Vol. 1294: M. Queffélec, Substitution Dynamical Systems – Spectral Analysis. XIII, 240 pages. 1987.

Vol. 1295: P. Lelong, P. Dolbeault, H. Skoda (Réd.), Séminaire d'Analyse P. Lelong – P. Dolbeault – H. Skoda. Seminar, 1985/1986. VII, 283 pages. 1987.

Vol. 1296: M.-P. Malliavin (Ed.), Séminaire d'Algèbre Paul Dubreil et Marie-Paule Malliavin. Proceedings, 1986. IV, 324 pages. 1987.

Vol. 1297: Zhu Y.-l., Guo B.-y. (Eds.), Numerical Methods for Partial Differential Equations. Proceedings. XI, 244 pages. 1987.

Vol. 1298: J. Aguadé, R. Kane (Eds.), Algebraic Topology, Barcelona 1986. Proceedings. X, 255 pages. 1987.

Vol. 1299: S. Watanabe, Yu.V. Prokhorov (Eds.), Probability Theory and Mathematical Statistics. Proceedings, 1986. VIII, 589 pages. 1988.

Vol. 1300: G.B. Seligman, Constructions of Lie Algebras and their Modules. VI, 190 pages. 1988.

Vol. 1301: N. Schappacher, Periods of Hecke Characters. XV, 160 pages. 1988.

Vol. 1302: M. Cwikel, J. Peetre, Y. Sagher, H. Wallin (Eds.), Function Spaces and Applications. Proceedings, 1986. VI, 445 pages. 1988.

Vol. 1303: L. Accardi, W. von Waldenfels (Eds.), Quantum Probability and Applications III. Proceedings, 1987. VI, 373 pages. 1988.

Vol. 1304: F.Q. Gouvêa, Arithmetic of p-adic Modular Forms. VIII, 121 pages. 1988.

Vol. 1305: D.S. Lubinsky, E.B. Saff, Strong Asymptotics for Extremal Polynomials Associated with Weights on $\mathbb{R}$. VII, 153 pages. 1988.

Vol. 1306: S.S. Chern (Ed.), Partial Differential Equations. Proceedings, 1986. VI, 294 pages. 1988.

Vol. 1307: T. Murai, A Real Variable Method for the Cauchy Transform, and Analytic Capacity. VIII, 133 pages. 1988.

Vol. 1308: P. Imkeller, Two-Parameter Martingales and Their Quadratic Variation. IV, 177 pages. 1988.

Vol. 1309: B. Fiedler, Global Bifurcation of Periodic Solutions with Symmetry. VIII, 144 pages. 1988.

Vol. 1310: O.A. Laudal, G. Pfister, Local Moduli and Singularities. V, 117 pages. 1988.

Vol. 1311: A. Holme, R. Speiser (Eds.), Algebraic Geometry, Sundance 1986. Proceedings. VI, 320 pages. 1988.

Vol. 1312: N.A. Shirokov, Analytic Functions Smooth up to the Boundary. III, 213 pages. 1988.

Vol. 1313: F. Colonius, Optimal Periodic Control. VI, 177 pages. 1988.

Vol. 1314: A. Futaki, Kähler-Einstein Metrics and Integral Invariants. IV, 140 pages. 1988.

Vol. 1315: R.A. McCoy, I. Ntantu, Topological Properties of Spaces of Continuous Functions. IV, 124 pages. 1988.

Vol. 1316: H. Korezlioglu, A.S. Ustunel (Eds.), Stochastic Analysis and Related Topics. Proceedings, 1986. V, 371 pages. 1988.

Vol. 1317: J. Lindenstrauss, V.D. Milman (Eds.), Geometric Aspects of Functional Analysis. Seminar, 1986–87. VII, 289 pages. 1988.

Vol. 1318: Y. Felix (Ed.), Algebraic Topology – Rational Homotopy. Proceedings, 1986. VIII, 245 pages. 1988

Vol. 1319: M. Vuorinen, Conformal Geometry and Quasiregular Mappings. XIX, 209 pages. 1988.

Vol. 1320: H. Jürgensen, G. Lallement, H.J. Weinert (Eds.), Semigroups, Theory and Applications. Proceedings, 1986. X, 416 pages. 1988.

Vol. 1321: J. Azéma, P.A. Meyer, M. Yor (Eds.), Séminaire de Probabilités XXII. Proceedings. IV, 600 pages. 1988.

Vol. 1322: M. Métivier, S. Watanabe (Eds.), Stochastic Analysis. Proceedings, 1987. VII, 197 pages. 1988.

Vol. 1323: D.R. Anderson, H.J. Munkholm, Boundedly Controlled Topology. XII, 309 pages. 1988.

Vol. 1324: F. Cardoso, D.G. de Figueiredo, R. Iório, O. Lopes (Eds.), Partial Differential Equations. Proceedings, 1986. VIII, 433 pages. 1988.

Vol. 1325: A. Truman, I.M. Davies (Eds.), Stochastic Mechanics and Stochastic Processes. Proceedings, 1986. V, 220 pages. 1988.

Vol. 1326: P.S. Landweber (Ed.), Elliptic Curves and Modular Forms in Algebraic Topology. Proceedings, 1986. V, 224 pages. 1988.

Vol. 1327: W. Bruns, U. Vetter, Determinantal Rings. VII,236 pages. 1988.

Vol. 1328: J.L. Bueso, P. Jara, B. Torrecillas (Eds.), Ring Theory. Proceedings, 1986. IX, 331 pages. 1988.

Vol. 1329: M. Alfaro, J.S. Dehesa, F.J. Marcellan, J.L. Rubio de Francia, J. Vinuesa (Eds.): Orthogonal Polynomials and their Applications. Proceedings, 1986. XV, 334 pages. 1988.

Vol. 1330: A. Ambrosetti, F. Gori, R. Lucchetti (Eds.), Mathematical Economics. Montecatini Terme 1986. Seminar. VII, 137 pages. 1988.

Vol. 1331: R. Bamón, R. Labarca, J. Palis Jr. (Eds.), Dynamical Systems, Valparaiso 1986. Proceedings. VI, 250 pages. 1988.

Vol. 1332: E. Odell, H. Rosenthal (Eds.), Functional Analysis. Proceedings, 1986–87. V, 202 pages. 1988.

Vol. 1333: A.S. Kechris, D.A. Martin, J.R. Steel (Eds.), Cabal Seminar 81–85. Proceedings, 1981–85. V, 224 pages. 1988.

Vol. 1334: Yu.G. Borisovich, Yu. E. Gliklikh (Eds.), Global Analysis – Studies and Applications III. V, 331 pages. 1988.

Vol. 1335: F. Guillén, V. Navarro Aznar, P. Pascual-Gainza, F. Puerta, Hyperrésolutions cubiques et descente cohomologique. XII, 192 pages. 1988.

Vol. 1336: B. Helffer, Semi-Classical Analysis for the Schrödinger Operator and Applications. V, 107 pages. 1988.

Vol. 1337: E. Sernesi (Ed.), Theory of Moduli. Seminar, 1985. VIII, 232 pages. 1988.

Vol. 1338: A.B. Mingarelli, S.G. Halvorsen, Non-Oscillation Domains of Differential Equations with Two Parameters. XI, 109 pages. 1988.

Vol. 1339: T. Sunada (Ed.), Geometry and Analysis of Manifolds. Procedings, 1987. IX, 277 pages. 1988.

Vol. 1340: S. Hildebrandt, D.S. Kinderlehrer, M. Miranda (Eds.), Calculus of Variations and Partial Differential Equations. Proceedings, 1986. IX, 301 pages. 1988.

Vol. 1341: M. Dauge, Elliptic Boundary Value Problems on Corner Domains. VIII, 259 pages. 1988.

Vol. 1342: J.C. Alexander (Ed.), Dynamical Systems. Proceedings, 1986–87. VIII, 726 pages. 1988.

Vol. 1343: H. Ulrich, Fixed Point Theory of Parametrized Equivariant Maps. VII, 147 pages. 1988.

Vol. 1344: J. Král, J. Lukeš, J. Netuka, J. Veselý (Eds.), Potential Theory – Surveys and Problems. Proceedings, 1987. VIII, 271 pages. 1988.

Vol. 1345: X. Gomez-Mont, J. Seade, A. Verjovski (Eds.), Holomorphic Dynamics. Proceedings, 1986. VII, 321 pages. 1988.

Vol. 1346: O. Ya. Viro (Ed.), Topology and Geometry – Rohlin Seminar. XI, 581 pages. 1988.

Vol. 1347: C. Preston, Iterates of Piecewise Monotone Mappings on an Interval. V, 166 pages. 1988.

Vol. 1348: F. Borceux (Ed.), Categorical Algebra and its Applications. Proceedings, 1987. VIII, 375 pages. 1988.

Vol. 1349: E. Novak, Deterministic and Stochastic Error Bounds in Numerical Analysis. V, 113 pages. 1988.

Vol. 1350: U. Koschorke (Ed.), Differential Topology. Proceedir 1987. VI, 269 pages. 1988.

Vol. 1351: I. Laine, S. Rickman, T. Sorvali, (Eds.), Complex Analy Joensuu 1987. Proceedings. XV, 378 pages. 1988.

Vol. 1352: L.L. Avramov, K.B. Tchakerian (Eds.), Algebra – S Current Trends. Proceedings, 1986. IX, 240 Seiten. 1988.

Vol. 1353: R.S. Palais, Ch.-l. Terng, Critical Point Theory Submanifold Geometry. X, 272 pages. 1988.

Vol. 1354: A. Gómez, F. Guerra, M.A. Jiménez, G. López (E Approximation and Optimization. Proceedings, 1987. VI, 280 pa 1988.

Vol. 1355: J. Bokowski, B. Sturmfels, Computational Synthetic G metry. V, 168 pages. 1989.

Vol. 1356: H. Volkmer, Multiparameter Eigenvalue Problems Expansion Theorems. VI, 157 pages. 1988.

Vol. 1357: S. Hildebrandt, R. Leis (Eds.), Partial Differential Equat and Calculus of Variations. VI, 423 pages. 1988.

Vol. 1358: D. Mumford, The Red Book of Varieties and Schemes 309 pages. 1988.

Vol. 1359: P. Eymard, J.-P. Pier (Eds.), Harmonic Analysis. Proce ings, 1987. VIII, 287 pages. 1988.

Vol. 1360: G. Anderson, C. Greengard (Eds.), Vortex Methc Proceedings, 1987. V, 141 pages. 1988.

Vol. 1361: T. tom Dieck (Ed.), Algebraic Topology and Transform Groups. Proceedings, 1987. VI, 298 pages. 1988.

Vol. 1362: P. Diaconis, D. Elworthy, H. Föllmer, E. Nelson, G Papanicolaou, S.R.S. Varadhan. École d'Été de Probabilités de S Flour XV–XVII, 1985–87. Editor: P.L. Hennequin. V, 459 pa 1988.

Vol. 1363: P.G. Casazza, T.J. Shura. Tsirelson's Space. VIII, pages. 1988.

Vol. 1364: R.R. Phelps, Convex Functions, Monotone Operators Differentiability. IX, 115 pages. 1989.

Vol. 1365: M. Giaquinta (Ed.), Topics in Calculus of Variati Seminar, 1987. X, 196 pages. 1989.

Vol. 1366: N. Levitt, Grassmannians and Gauss Maps in PL-Topol V, 203 pages. 1989.

Vol. 1367: M. Knebusch, Weakly Semialgebraic Spaces. XX, pages. 1989.

Vol. 1368: R. Hübl, Traces of Differential Forms and Hochsc Homology. III, 111 pages. 1989.

Vol. 1369: B. Jiang, Ch.-K. Peng, Z. Hou (Eds.), Differential Geom and Topology. Proceedings, 1986–87. VI, 366 pages. 1989.

Vol. 1370: G. Carlsson, R.L. Cohen, H.R. Miller, D.C. Ravenel (E Algebraic Topology. Proceedings, 1986. IX, 456 pages. 1989.

Vol. 1371: S. Glaz, Commutative Coherent Rings. XI, 347 pa 1989.

Vol. 1372: J. Azéma, P.A. Meyer, M. Yor (Eds.), Séminaire de Prob tés XXIII. Proceedings. IV, 583 pages. 1989.

Vol. 1373: G. Benkart, J.M. Osborn (Eds.), Lie Algebras, Ma 1987. Proceedings. V, 145 pages. 1989.

Vol. 1374: R.C. Kirby, The Topology of 4-Manifolds. VI, 108 1989.

Vol. 1375: K. Kawakubo (Ed.), Transformation Groups. Proceedings, 1 VIII, 394 pages, 1989.

Vol. 1376: J. Lindenstrauss, V.D. Milman (Eds.), Geometric Aspect Functional Analysis. Seminar (GAFA) 1987–88. VII, 288 pages. 19

Vol. 1377: J.F. Pierce, Singularity Theory, Rod Theory, and Symme Breaking Loads. IV, 177 pages. 1989.

Vol. 1378: R.S. Rumely, Capacity Theory on Algebraic Curves. III, pages. 1989.

Vol. 1379: H. Heyer (Ed.), Probability Measures on Groups Proceedings. 1988. VIII, 437 pages. 1989